W9-BTI-449

Biological Magnetic Resonance
Volume 20

Protein NMR for
the Millennium

Biological Magnetic Resonance
Volume 20

Protein NMR for the Millennium

Edited by

N. Rama Krishna
University of Alabama at Birmingham
Birmingham, Alabama

and

Lawrence J. Berliner
University of Denver
Denver, Colorado

KLUWER ACADEMIC / PLENUM PUBLISHERS
NEW YORK, BOSTON, DORDRECHT, LONDON, MOSCOW

ISBN: 0-306-47448-4

© 2003 Kluwer Academic / Plenum Publishers, New York
233 Spring Street, New York, New York 10013

http://www.wkap/nl

10 9 8 7 6 5 4 3 2 1

A C.I.P. record for this book is available from the Library of Congress

Printed in the United States of America

Contributors

Charles R. Babu • Johnson Research Foundation and Department of Biochemistry & Biophysics, University of Pennsylvania, Philadelphia, PA 19104-6059

John L. Battiste • Department of Biological Chemistry and Molecular Pharmacology, Harvard Medical School, Boston, MA 02115

Thorsten Biet • Institute of Chemistry, Medical University of Luebeck, Luebeck, Germany

Ulrich K. Blaschke • Laboratories of Synthetic Protein Chemistry, The Rockefeller University, 1230 York Avenue, New York, NY 10021

Florence Cordier • Department of Structural Biology, Biozentrum, University of Basel, 4056 Basel, Switzerland

David Cowburn • Laboratory of Physical Biochemistry, The Rockefeller University, 1230 York Avenue, New York, NY 10021, and New York Structural Biology Center, c/o Box 163, 1230 York Avenue, New York, NY 10021

Andrew J. Dingley • Institute of Physical Biology, Heinrich-Heine-Universität, 40225 Düsseldorf, Germany, and Institute of Structural Biology, IBI-2, Forschungszentrum Jülich, 52425 Jülich, Germany

Peter F. Flynn • Johnson Research Foundation and Department of Biochemistry & Biophysics, University of Pennsylvania, Philadelphia, PA 19104-6059

David Fushman • Dept. of Chemistry and Biochemistry, University of Maryland, College Park, MD 20742

Christian Griesinger ● Max Planck Institute of Biophysical Chemistry, Group 030, Am Fassberg 11, 37077 Göttingen, Germany

Angela M. Gronenborn ● Laboratory of Chemical Physics, NIDDK, NIH, Bethesda, MD

John D. Gross ● Department of Biological Chemistry and Molecular Pharmacology, Harvard Medical School, Boston, MA 02115

Stephan Grzesiek ● Department of Structural Biology, Biozentrum, University of Basel, 4056 Basel, Switzerland

Lars Herfurth ● Institute of Chemistry, Medical University of Luebeck, Luebeck, Germany

Ann McDermott ● Columbia University, Department of Chemistry, New York, NY

Jens Meiler ● Max Planck Institute of Biophysical Chemistry, Group 030, Am Fassberg 11, 37077 Göttingen, Germany

Mark J. Milton ● Johnson Research Foundation and Department of Biochemistry & Biophysics, University of Pennsylvania, Philadelphia, PA 19104-6059

Tom W. Muir ● Laboratories of Synthetic Protein Chemistry, The Rockefeller University, 1230 York Avenue, New York, NY 10021

Jennifer J. Ottesen ● Laboratory of Synthetic Protein Chemistry, The Rockefeller University, 1230 York Avenue, New York, NY 10021

Konstantin V. Pervushin ● Laboratorium für Physikalische Chemie, Eidgenössische Technische Hochschule CH-8092 Zürich, Switzerland

Thomas Peters ● Institute of Chemistry, Medical University of Luebeck, Luebeck, Germany

Wolfgang Peti ● Max Planck Institute of Biophysical Chemistry, Group 030, Am Fassberg 11, 37077 Göttingen, Germany

Gerhard Wagner ● Department of Biological Chemistry and Molecular Pharmacology, Harvard Medical School, Boston, MA 02115

A. Joshua Wand ● Johnson Research Foundation and Department of Biochemistry & Biophysics, University of Pennsylvania, Philadelphia, PA 19104-6059

Preface

Volume 20 is the third in a special topic series devoted to the latest developments in protein NMR under the Biological Magnetic Resonance series. Previous volumes in this series included 16 (*Modern Techniques in Protein NMR*) and 17 (*Structure Computation and Dynamics in Protein NMR*). Once again, we are indeed honored to have brought together in Volume 20 some of the world's foremost experts who have provided broad leadership in advancing the protein NMR field. This volume is broadly divided into three major sections: (*I*) *Toward Larger Proteins in Solution and Solid State;* (*II*) *Structure Refinement*; and (*III*) *NMR Methods for Screening Bioactive Ligands.*

The first section is entirely devoted to recent advances in further pushing the limits of protein size amenable to solution and solid state NMR methods. The opening chapter of Volume 20 deals with Transverse Relaxation Optimized Spectroscopy (TROSY) by Konstantin Pervushin. TROSY is one of the most significant developments that has extended the limit of molecular size that can be studied by solution NMR to in excess of 100 kDa. It exploits the line-narrowing resulting from cross-correlation effects between chemical shift anisotropy relaxation at high magnetic fields and the dipolar relaxation. Konstantin Pervushin describes the theory of TROSY, the pulse sequences, and some applications of this technique. The next two chapters deal with advancements in the study of multidomain proteins. Chapter 2 by David Cowburn, Tom Muir and their collaborators deals with strategies for segmental isotopic labeling in multidomain proteins. In Chapter 3, Fushman and Cowburn describe techniques based on relaxation to characterize inter-domain orientations in solution state. They also give an example where a difference in the interdomain orientations has been observed between solution and crystal structures. In Chapter 4 Gerhard Wagner and his associates describe the site-directed spin labeling approach to study the global folds of large proteins. In Chapter 5 Ann McDermott summarizes the recent advances in the study of uniformly isotopically labeled proteins by solid state NMR methods. A novel approach for achieving narrowing of NMR lines and, hence, extend the size limit of proteins, has been described by Joshua Wand and his collaborators in Chapter 6. Their approach utilizes the combined use of low viscosity solvents and encapsulation of proteins in reverse micelles; an interesting extension of this method is to the study of membrane proteins.

The section on *Structure Refinement* starts with a chapter by Christian Griesinger and his collaborators on the use of angular restraints derived from residual dipolar couplings in proteins and oligosaccharides weakly aligned in high magnetic fields. Griesinger also discusses the impact of dipolar couplings in the area of structural genomics. This is followed by Chapter 8, wherein Angela Gronenborn discusses methods for protein structure refinement based on residual dipolar couplings. Next, Stephan Grzesiek and his associates describe developments involving the detection of scalar couplings across hydrogen bonds in proteins and nucleic acids, and the use of these couplings as structural constraints. The third section deals with a review by Thomas Peters and his associates on the various NMR techniques for the screening of the binding of ligands to target proteins and the further characterization of bioactive ligands.

We are extremely proud of this compilation of excellent contributions from leading investigators describing significant recent advances in the biomolecular NMR field. Because the new publisher of this series has required the authors to prepare their chapters as camera-ready manuscripts using a template supplied by the publisher, there are some inevitable stylistic variations from chapter to chapter. The editors have tried their best to correct these variations, and they take full responsibility for any irregularities that might still remain in the finished volume. As always, the editors welcome suggestions for topics to be covered in future volumes.

N. Rama Krishna
Lawrence J. Berliner

Contents

Chapter 3

Characterization of Inter-Domain Orientations in Solution Using the NMR Relaxation Approach

David Fushman and David Cowburn

Chapter 4

Global Fold Determination of Large Proteins using Site-Directed Spin Labeling

John L. Battiste, John D. Gross, and Gerhard Wagner

Chapter 5

Solid State NMR Studies of Uniformly Isotopically Enriched Proteins

Ann McDermott

Chapter 6

NMR Spectroscopy of Encapsulated Proteins Dissolved in Low Viscosity Fluids

A. Joshua Wand, Charles R. Babu, Peter F. Flynn, and Mark J. Milton

Section II. Structure Refinement

Chapter 7

Angular Restraints from Residual Dipolar Couplings for Structure Refinement

Christian Griesinger, Jens Meiler, and Wolfgang Peti

Chapter 8

Protein Structure Refinement using Residual Dipolar Couplings

Angela M. Gronenborn

Chapter 9

Hydrogen Bond Scalar Couplings – A New Tool In Biomolecular NMR

Stephan Grzesiek, Florence Cordier, and Andrew J. Dingley

Section III. NMR Methods for Screening Bioactive Ligands

Chapter 10

NMR Methods for Screening the Binding of Ligands to Proteins – Identification and Characterization of Bioactive Ligands

Thomas Peters, Thorsten Biet, and Lars Herfurth

I

Toward Larger Proteins in Solution and Solid State

Chapter 1

Transverse Relaxation Optimized Spectroscopy

Konstantin V. Pervushin
Laboratorium für Physikalische Chemie
Eidgenössische Technische Hochschule
CH-8092 Zürich, Switzerland

Abstract: TROSY represents a spectroscopic contribution in a constant pursuit of NMR structures of ever larger proteins, nucleic acids and their complexes. As it was theoretically predicted and proven in practice the most efficient application of TROSY is achieved in combination with the selective or uniform deuteration of biomolecules, in particular proteins. In this way biomolecular particles with the molecular sizes exceeding 100 kDa have now become amenable for high resolution NMR studies. Due to its simplicity and flexibility the TROSY principle has also been integrated into many other NMR applications not necessary targeting extremely large biomolecues. TROSY-based NMR experiments have been developed for the detection and characterization of weak interactions such as scalar couplings across hydrogen bonds and residual dipolar couplings. The use of TROSY in Nuclear Overhauser spectroscopy might significantly facilitate the spectral analysis by suppressing the strong diagonal peaks benefiting studies of a wide range of biomolecules.

1.   TROSY SCOPE

The non-equivalence of certain magnetization transfer pathways in NMR experiments viewed from the stand point of the decay of evolving coherences known as nuclear spin relaxation was recognized at the very onset of the multidimensional multinuclear NMR (Griffey and Redfield, 1987; Gueron *et al.*, 1983; Rance, 1988; Shimizy, 1964). Although for relatively small biomolecules with M.W. of a few kDa at low magnetic field strength the difference in relaxation could be safely ignored, it is not the case for larger systems at higher magnetic fields. The ever lasting demand for larger structures and commercial availability of high field spectrometers pushed high resolution NMR into a realm of proteins and protein complexes of at least an order of magnitude larger where these differences become increasingly more stricking (Clore and Gronenborn, 1998; Pervushin et al., 1997; Wider and Wüthrich, 1999; Wüthrich, 1998). The transverse relaxation optimization spectroscopy, TROSY, emerged as a spectroscopic methodology, which concerns itself with the analysis of transverse relaxation of spin coherences involved and construction of optimal magnetization transfer pathways where the coherences with the most preferable relaxation properties are employed.

The original TROSY (Pervushin et al., 1997; Pervushin et al., 1999) was referred to the use of interference between dipole–dipole coupling (DD) and chemical shift anisotropy (CSA) (Goldman, 1984; Shimizy, 1964) or between two remote CSA interactions (Kumar and Kumar, 1996) to suppress or at least reduce transverse relaxation in two spin 1/2 systems such as ^1H– ^{13}C and ^1H–^{15}N moieties in proteins and nucleic acids (Brutscher et al., 1998; Meissner and Sørensen, 1999a; Pervushin et al., 1997; Pervushin et al., 1998b; Pervushin et al., 1999; Yang and Kay, 1999c). Since then new TROSY-type experiments have been proposed which rely on DD/CSA interference in the ^{13}C–^{13}C spin systems (Riek et al., 2000) or interference between nuclear spin DD and CSA interactions and the electron Curie spin in paramagnetic proteins (Boisbouvier et al., 1999b; Madhu et al., 2000). Those interactions are modulated by rotational molecular motions of biomolecules in study and thus impose a dominant impact on the size limit for biomacromolecular structures that can be studied by NMR spectroscopy in solution (Wider and Wüthrich, 1999).

A dynamic exchange between different conformational or chemical states with the characteristic time scale of microseconds to milliseconds frequently represents yet another significant transverse relaxation mechanism in biomolecules (Kaplan and Fraenkel, 1980; Sandstrom, 1982). The presence of the slow motional processes which cause a modulation of the isotropic chemical shifts of two nuclei involved in multiple-quantum coherences can be manifested in significantly different relaxation of the zero- and double-quantum coherences (Kloiber and Konrat, 2000; Rance, 1988; Tessari and Vuister, 2000). The TROSY principle is now extended to the use of differential conformational exchange-induced transverse relaxation (CSX) in the ZQ and DQ coherences in order to optimize multinuclear NMR experiments. This type of TROSY optimization is independent of the molecular size and can be applied with equal efficiency to a broad range of biomolecules.

Combined with the extensive deuteration TROSY is capable to overcome a key obstacle opposing solution NMR of biomolecules above 100 kDa. Thus, with the use of TROSY the backbone ^{13}C, ^{15}N and $^1H^N$ resonances were assigned in the 42 kDa single polypeptide chain E. coli maltose binding protein/β-cyclodextrin complex at 4°C, which served as a model for even larger proteins at ambient temperature (Yang and Kay, 1999c). Many large systems may give spectra of manageable complexity, such as symmetric multimeric proteins, selectively isotopically labeled proteins in supramolecular complexes or membrane proteins solubilized in detergents. For example, TROSY was successfully used to establish the backbone resonance assignment of a uniformly $^2H,^{15}N,^{13}C$ labeled homooctameric $S.$ $aureus$ 7,8-dihydroneopterin aldolase (Salzmann et $al.$, 2000) with molecular weight of 110 kDa. The backbone resonances were assigned in the TROSY spectra of the $^2H,^{15}N,^{13}C$ labeled 171 residue transmembrane domain of $E.coli$ outer membrane protein A solubilized in 1,2-dicaproyl-glycero-3-phosphocholine micelles (Pervushin et $al.$, 2000a). The cross relaxation-enhanced polarization transfer (CRINEPT) technique optimized for large systems promises yet another several fold increase in the molecular size limit for NMR studies (Riek et $al.$, 1999).

For large single polypeptide chain proteins the complexity of the spectra might be reduced using a combination of TROSY and segmental isotope labeling (see Chapter 2). Even if the detailed 3D structure is not of prime interest, the TROSY-type direct correlation experiments might be useful in chemical shift mapping of the intermolecular contacts in very large aggregates. For example, on the surface of the NMR structure of the 23 kDa two-domain periplasmic chaperone FimC from $E.$ $coli$ the contact sites with the 28 kDa mannose-binding type-1 pilus subunit FimH were identified by ^{15}N and ^1H NMR chemical shift mapping, using TROSY (Pellecchia et $al.$, 1999). NOE steady-state cross-relaxation combined with TROSY allowed identification of the binding surface of protein A with the Fc portion of immunoglobulin G (Takahashi et $al.$, 2000). TROSY was used to extend applicability of the conventional ^{15}N and ^{13}C relaxation measurements (Wagner, 1993a) for studies of intramolecular dynamics in larger proteins (Loria et $al.$, 1999b) or nucleic acids (Boisbouvier et $al.$, 1999a; Fiala et $al.$, 2000). In the following sections the theoretical background of TROSY is described and the experimental applications are illustrated with a number of large proteins with molecular masses of up to 110 kDa.

2.   TROSY: HOW DOES IT WORK?

We consider a system of two scalar coupled spins 1/2, I and S, with a scalar coupling constant J_{IS}, which is located in a rigid molecule. The relaxation rates of the individual multiplet components of the spins I and S in a single quantum spectrum may then be widely different due to the effect of interference between IS dipolar coupling and anisotropic chemical shift on transverse relaxation (Pervushin, 2000). Detailed descriptions of the interference effects on transverse relaxation in coupled spin systems are available (Blicharski, 1967; Blicharski, 1972; Goldman, 1984; Gueron et al., 1983; Mayne and Smith, 1996; Vold and Vold, 1975; Vold and Vold, 1978).

A convenient functional form of the transverse relaxation equations can be obtained by using single-transition and zero- and double-quantum basis operators (Ernst et al., 1987; Sørensen et al., 1983). In the slow-tumbling limit for an isolated IS spin system in the absence of rf pulses and rf field inhomogeneities only terms in $J(0)$ need to be retained resulting in an uncoupled system of differential equations with the diagonal form of the first-order relaxation matrix given by Eq. 1.

$$\frac{d}{dt}\begin{bmatrix}\langle I_{13}^{\pm}\rangle\\ \langle I_{24}^{\pm}\rangle\\ \langle S_{12}^{\pm}\rangle\\ \langle S_{34}^{\pm}\rangle\\ \langle DQ^{\pm}\rangle\\ \langle ZQ^{\pm}\rangle\end{bmatrix} = -diag\left(i\begin{bmatrix}\pm\omega_I^{13}\\ \pm\omega_I^{24}\\ \pm\omega_S^{12}\\ \pm\omega_S^{34}\\ \pm\omega^{DQ}\\ \pm\omega^{ZQ}\end{bmatrix} + \frac{8}{5}\tau_c\begin{bmatrix}p^2-2C_{p\delta_I}p\delta_I+\delta_I^2\\ p^2+2C_{p\delta_I}p\delta_I+\delta_I^2\\ p^2-2C_{p\delta_S}p\delta_S+\delta_S^2\\ p^2+2C_{p\delta_S}p\delta_S+\delta_S^2\\ \delta_I^2+2C_{\delta_I\delta_S}\delta_I\delta_S+\delta_S^2\\ \delta_I^2-2C_{\delta_I\delta_S}\delta_I\delta_S+\delta_S^2\end{bmatrix} + \begin{bmatrix}R_I^{csx}\\ R_I^{csx}\\ R_S^{csx}\\ R_S^{csx}\\ R_{DQ}^{csx}\\ R_{ZQ}^{csx}\end{bmatrix}\right)\bullet\begin{bmatrix}\langle I_{13}^{\pm}\rangle\\ \langle I_{24}^{\pm}\rangle\\ \langle S_{12}^{\pm}\rangle\\ \langle S_{34}^{\pm}\rangle\\ \langle DQ^{\pm}\rangle\\ \langle ZQ^{\pm}\rangle\end{bmatrix}$$

[1]

where ω_S and ω_I are the Larmor frequencies of the spins S and I, $p = \gamma_I\gamma_S h/r_{IS}^3$, $\delta_S = \gamma_S B_0\Delta\sigma_S$ and $\delta_I = \gamma_I B_0\Delta\sigma_I$, where γ_I and γ_S are the gyromagnetic ratios of I and S, h is the Plank constant, r_{IS} the distance between S and I, B_0 the polarizing magnetic field, and $\Delta\sigma_S$ and $\Delta\sigma_I$ are the differences between the axial and the perpendicular principal components of the axially symmetric chemical shift tensors of the spins S and I, respectively. $C_{kl} = 0.5(3\cos^2\Theta_{kl} -1)$ and Θ_{kl} the angle between the unique tensor axes of the interactions k and l. R^{csx} are the relaxation rates due to the conformational exchange (see section 4). The single-transition basis operators $I_{13}^{\pm}$, $I_{24}^{\pm}$, $S_{12}^{\pm}$ and $S_{34}^{\pm}$, which refer to the transitions 3–1, 4–2, 2–1 and 4–3, respectively, in the standard energy-level diagram for a system of two spins 1/2 shown in Fig. 1a, are associated with the corresponding resonance frequencies given by Eqs. 2–5 and can be identified with particular multiplet components in the single-quantum spectrum (Fig. 1c and d).

$$I_{13}^{\pm}= I^{\pm}/2 - I^{\pm}S_z, \quad \text{precession frequency} \quad \omega_I^{13} = \omega_I - \pi J_{IS}, \quad [2]$$

$$I_{24}^{\pm}= I^{\pm}/2 + I^{\pm}S_z \quad\quad\quad\quad\quad\quad\quad\quad \omega_I^{24} = \omega_I + \pi J_{IS}, \quad [3]$$

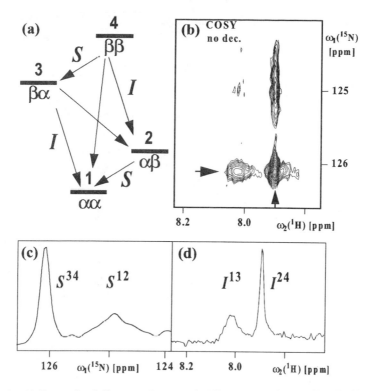

Figure 1. (a) Energy level diagram of a two spin 1/2 system IS showing the identification of components of the 2D multiplet with off-diagonal density matrix elements expressed via single transition basis operators, I^{13}, I^{24}, S^{12} and S^{34}, connecting eigenstates j and k of the static Hamiltonian superoperator. The diagram is not drawn to scale. (b) 2D cross-peak from the [^{15}N,^{1}H]-HSQC spectrum (Bodenhausen and Ruben, 1980) of a backbone amide moiety of u- [^{15}N,^{2}H]-aldolase, 110 kDa at 20°C measured without ^{1}H and ^{15}N decoupling. (c) and (d) 1D slices taken along ω_1 and ω_2 dimensions at the positions indicated in (b).

$$S^{\pm}_{12}= S^{\pm}/2 - I_z S^{\pm} \qquad\qquad \omega_S^{12} = \omega_S - \pi J_{IS}, \qquad [4]$$

$$S^{\pm}_{34}= S^{\pm}/2 + I_z S^{\pm} \qquad\qquad \omega_S^{34} = \omega_S + \pi J_{IS} \qquad [5]$$

The zero- and double-quantum basis operators $ZQ^{\pm}$ and $DQ^{\pm}$, which refer to the transitions 3–2 and 4–1, respectively, in the standard energy-level diagram (Fig. 1a) and are associated with the corresponding resonance frequencies given by Eqs. 6–7.

$$DQ^{\pm} = I^{\pm}S^{\pm}, \qquad \text{precession frequency:} \qquad \omega^{DQ} = \omega_I + \omega_S \qquad [6]$$

$$ZQ^{\pm} = I^{\pm}S^{-/+} \qquad\qquad\qquad\qquad \omega^{ZQ} = \omega_I - \omega_S \qquad [7]$$

Eqs. 2–7 show that whenever CSA and DD coupling are comparable in magnitude and the principal symmetry axis of the chemical shift tensor and the vector r_{IS} are almost collinear as is the case for amide groups in the protein backbone, the resonances at frequencies ω_S^{34} and ω_I^{24} may exhibit slow transverse relaxation even for very large molecules (Pervushin et al., 1997; Pervushin et al., 1998c). A 2D NMR pulse sequence which correlates exclusively the 4–3 transition of the spin S with the 4–2 transition of the spin I represents a Single-Quantum Transverse Relaxation Optimized experiment, SQ-TROSY, which utilizes the single-quantum transitions to record chemical shifts in both spectral dimensions (Pervushin et al., 1998c). Correspondingly, an NMR experiment connecting the zero-quantum transition of the spin S with preferable transverse relaxation properties with the 4–2 transition of the spin I represents a Zero-Quantum TROSY, ZQ-TROSY (Pervushin et al., 1999).

To quantitatively evaluate the transverse relaxation optimization effect and contributions from other mechanisms of relaxation we consider a specific example where I and S are identified as the ${}^1H^N$ and ${}^{15}N$ spins in a ${}^{15}N-{}^1H$ moiety. The internuclear distances and CSA tensors in various model compounds including amide moieties in short peptides were extensively studied by solid state NMR, so that we use the data obtained there: the internuclear distance $r({}^{15}N-{}^1H^N) = 1.04$ Å for the peptide backbone amide moieties (Roberts et al., 1987); $\Delta\sigma({}^1H^N) = \sigma_{zz} - (\sigma_{xx} + \sigma_{yy})/2 = 15$ ppm, is axially symmetric with the angle between the zz axis and the ${}^{15}N-{}^1H^N$ bond = 10° (Gerald et al., 1993; Ramamoorthy et al., 1997); $\Delta\sigma({}^{15}N) = \sigma_{zz} - (\sigma_{xx} + \sigma_{yy})/2 = -155$ ppm, is axially symmetric with the angle between zz axis and ${}^{15}N-{}^1H^N$ bond = 15° (Wu et al., 1995).

The above parameters of the protein backbone amide moiety along with DD contributions from remote protons are used to calculate dependence of the TROSY ${}^1H^N$ and ${}^{15}N$ linewidths on the polarizing magnetic field strength expressed in the frequency units for three hypothetical proteins with molecular weights in the range 50 to 800 kDa. Figure 2 shows that at ${}^1H$ frequencies in the range of 900 to 1100 MHz significant attenuation of transverse relaxation can be achieved simultaneously for the TROSY component of the ${}^1H^N$ and ${}^{15}N$ multiplets observed for the ${}^{15}N-{}^1H$ moieties. For a comparison, predicted ${}^1H^N$ and ${}^{15}N$ linewidth for the conventional [${}^{15}N,{}^1H$]-HSQC experiment is plotted showing that the conventional scheme is not applicable for proteins with molecular weights above 150 kDa.

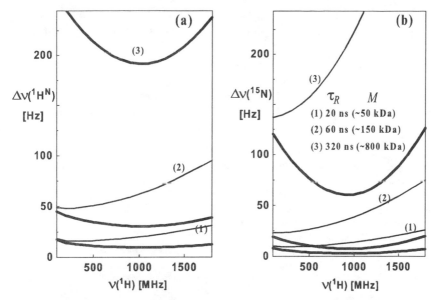

Figure 2. Frequency dependence from 100-1800 MHz of the full resonance linewidth at half height for amide groups in [^{15}N,^{1}H]-SQ-TROSY (bold lines) (Pervushin *et al.*, 1997; Pervushin *et al.*, 1998c) and in [^{15}N,^{1}H]-HSQC (thin lines) (Bodenhausen and Ruben, 1980) experiments calculated for three correlation times of τ_c = 20, 60 and 320 ns, which represent spherical proteins with molecular weights of 50,000, 150,000 and 800,000 M_r (M_r = molecular mass). (a) ^{1}H^{N} linewidth. (b) ^{15}N linewidth. The following parameters were used: r_{HN} = 1.04A, $\Delta\sigma_N$ = - 155 ppm, $\Delta\sigma_H$ = 15 ppm, Θ_N = 15°, Θ_H = 10°, remote protons r (^{1}H^{N}i-H$^{\alpha}$i) = 0.22 nm, r (^{1}H^{N}i- H$^{\alpha}$i+1) = 0.28 nm, r (^{1}H^{N}i-^{1}H^{N}i+1) = 0.4 nm, r (^{1}H^{N}i-^{1}H^{N}i-1) = 0.4 nm, r (^{1}H^{N}i-H$^{\beta2}$i) = 0.3 nm, r (^{1}H^{N}i-H$^{\beta3}$i) = 0.3 nm.

For ^{1}H^{N} in both ZQ$^{\pm}$ and $I^{\pm}_{24}$ states, the contributions from other sources of relaxation are dominated by DD interactions with remote protons I_k at distance r_k and appear as an additive term to the corresponding TROSY relaxation rates given by Eq. 1. Uniform or partial replacement of nonlabile protons with deuterons significantly reduces transverse relaxation by scaling down dipole dipole interactions with remote hydrogens (Gardner and Kay, 1998; Grzesiek *et al.*, 1995b; LeMaster, 1994; Shan *et al.*, 1996; Yamazaki *et al.*, 1994a). Advantages are demonstrated in selective protonation of peripheral side-chain groups such as the methyl groups of Val, Leu, Ile in otherwise ^{15}N, ^{13}C, ^{2}H-labeled proteins (Gardner and Kay, 1998; Gardner *et al.*, 1996; Gardner *et al.*, 1997; Gardner *et al.*, 1998; Goto *et al.*, 1999; Rosen *et al.*, 1996) or when proton labels are retained in the amino acids of a given type in an otherwise deuterated background (Kelly *et al.*, 1999; McCallum *et al.*, 1999). In all cases transverse relaxation of the amide protons and remaining aliphatic and aromatic protons is substantially reduced which is frequently a prerequisite of TROSY NMR studies of large proteins.

Figure 3 (overleaf). Experimental schemes for (a) 2D Single-Quantum [^{15}N,^{1}H]-TROSY using single transition-to-single transition polarization transfer (box identified with ST2-PT) (Pervushin *et al.*, 1998c) and (b) 2D Zero-Quantum [^{15}N,^{1}H]-TROSY (Pervushin *et al.*, 1999). On the lines marked ^{1}H and ^{15}N, narrow and wide bars stand for non-selective 90° and 180° radio-frequency pulses, respectively. The delay $\tau = 2.7$ ms. The line marked PFG indicates the pulsed magnetic field gradients applied along the z-axis: in (a) G_1, amplitude 30 G/cm, duration 1 ms; G_2, 40 G/ cm, 1 ms; G_3, 48 G/cm, 1 ms; G_N, -60 G/cm, 0.75 ms; G_H, 60 G/cm, 0.076 ms; in (b) G_1, 50 G/ cm, 1 ms; G_N, -60 G/cm, 0.75 ms; G_H, 60 G/cm, 0.076 ms; in (c) G_1, 30 G/cm, 1 ms; G_2, 5 G/ cm, 0.5·t_1; G_3, -5 G/cm, 0.5·t_1; G_4, 40 G/cm, 1 ms; G_N, -60 G/cm, 0.75 ms; G_H, 60 G/cm, 0.076 ms. (a) The following two-step phase cycling scheme was used: $\psi_1 = \{y, -x\}$; $\psi_2 = \{-y\}$; $\psi_3 = \{y\}$; $\psi_4 = \{-y\}$; ψ_5(receiver) = $\{y, -x\}$; x on all other pulses. To obtain a complex interferogram a second FID is recorded for each t_1 delay, with $\psi_1 = \{y, x\}$; $\psi_2 = \{y\}$; $\psi_3 = \{-y\}$; $\psi_4 = \{y\}$, and G_N inverted. The use of ST2-PT thus results in a 2D [^{1}H,^{15}N]-correlation spectrum that contains only the most slowly relaxing component of the 2D ^{15}N–^{1}H multiplet. The data are processed as described by (Kay *et al.*, 1992). (b) The phases for the rf-pulses are: $\phi_1 = \{-x\}$; $\phi_2 = \{x\}$; $\phi_3 = \{x, -x, -y, y\}$; $\psi_1 = \{-x, x, -y, y\}$; $\psi_2 = \{y, -y, x, -x\}$; $\psi_3 = \{y\}$; x on all other pulses. To obtain a complex interferogram, a second FID is recorded for each t_1 delay, with the following different phases: $\phi_1 = \{x\}$, $\phi_2 = \{-x\}$, $\phi_3 = \{x, -x, y, -y\}$, $\psi_3 = \{-y\}$. After Fourier transformation in the ω_2-dimension the complex interferogram is multiplied by $\exp[-i\Omega_H t_1]$, where Ω_H is the offset in the ω_2-dimension relative to the ^{1}H carrier frequency in rad s^{-1}. Further data processing follows Kay *et al.* (1992). Water saturation is minimized by keeping the water magnetization along the +z-axis during the entire experiment, which is achieved by the application of the water-selective 90° rf-pulses indicated by curved shapes on the ^{1}H line (Grzesiek and Bax, 1993; Piotto *et al.*, 1992). The use of the gradients G_N and G_H (broken lines) allows the recording of the pure phase absorption spectrum without any cycling of the pulse phases. This may be attractive, for example, to minimize the recording time of the experiment, but the first point of the FIDs generated in t_1 by transformation of t_2 must then be back-predicted in both dimensions. Alternatively, when the two-step phase cycle is employed, the gradients G_N and G_H are usually not applied.

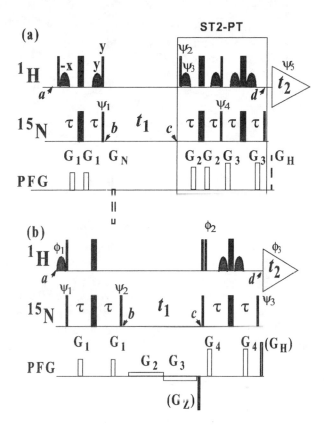

3. DIRECT HETERONUCLEAR CHEMICAL SHIFT CORRELATIONS

The evolution of the density operator in a variant of the single-quantum [^{15}N,^{1}H]-TROSY (Andersson *et al.*, 1998a; Brutscher *et al.*, 1998; Czisch and Boelens, 1998; Meissner *et al.*, 1998; Pervushin *et al.*, 1998c; Rance *et al.*, 1999; Weigelt, 1998; Zhu *et al.*, 1999) the experimental scheme of which is shown in Fig. 3a, can be schematically represented as:

$$uI_z + vS_z \rightarrow 0.5(u - v)S^+_{12}\, exp(-i\omega_S^{12}t_1) \rightarrow 0.5(u - v)I^-_{13}\, exp(-i\omega_S^{12}t_1) \quad [8]$$

$$uI_z + vS_z \rightarrow 0.5(u + v)S^+_{34}\, exp(-i\omega_S^{34}t_1) \rightarrow 0.5(u + v)I^-_{24}\, exp(-i\omega_S^{34}t_1) \quad [9]$$

The first arrow designates coherence transfer from ^{1}H to ^{15}N and ^{15}N chemical shift evolution during the delay t_1, and the second arrow represents the coherence transfer from ^{15}N to ^{1}H. The constant factors u and v reflect the relative magnitude of the steady-state ^{1}H and ^{15}N magnetization, respectively, which are determined by the respective gyromagnetic ratios, the spin–lattice relaxation rates and the delay between individual data recordings. Since the $S^\pm_{34}$ operators are transferred to observable magnetization, both the ^{1}H and ^{15}N steady-state magnetizations contribute to the signal, which is then proportional to $0.5(u - v)$.

When both pathways indicated by Eqs. 8 and 9 are retained (Andersson *et al.*, 1998a), two diagonally shifted signals are observed representing two out of the four ^{15}N–^{1}H multiplet components in the resulting [^{15}N,^{1}H]-correlation spectrum (see Fig. 1). The undesired polarization transfer pathway, $S^\pm_{12}$ –> $\varGamma_{13}$, is suppressed either by 2-step cycling of the phases ψ_1 and ψ_5 (Fig. 3a), or by application of PFGs during t_1 and at time point d (Fig. 3a). The remaining anti-echo polarization transfer pathway, $S^\pm_{34}$ –> $\varGamma_{24}$, connects a single transition of the spin S with a single transition of the spin I, and in alternate scans with inversion of the rf- phases ψ_2 and ψ_4, the corresponding echo transfer, S^-_{34} –> $\varGamma_{24}$, is recorded. Lately variants of the ST2-PT polarization transfer scheme were introduced by Yang and Kay (1999a) and Schulte-Herbruggen *et al.* (1999).

The use of the TROSY-type ^{15}N,^{1}H correlation experiments is demonstrated with the homooctameric *S. aureus* 7,8-dihydroneopterin aldolase (DHNA) (Hennig *et al.*, 1998) with a molecular weight of 110 kDa for the unlabeled protein. DHNA represents an oligomeric protein of high molecular weight which yields a [^{15}N,^{1}H]-correlation spectrum with slightly more than 100 resonances (Fig. 4) due to the degeneracy of the resonances stemming from the identical structural subunits (Hennig *et al.*, 1998; Wüthrich, 1998). For this exceptionally large protein as judged using conventional NMR methods (Wagner, 1993b), a dramatic improvement in the linewidth was obtained using TROSY. The detailed comparison of the line shapes in both spectral dimension for three selected residues, Arg 90, Ile 64 and Glu 119, located in the α-helical, loop and β-sheet secondary structure elements of DHNA, respectively, is shown in Fig. 5. The stronger TROSY effect is observed for the amide moieties located in β-sheet secondary structure, which might be explained by the fact that in β-sheets somewhat larger values of the ^{1}H^N CSA anisotropy tensor with the almost parallel orientation of the unique axis to the H-N vector are usually found (Tessari *et al.*, 1997a; Tessari *et al.*, 1997b; Tjandra and Bax, 1997b).

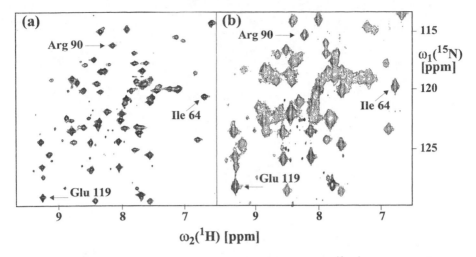

$\omega_2(^1H)$ [ppm]

Figure 4. Contour plots of a region containing most of the backbone ^{15}N–^1H resonances from $^{15}N,^1H$-correlation spectra of a 1 mM solution of uniformly ^{15}N-labeled 7,8-dihydroneopterin aldolase (DHNA) *S. aureus* (Hennig *et al.*, 1998) with a molecular weight of 110 kDa for the unlabeled protein recorded using (a) SQ-$[^{15}N,^1H]$-TROSY of Fig. 3a and (b) conventional $[^{15}N,^1H]$-FHSQC (Andersson *et al.*, 1998b). Both spectra were measured at a proton resonance frequency of 750 MHz at 25°C and pH=6.0. The arrows marked according to the residue specific assignment identify cross-peaks for which cross-sections are shown in Fig. 5.

For the zero-quantum $[^{15}N,^1H]$-TROSY experiment the two equivalent magnetization transfer pathways are given by Eqs. 10 and 11:

$$u I_z^{13} + v S_z \rightarrow (u - 0.5v) ZQ\_ exp[(-R^{ZQ} - i\,\Omega^{ZQ})t_1] \rightarrow$$
$$(u - 0.5v)\, I^-{}_{24}\, exp[(-R^{24} + i\,\omega^{24})t_2] \qquad\qquad [10]$$

$$u I_z^{24} + v S_z \rightarrow (u + 0.5v) DQ_+ exp[(-R^{DQ} + i\,\Omega^{DQ})t_1] \rightarrow$$
$$(u + 0.5v)\, I^-{}_{13}\, exp[(-R^{13} + i\,\omega^{13})t_2] \qquad\qquad [11]$$

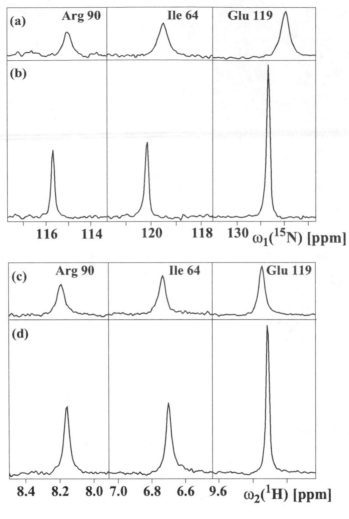

Figure 5. (a) and (c), cross sections of SQ-$[^{15}N,^1H]$-TROSY and, (b) and (d), $[^{15}N,^1H]$-FHSQC spectra of Fig. 4 taken along the $\omega_1(^{15}N)$ and $\omega_2(^1H)$ dimensions. The letters refer to the corresponding cross-peaks indicated in Fig. 4.

The initial steady-state ^1H polarization at time point a (Fig. 3b), I_z, is the sum of the single- transition polarization operators, $I_z = I_z^{13} + I_z^{24}$, which can be considered separately. R^{24} and R^{13} are the previously described relaxation rates of the individual single-quantum transitions 4– 2 and 3–1 (Fig. 1). R^{ZQ} and R^{DQ} are the relaxation rates of the zero-quantum and double-quantum coherences given by Eq. 1.

For backbone ^{15}N–^{1}H amide moieties in proteins, where δ_S can be found to be approximately equal to δ_I (see Eq. 1) and the unique axes of the ^{1}H and ^{15}N CSA tensors are nearly parallel, transverse relaxation due to ^{1}H CSA and ^{15}N CSA interactions is largely suppressed for the ZQ coherence. Since ^{15}N–^{1}H DD relaxation is negligible for the ZQ coherence (Grzesiek *et al.*, 1995a; Kumar and Kumar, 1995) and the acquired signal comes from the relaxation-optimized single-transition 4–2, Eq. 10 represents a ZQ-based method to obtain TROSY-type 2D [^{15}N,^{1}H]-correlation spectra with transverse relaxation optimized in both dimensions. The unfavorable pathway of Eq. 11 is suppressed by phase cycling, PFGs or a combination of both (Fig. 3b).

The ZQ-[^{15}N,^{1}H]-TROSY experiment contains a novel type of the sensitivity-enhanced quadrature detection scheme that requires only a single INEPT element for the transfer of both precessing orthogonal components of the ZQ coherence to the observable signal (Pervushin *et al.*, 1999). When compared with the serial arrangement of two INEPT elements in the standard ST2-PT enhancement scheme (Pervushin *et al.*, 1998c) this resulted in a smaller number of ^{1}H and ^{15}N pulses and an overall shorter pulse sequence, with a concomitant improvement in sensitivity. With the use of the PFGs for the coherence pathway selection according to Eq. 10, only a one-step phase cycle is required for the collection of a phase-sensitive, sensitivity- enhanced 2D spectrum. These features make 2D ZQ-[^{15}N,^{1}H]-TROSY an attractive building block in more complex NMR experiments (see below).

In aromatic spin systems the relaxation mechanisms of interest are due to ^{13}C–^{1}H dipole–dipole coupling and ^{13}C chemical shift anisotropy. To optimally suppress relaxation, the potentially competing interactions must be comparable in magnitude. Aromatic ^{13}C–^{1}H groups have favorable CSA for the implementation of TROSY-type NMR experiments. For a carbon spin in a six-membered aromatic ring the most highly shielded direction, σ_{33}, is perpendicular to the plane of the ring, and the least-shielded orientation, σ_{11}, is directed approximately along the ^{13}C–^{1}H bond (Veeman, 1984), with average values of $\sigma_{11} = 225$ ppm, $\sigma_{22} = 149$ ppm and $\sigma_{33} = 15$ ppm. In nucleic acids similar values are calculated for the base carbons (Fiala *et al.*, 2000; Riek *et al.*, 2000). In TROSY experiments these large values and the favorable orientation of the ^{13}C CSA tensor in aromatic ^{13}C–^{1}H groups provide efficient compensation of relaxation due to ^{13}C–^{1}H dipolar coupling during the ^{13}C chemical shift evolution. In contrast, the small CSA values measured for aromatic protons advise against the use of the TROSY method during proton chemical shift evolution periods. Therefore, only the aromatic carbon chemical shifts can be recorded with the TROSY method, with the use of a constant-time evolution period to reduce resonance overlap and eliminate the effect of $^1J(^{13}C,^{13}C)$ couplings (Vuister and Bax, 1992), whereas broad-band ^{13}C-decoupling is applied during proton signal acquisition (Pervushin *et al.*, 1998b). With this strategy the advantage of

TROSY-type experiments is in improved sensitivity, whereas, in contrast to the situation with ^{15}N–^1H groups (Pervushin *et al.*, 1997), the resonance line shapes and hence the spectral resolution are not affected (Sørensen *et al.*, 1997).

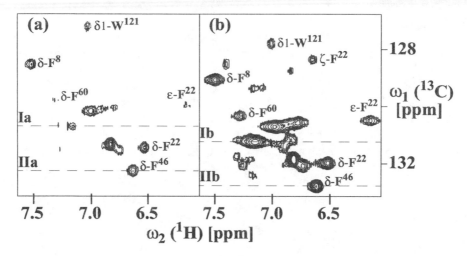

Figure 6.   Comparison of two direct ^1H–^{13}C correlation experiments performed with uniformly ^{13}C-labeled cyclophilin A (Clubb *et al.*, 1994; Wüthrich *et al.*, 1991) in ^2H_2O solution. Contour plots of a spectral region that contains resonances of phenylalanyl residues are shown. (a) 2D ct-[^{13}C,^1H]-COSY (Vuister and Bax, 1992). (b) 2D ct-TROSY-[^{13}C,^1H] (Pervushin *et al.*, 1998b). The dashed lines identify the locations of the cross sections along ω_2 that are shown in Fig. 7. In (a) and (b) the assignments of selected cross-peaks are indicated.

The potential of the TROSY approach for NMR experiments with aromatic spin systems in uniformly ^{13}C-labeled proteins are first illustrated by a comparison of the use of 2D ct- [^{13}C,^1H]-TROSY and ct-[^{13}C,^1H]-COSY (Vuister and Bax, 1992). The key difference between the two experiments is that in TROSY the evolution of the *IS* spin system due to the $^1J_{IS}$ scalar coupling is not refocused during t_1, which preserves the differences in DD–CSA interference for the individual multiplet components during this period. Fig. 6, a and b show small regions from the ^1H–^{13}C correlation spectra which contain resonances of aromatic groups of the uniformly ^{13}C-labeled 18 kDa protein cyclophilin A (Clubb *et al.*, 1994; Wüthrich *et al.*, 1991). The spectra were measured at 10°C, where the correlation time of the cyclophilin A was estimated to be $\tau_c = 16$ ns based on ^{15}N relaxation measurements (Ottiger *et al.*, 1997). In the TROSY spectrum (Fig. 6b) only the downfield component of the ^{13}C doublets is observed, so that the peak positions along ω_1 differ from those observed for the same C–H fragment in Fig. 6a by about 80 Hz. All the peaks are more intense in Fig. 6b, and a number of signals can be identified in the 2D ct-[$^{13}C,^1H$]-TROSY spectrum that are completely obscured by noise in the conventional experiment. A quantitative comparison of the relative sensitivity of the two experiments can be obtained from cross-sections taken along the ω_2 dimension (Fig. 7). On average, a six-fold signal-to-noise enhancement was obtained for the TROSY-type experiment. For the individual peaks the enhancement varies from 4 to 10, which reflects different local mobility of the individual aromatic rings. The largest enhancements were obtained for the aromatic rings in the core of the protein, which have the shortest transverse ^{13}C relaxation times.

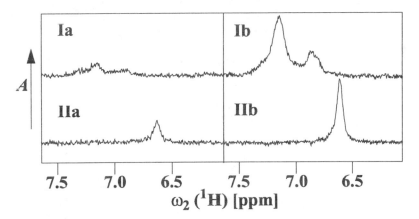

Figure 7. Cross sections through the spectra of Fig. 6. The numbering Ia, IIa, *etc* identifies the cross sections along the corresponding dashed lines in Fig. 6.

The direct correlation experimental schemes were used as building blocks in more complicated experiments, *e.g.*, for the measurement of the ^{13}C relaxation rates in RNA and DNA bases (Boisbouvier *et al.*, 1999a), the accurate measurement of the ^{13}C-^1H one-bond coupling constants in the two H$^{(\alpha)}$- and H$^{(\beta)}$-TROSY spectra (Brutscher *et al.*, 1998), for recording high resolution 3D ^{13}C-edited NOESY spectra (Brutscher *et al.*, 1998) of a 33-mer

RNA molecule and an RNA-theophylline complex or for establishing the sugar-to-base connectivity in the NMR spectra of isotopically labeled oligonucleotides (Fiala *et al.*, 2000; Riek *et al.*, 2000). Another application of $[^{13}C,^{1}H]$-TROSY is to construct experiments which correlate the ^{1}H and ^{13}C resonances of a given $^{13}C-^{1}H$ moiety with the resonances of the directly bound $^{13}C-^{1}H$ fragments (Meissner and Sørensen, 1999a; Pervushin *et al.*, 1998b). Such experiments enable the ^{1}H and ^{13}C resonance assignment of complete aromatic spin systems in uniformly ^{13}C-labeled proteins. Fig. 8b shows the same spectral region as in Fig. 6b from the 2D ct-TROSY-(H)C(C)H-COSY spectrum. Here each aromatic ^{13}C resonance is correlated with the resonance of the directly attached proton with positive sign and with the resonances of one or two protons separated by two bonds with negative sign. A comparison with the spectrum in Fig. 8a, which was measured with the corresponding conventional experimental scheme (Ikura *et al.*, 1991), demonstrates that for both the direct peaks and the relay peaks the amplitudes are enhanced by a factor of 4 to 10 for individual lines using TROSY, as can be seen from a comparison of the cross sections taken along the ω_2 dimension (Fig. 9). In the case of Phe 22, which is located in the core of the protein and has short ^{13}C transverse relaxation times, the connectivities identified in Fig. 8b (three dotted lines on the right), are not even visible in Fig. 8a.

4.   SUPPRESSION OF CONFORMATIONAL EXCHANGE LINE BROADENING

The suppression of CSX relaxation is based on the constructive use of the interference effect between two CSX interactions of the individual spins I and S resulting in the different relaxation rates of the ZQ and DQ coherences (Kloiber and Konrat, 2000; Rance, 1988; Tessari and Vuister, 2000). For the two-sites jump model with the populations P_1, P_2 and the jump rate $1/(2\tau_{ex}) = \kappa_1 P_1 = \kappa_2 P_2$ the relaxation rates are given by Eqs. 12 and 13.

$$R^{csx}_{ZQ} = 2P_1P_2\tau_{ex}(\Delta\delta\omega_I^{12} - \Delta\delta\omega_S^{12})^2 \qquad [12]$$

$$R^{csx}_{DQ} = 2P_1P_2\tau_{ex}(\Delta\delta\omega_I^{12} + \Delta\delta\omega_S^{12})^2 \qquad [13]$$

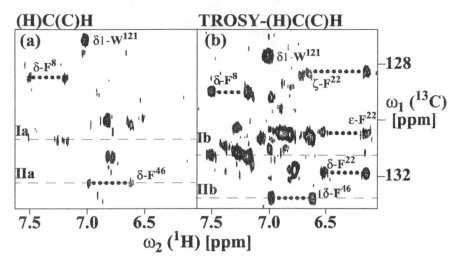

Figure 8. Comparison of two relayed ^{1}H–^{13}C correlation experiments performed with uniformly ^{13}C-labeled cyclophilin A (Clubb *et al.*, 1994; Wüthrich *et al.*, 1991) in ^{2}H$_2$O solution. Contour plots of a spectral region that contains resonances of phenylalanyl residues are shown. (a) 2D ct-(H)C(C)H-COSY (Ikura *et al.*, 1991). (b) 2D ct-TROSY-(H)C(C)H-COSY (Pervushin *et al.*, 1998b) spectrum. The dashed lines identify the locations of the cross sections along ω_2 that are shown in Fig. 9. The assignments of selected cross-peaks are indicated. The dotted horizontal lines indicate three-bond ^{1}H–^{13}C connectivities in the aromatic rings.

$\Delta\delta\omega_I^{12}$ is the difference in the chemical shifts between site 1 and 2. Note, that for $|\Delta\delta\omega_I^{12}| \simeq |\Delta\delta\omega_S^{12}|$, total CSX relaxation of one of the two (zero- or double-quantum) coherences is greatly reduced. Thus, for the case of the exchange process between two conformations, the complete cancellation of CSX transverse relaxation can be achieved provided that variations in the chemical shifts of the spins I and S between two exchanging conformations are comparable. Note, that the population factors and the exchange rate do not impose any limitation on the efficiency of the relaxation compensation. In contrast to the TROSY effect utilizing DD/CSA or CSA/CSA interference, any of the ZQ and DQ coherences may exhibit preferable relaxation properties depending on the relative configuration of the two exchanging sites. Some deterioration of the relaxation compensation effect is expected if the exchange process proceeds via a large number of exchanging sites (Pervushin, 2001).

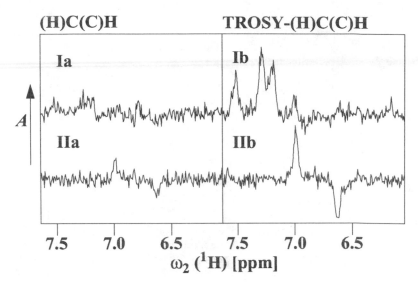

Figure 9.   Cross sections through the spectra of Fig. 8. The numbering Ia, IIa, *etc* identifies the cross sections along the corresponding dashed lines in Fig. 8.

The basic experimental scheme utilizing the CSX suppression is a variant of the $[^{15}N,^{1}H]$-ZQ-TROSY of Fig. 3b where both magnetization transfer pathways represented by Eqs. 10 and 11 are retained resulting into ZQ- and DQ- $[^{15}N,^{1}H]$-correlation subspectra. Figure 10 a and b compare the ZQ- and DQ- $[^{15}N,^{1}H]$-correlation subspectra of the 14-mer DNA duplex at 30°C. The partial labeling of the selected nucleotides indicated on the schematic drawings of the DNA duplex (Fig. 10, top) enables observation of apparent doubling of the $[^{15}N,^{1}H]$-correlation cross-peak of T^{21}. The two cross-peaks assigned to T^{21} might represent two distinct conformational states with the relative populations of 0.3 and 0.7 as it is derived from the relative peak intensities. The increase of the exchange rate at 37°C results in the collapse of these resonances into a broad peak (Fig. 10 c and d). At the elevated temperature the presence of conformational exchange-induced transverse relaxation is detected by a comparison of the signal intensities in the ZQ- and DQ- subspectra using the 1D slices along the ω_2 dimension shown on the inserts in Fig. 1. This experiment indicates that conformational exchange-induced line broadening can be a dominant source of transverse relaxation for the ^{1}H and ^{15}N spins. The suppression of conformational exchange-induced transverse relaxation in the DQ-[^{15}N,^{1}H]-correlation subspectrum results in the significant increase of the signal intensity of the [^{15}N,^{1}H]-correlation cross-peak of T^{21}, which can be used in construction of new of experiments. For example 2D $^{h1}J_{HN}$,$^{h2}J_{NN}$-quantitative [^{15}N,^{1}H]-TROSY to measure $^{h1}J_{HN}$ and $^{h2}J_{NN}$ scalar coupling constants across hydrogen bonds in nucleic acids and 2D ($^{h2}J_{NN}$+$^{h1}J_{NH}$)-correlation-[^{15}N,^{1}H]-TROSY to correlate ^{1}H^N chemical shifts of bases with the chemical shifts of the tertiary ^{15}N spins across hydrogen bonds using the sum of the trans- hydrogen bond coupling constants were proposed (Pervushin, 2001).

5.   BACKBONE RESONANCE ASSIGNMENT

Assignment of the backbone NMR resonances of ^{15}N,^{13}C-labeled proteins is based on a strategy in which chemical shifts from one residue are linked to those of sequential residues using a series of experiments which correlate ^{1}H, ^{15}N and ^{13}C chemical shifts (Bax, 1994; Cavanagh et al., 1996; Wider, 1998). Size limitations for the use of these triple-resonance experiments arise from fast transverse relaxation of the heteronuclei, especially ^{13}C$^\alpha$, which is efficiently relaxed by DD interactions with the attached ^{1}H$^\alpha$. Deuteration significantly decreases the ^{13}C$^\alpha$ relaxation rate and extends the accessible molecular size range (Gardner and Kay, 1998; Kay and Gardner, 1997; LeMaster, 1994), but transverse ^{15}N relaxation during coherence transfer steps is only little affected by deuteration of aliphatic protons (Salzmann et al., 1998). In the case of highly deuterated proteins, a family of HNC-type experiments was developed (Farmer and Venters, 1995; Farmer and Venters, 1996; Shan et al., 1996; Yamazaki et al., 1994a; Yamazaki et al., 1994b), which still were not able to surpass a 50-60 kDa molecular size limit for the backbone resonance assignment due to strong ^{15}N transverse relaxation even in deuterated proteins. The implementation of TROSY in a suite of triple-resonance experiments yielded substantial reduction of the ^{15}N transverse relaxation rates with a concomitant dramatic increase of sensitivity, enabling the backbone resonance assignment of large proteins (Salzmann et al., 1998).

Since the triple-resonance experiments use coherence transfers along a network of scalar coupled ^{15}N, ^{13}C and ^{1}H spins (Grzesiek and Bax, 1992; Yamazaki et al., 1994b), the coherence transfer from ^{15}N to either ^{13}C$^\alpha$ or ^{13}C' require long delays due to the small ^{1}J(^{15}N,^{13}C$^\alpha$) and ^{1}J(^{15}N,^{13}C') coupling constants. The ^{15}N chemical shift evolution is recorded in a constant-time (ct) fashion during the magnetization transfer (Bax, 1994; Cavanagh et al., 1996; Wider, 1998), and relaxation of ^{15}N due to DD coupling with the directly bound $^1H^N$ and to ^{15}N CSA results in severe loss of signal. Using the TROSY principle, transverse relaxation during these critical ^{15}N evolution periods can be efficiently suppressed. Similarly, $^1H^N$ transverse relaxation during detection due to $^1H^N$ CSA and DD coupling with ^{15}N can be additionally suppressed by TROSY.

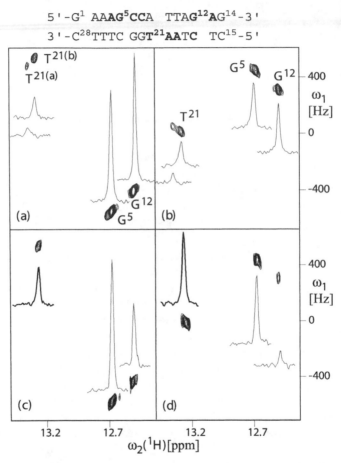

$5'-G^1$ AAAG^5CCA TTA$G^{12}AG^{14}-3'$

$3'-C^{28}$TTTC GGT^{21}AATC T$C^{15}-5'$

Figure 10. Detection of conformational exchange-induced transverse relaxation in the BS2 DNA duplex. (a) and (c), contour plots and cross-sections along $\omega_2(^1H)$ through the individual cross peaks from [$^{15}N,^1H$]-Zero-Quantum TROSY spectra of the partially $^{13}C,^{15}N$-labeled DNA duplex (see insert on the top, underlined letters identify the nucleotides that contain ^{15}N in the partially labelled duplex) at T=30°C and 37°C, respectively. (b) and (d), corresponding contour plots from [$^{15}N,^1H$]-Double-Quantum TROSY spectra extracted from the same data set. Chemical shifts relative to DSS in ppm are indicated in the $\omega_2(^1H)$ dimension and shifts in Hz relative to the centre of the spectrum are indicated in the ω_1 dimension.

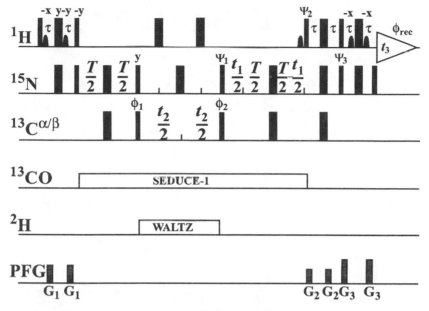

Figure 11. Experimental scheme for the [^{15}N,^{1}H]-TROSY-HNCA experiment (Eletsky *et al.*, 2001; Salzmann *et al.*, 1998; Salzmann *et al.*, 1999c). The radio-frequency pulses on ^{1}H, ^{15}N, ^{13}C$^\alpha$, ^{13}C$'$, ^{2}H and ^{1}H$^\alpha$ are applied at 4.7, 118, 56, 177, 3.6 and 4.7 ppm, respectively. Narrow and wide black bars indicate non-selective 90° and 180° pulses. Sine bell shapes on the lines marked ^{1}H and ^{1}H$^\alpha$ indicate selective 90° pulses. On the line marked ^{13}C$'$ three selective 180° pulses are applied off-resonance with a duration of 120 μs and a Gaussian shape. The line marked PFG indicates the durations and amplitudes of pulsed magnetic field gradients applied along the z-axis: G_1: 800 μs, 15 G/cm; G_2: 800 μs, 9 G/cm; G_3: 800 μs, 22 G/cm. The delays are τ = 2.7 ms and T = 21.8 ms. The phase cycle is: ψ_1 = {y, -y, -x, x}; ψ_2 = {-y}; ψ_3 = {-y}; ϕ_1 = {4x, 4(-x)}; ϕ_2 = {x}; ϕ_{rec} = {y, -y, -x, x, -y, y, x, -x}, with all other radio-frequency pulses applied with phase x. A phase-sensitive spectrum in the ^{15}N(t_1) dimension is obtained by recording a second FID for each t_1 value, with ϕ_1 = {y, -y, x, -x}, ϕ_3 = {y} and ϕ_4 = {y}, and data processing as described by Kay *et al.* (1992). Quadrature detection in the ^{13}C$^\alpha$(t_2) dimension is achieved by the States-TPPI method (Marion *et al.*, 1989) applied to the phase ϕ_2. The use of water flip-back pulses (Piotto *et al.*, 1992) ensures that the water magnetization stays aligned along +z throughout both the constant-time period, T, and the data acquisition period, t_3. For ^{2}H-decoupling, WALTZ-16 (Shaka *et al.*, 1983a; Shaka *et al.*, 1983b) was used with a field strength of 2.5 kHz. ^{13}C$'$ decoupling is performed using off-resonance SEDUCE-1 (McCoy and Mueller, 1992) with a field strength of 0.83 kHz.

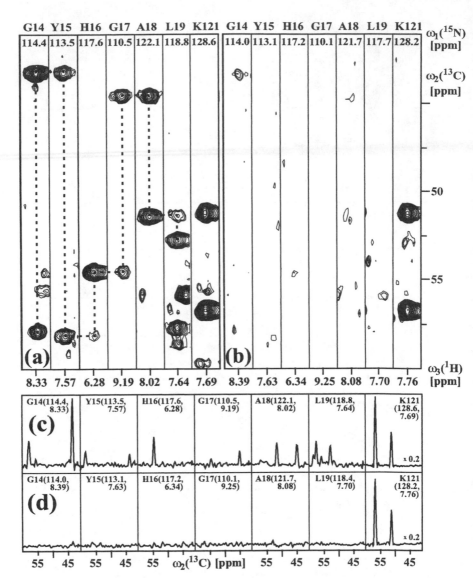

Figure 12. Comparison of corresponding [$\omega_2(^{13}C)$, $\omega_3(^1H)$] strips from two 3D HNCA experiments recorded with a 0.5 mM solution of the uniformly $^2H,^{13}C,^{15}N$-labeled octameric 110 kDa *S. aureus* DHNA (Hennig *et al.*, 1998): (a) [$^{15}N,^1H$]-TROSY- HNCA (Salzmann *et al.*, 1998; Salzmann *et al.*, 1999c). (b) Conventional HNCA (Grzesiek and Bax, 1992; Yamazaki *et al.*, 1994b) adapted to the scheme employed for (a) by using a WATERGATE sequence (Piotto *et al.*, 1992). The strips were taken at the ^{15}N chemical shifts of the residues 14 to 19 and the C- terminal Lys 121, are centered about the corresponding amide proton chemical shifts and have a width of 131 Hz along $\omega_3(^1H)$. In (a) the dashed line indicates sequential connectivities that could be reliably identified.

Figure 11 shows the experimental scheme of the TROSY-type HNCA experiment (Salzmann *et al.*, 1998; Weigelt, 1998), which correlate the ^{15}N–$^1H(i)$ group with $^{13}C^\alpha(i)$ and $^{13}C^\alpha(i\text{-}1)$, respectively, and can serve as a template for other TROSY HNC-type triple resonance experiments (Konrat *et al.*, 1999; Lohr *et al.*, 2000; Loria *et al.*, 1999a; Salzmann *et al.*, 1999a; Salzmann *et al.*, 1999b; Yang and Kay, 1999b; Yang and Kay, 1999c). In only partially deuterated proteins the evolution of the ^{13}C spins under the scalar coupling to the directly attached hydrogen spin in both ^{13}C-^1H and ^{13}C-^2H moieties has to be simultaneously refocused. This is achieved by an application of the conventional broad band ^2H decoupling and two 180° ^1H hard pulses as it is shown in Fig. 11. Such a combination interchanges ^{15}N magnetization modes twice, thus resulting in no mixing of the TROSY and 'anti-TROSY' magnetization transfer pathways. The characteristic feature of this experiment is the detection of the signal via the ^1H single transition operator using the ST2-PT element (Pervushin *et al.*, 1998c; Salzmann *et al.*, 1999c). For very large molecules an alternative detection scheme replacing the ST2-PT element was proposed by Yang and Kay (1999a) although the concatenation of the $^{13}C/^{15}N$ magnetization transfer with the ST2-PT element (Loria *et al.*, 1999a; Salzmann *et al.*, 1999c) yielded similar signal enhancement to the scheme of Yang and Kay (1999a).

The TROSY HNC-type suite of triple resonance experiments was applied to achieve the sequential resonance assignment of a uniformly $^2H,^{13}C,^{15}N$-labeled fragment of 7,8-dihydroneopterin aldolase (DHNA) *S. aureus* (Hennig *et al.*, 1998) with a molecular weight of 110 kDa for the unlabeled protein (Salzmann *et al.*, 2000). Using a conventional scheme (Grzesiek and Bax, 1992; Yamazaki *et al.*, 1994b) the HNCA connectivities were unambiguously detected for only 7 out of the 121 residues of DHNA, and only the apparently highly flexible C-terminal Lys 121 gave strong sequential and intraresidual peaks (Fig. 12). Using [$^{15}N,^1H$]- TROSY-HNCA (Salzmann *et al.*, 1999c) a 20- to 50-fold sensitivity gain was achieved (Fig. 12, c and d), and for a large part of the polypeptide chain the peaks were sufficiently well resolved to establish the sequential connectivities, which are indicated by dotted lines.

To resolve some ambiguities in $^{13}C^\alpha$ assignment, 3D [$^{15}N,^1H$]-TROSY-HNCACB (Salzmann *et al.*, 1999b) and 3D [^{13}C]-ct-[$^{15}N,^1H$]-TROSY-HNCA (Salzmann *et al.*, 1999a) experiments can be used. The backbone $^{13}C'$ chemical shift assignment are usually obtained with the [$^{15}N,^1H$]-TROSY-HNCO measurement (Loria *et al.*, 1999a; Salzmann *et al.*, 1998; Yang and Kay, 1999b). The problem of signal overlap in large proteins is addressed by introducing the constant-time $^{13}C^\alpha$ evolution in to the [$^{15}N,^1H$]-TROSY-HNCA experiment (Salzmann *et al.*, 1999a) or by increasing the dimensionality of the TROSY HNC-type experiments (Konrat *et al.*, 1999). For example, the experiments, 4D-HNCACO and 4D-HNCOCA, establish correlations of the form ($^{13}C^{\alpha}(i,i\text{-}1)$, $^{13}C'(i,i\text{-}1)$, $^{15}N(i)$, $^{1}H^{N}(i)$) and ($^{13}C^{\alpha}(i\text{-}1)$, $^{13}C'(i\text{-}1)$, $^{15}N(i)$, HN(i)), respectively, using TROSY (Yang and Kay, 1999c). Another 4D TROSY-based triple resonance experiment, 4D-HNCO(i-1)CA(i), correlating intra-residue ^{1}H, ^{15}N, $^{13}C^{\alpha}$ chemical shifts with the carbonyl $^{13}C'$ shift of the preceding residue can be recommended (Konrat *et al.*, 1999) to complement the 4D TROSY-HNCOCA and 4D HNCACO experiments. The utility of the experiments was demonstrated by an application to a complex of the maltose binding protein and β-cyclodextrin with 46 ns correlation time. Approximately 95% of the expected intra- and interresidue correlations were observed with more than factor of 30 average signal-to-noise values. The methodology promises to be particularly powerful for applications to high molecular weight complexes comprised of a labeled fragment and unlabeled components or proteins with segmental labeling (Cotton *et al.*, 1999; Muir *et al.*, 1998; Otomo *et al.*, 1999a; Otomo *et al.*, 1999b; Xu *et al.*, 1999; Yamazaki *et al.*, 1998).

6.   NOE SPECTROCOPY OF LARGE BIOMOLECULES

Overhauser effect (NOE) spectroscopy (NOESY) yields ^{1}H–^{1}H distance constraints that enable *de novo* determination of three-dimensional structures of biological macromolecules by NMR in solution (Wüthrich, 1986). In initial implementations of TROSY in NOE experiments, [$^{13}C,^{1}H$]-TROSY and [$^{15}N,^{1}H$]-TROSY were used in 3D ^{13}C-resolved and ^{15}N-resolved [$^{1}H,^{1}H$]-NOESY to resolve homonuclear ^{1}H–^{1}H NOEs along the heteronuclear dimension (Brutscher *et al.*, 1998; Zhu *et al.*, 1999). In these NOE experiments the conventional [$^{13}C,^{1}H$]- or [$^{15}N,^{1}H$]- HSQC or HMQC schemes are substituted by the corresponding single-quantum TROSY-type chemical shift correlation schemes (Pervushin *et al.*, 1998c), whereby the desired transverse relaxation optimization is achieved along the ^{13}C frequency axis (Brutscher *et al.*, 1998), or in the ^{15}N and $^{1}H^{N}$ dimensions (Zhu *et al.*, 1999), respectively.

In a more sophisticated 3D ^{15}N-resolved NOE experiment, the 3D NOESY-[$^{1}H,^{15}N,^{1}H$]- ZQ-TROSY, optimization of transverse relaxation in all three spectral dimensions is achieved (Pervushin *et al.*, 1999). A key element of 3D NOESY-[$^{1}H,^{15}N,^{1}H$]-ZQ-TROSY is that the ^{15}N and ^{1}H chemical shifts are correlated by the sensitivity-enhanced 2D ZQ-[$^{15}N,^{1}H$]-TROSY steps which use the remote cross-correlation between ^{1}H chemical shift anisotropy (CSA) and ^{15}N CSA interactions to reduce the transverse relaxation rate of the [$^{15}N,^{1}H$]-zero-quantum coherence (Pervushin *et al.*, 1999). An additional important feature is artifact-free suppression of the diagonal peaks, which are typically much stronger than the cross-peaks and could interfere with cross peak integration in conventional NOESY spectra (Meissner and Sørensen, 1999b; Meissner and Sørensen, 2000; Pervushin *et al.*, 1999).

Fig. 13a shows a spectrum of the protein DHNA recorded with the 3D NOESY- [^{1}H,^{15}N,^{1}H]-ZQ-TROSY experiment (Pervushin *et al.*, 1999) in the form of 2D [ω_1(^{1}H)/ ω_3(^{1}H^N)] strips, where the diagonal peaks are identified by the arrows. As a reference for evaluation of the TROSY experiment, a similar presentation of the same regions of a conventional 3D ^{15}N-resolved [^{1}H,^{1}H]-NOESY spectrum (Talluri and Wagner, 1996) of DHNA is displayed in Fig. 13b. Along both ^{1}H frequency axes, corresponding peak positions in the two spectra differ by approximately 45 Hz. Comparison of the two spectra shows that an important advantage of the 3D NOESY-[^{1}H,^{15}N,^{1}H]-ZQ-TROSY experiment results from the fact that the diagonal peaks are either completely suppressed or have very small residual negative intensity (arrows in Fig. 13) so that the cross peaks near the diagonal become amenable for detailed analysis. The two spectra further demonstrate that significantly narrower cross-peak linewidths were achieved when using TROSY, which in turn also yielded higher signal amplitudes.

The TROSY-type NOESY experiments (Brutscher *et al.*, 1998; Pervushin *et al.*, 1999; Zhu *et al.*, 1999) were used to supported the sequential assignment obtained from the 3D triple resonance spectra and to characterize the secondary structure of DHNA (Salzmann *et al.*, 2000). In addition, by reference to the crystal structure of DHNA (Hennig *et al.*, 1998), NOE contacts between amide protons across β strands within one subunit, $d_{NN}(i,j)$ could be assigned as well as inter-subunit contacts.

7. OTHER APPLICATIONS

The recent discovery of spin-spin scalar couplings across hydrogen bonds initially in nucleic acids (Dingley and Grzesiek, 1998) and subsequently in proteins (Cornilescu *et al.*, 1999a) opened a possibility for the direct detection of the hydrogen bonds in biomolecules as it was previously achieved for small organic compounds (Platzer *et al.*, 1992). (See Chapter 9). Compared with conventional NMR correlation spectroscopy (Cavanagh *et al.*, 1996), [^{15}N,^{1}H]- TROSY (Pervushin *et al.*, 1997; Pervushin *et al.*, 1998c) yielded about 70% and 30% reduction of the ^{15}N and ^{1}H linewidths, respectively, in the signals of the guanosine ^{15}N$_1$-^{1}H and thymidine ^{15}N$_3$-^{1}H imino groups in ^{15}N-labeled DNA (Pervushin *et al.*, 1998a). The reduced TROSY linewidths then allowed the direct observation of scalar couplings across hydrogen bonds involved in base pairing either by direct measurement in resolved multiplet fine structures (Pervushin *et al.*, 2000b; Pervushin *et al.*, 1998a) or in more complex coherence transfer experiments (Dingley and

Grzesiek, 1998; Dingley *et al.*, 1999; Majumdar *et al.*, 1999a; Majumdar *et al.*, 1999b; Pervushin *et al.*, 2000b; Pervushin *et al.*, 1998a; Wohnert *et al.*, 1999). The variability of such couplings observed so far indicates that they may become sensitive new parameters for detection of hydrogen bond formation and associated subtle conformational changes (Cornilescu *et al.*, 1999a; Cornilescu *et al.*, 1999b). Furthermore, in conjunction with quantum-chemical calculations (Dingley *et al.*, 1999; Scheurer and Bruschweiler, 1999), precise measurements of scalar couplings across hydrogen bonds can be expected to provide novel insights into the nature of hydrogen bonds in macromolecules.

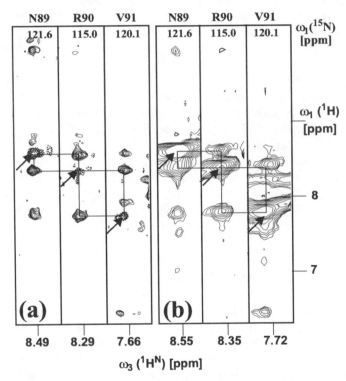

Figure 13. Comparison of corresponding spectral regions in a 3D NOESY-ZQ-[^{1}H,^{15}N,^{1}H]-TROSY spectrum (Pervushin *et al.*, 1999) and a conventional 3D ^{15}N-resolved [^{1}H,^{1}H]-NOESY spectrum recorded according to Talluri and Wagner (1996). Both experiments were recorded with the uniformly ^{2}H,^{13}C,^{15}N-labeled octameric 110 kDa *S. aureus* DHNA (Hennig *et al.*, 1998). (a) and (b): Contour plots of [ω_1(^{1}H),ω_3(^{1}H)] strips. NOE connectivities are shown with thin horizontal and vertical lines. At the top of each strip the sequence-specific assignment is indicated by the one-letter amino acid symbol and the residue number in the amino acid sequence. (c) and (d): Cross-sections taken along ω_1(^{1}H) at the ω_3(^{1}H) positions of the diagonal peaks in the strips (a) and (b), respectively. The positions of the diagonal peaks are indicated by arrows.

The use of TROSY allowed observation of the $^{h3}J_{NC'}$ scalar couplings across hydrogen bonds in larger, perdeuterated proteins (Wang et al., 1999). The TROSY experiments were applied to the uniformly $^{2}H, ^{13}C, ^{15}N$ labeled 30 kDa ribosome-inactivating protein MAP30. The $^{h3}J_{NC}$ interactions were found to be smaller than 1 Hz and have negative sign (Cornilescu et al., 1999b), but their detection in an TROSY-HNCO experiment was made possible through the use of TROSY (Wang et al., 1999). Despite the small size of the $^{h3}J_{NC'}$ scalar couplings sufficient magnetization was transferred across hydrogen bonds resulting in the high sensitivity. A direct correlation between backbone-backbone hydrogen-bond length in protein crystal structures and experimentally observed $^{h3}J_{NC'}$ couplings across such bonds has been established (Cornilescu et al., 1999b).

In isotropic solution, internuclear dipolar couplings average to zero as a result of rotational diffusion. By dissolving macromolecules in a dilute aqueous nematic discotic liquid-crystalline medium containing widely spaced magnetically oriented particles, a tunable degree of solute alignment with the magnetic field can be created while retaining the high resolution and sensitivity of the standard isotropic nuclear magnetic resonance (NMR) spectrum (see Chapter 7 and 8). Dipolar couplings between $^{1}H-^{1}H$, $^{1}H-^{13}C$, $^{1}H-^{15}N$, and $^{13}C-^{13}C$ pairs in such an oriented macromolecule no longer average to zero, and are readily measured (Tjandra and Bax, 1997a). TROSY-based HNCO-type 3D pulse schemes were developed for measuring $^{1}H-^{15}N$, $^{15}N -(C')-^{13}C$, $^{1}H-(C')-^{13}C$, $(C')-^{13}C-^{13}C^{\alpha}$ and $(HN)-^{1}H-^{13}C^{\alpha}$ dipolar couplings in $^{2}H, ^{15}N, ^{13}C$-labeled proteins (Yang et al., 1999). Data sets recorded on a complex of the $^{2}H, ^{15}N, ^{13}C$-labeled maltose binding protein and β-cyclodextrin as well as the $^{2}H, ^{15}N, ^{13}C$- labeled human carbonic anhydrase II demonstrated that precise dipolar couplings can be obtained from proteins in the 30-40 kDa molecular weight range. These couplings serve as powerful restraints for obtaining global folds of highly deuterated proteins (Yang et al., 1999).

A method to measure the one-bond $^{1}H-^{15}N$ coupling constants in the protein backbone has been proposed based on the 2D TROSY-type correlation experiments (Lerche et al., 1999). Two 2D subspectra corresponding to the two possible spin states of the coupling partner were measured to separate a scalar and a residual dipolar contributions. The new experiment was demonstrated for the ^{15}N-labeled protein chymotrypsin inhibitor 2 in a lipid bicelle mixture (Lerche et al., 1999). The upfield and downfield components of the HN multiplet were separated in the indirect dimension in a sensitivity-enhanced HSQC experiment (Cordier et al., 1999). The pulse sequence is similar to the generalized TROSY scheme; however, decoupling of the X- nucleus is used during detection. A detailed analysis of relaxation effects, precision and sensitivity of the method was presented. The approach was demonstrated in a 2D water flip-back $[^{15}N, ^{1}H]$-TROSY which measures $^1J_{HN}$ splitting in isotropic and oriented samples of the human ubiquitin and the hepatitis C protease (Cordier *et al.*, 1999).

Measurements of $^1J_{NC'}$ and $^2J_{NC'}$ coupling constants from a simplified 2D ^{15}N-^{1}H TROSY-type correlation spectrum were reported (Permi *et al.*, 1999). The multiplet components of the $^1J_{NC'}$ doublet in the indirect dimension and $^2J_{NC'}$ in the direct dimension were separated into two subspectra by spin-state-selective filters. The results were verified against the measurements of $^1J_{NC'}$ from spin-state-selective ^{13}C'-^{1}H correlation spectra recorded (Permi *et al.*, 1999).

8. CONCLUSIONS

The TROSY method represents a spectroscopic contribution in a constant pursuit of NMR structures of ever larger proteins, nucleic acids and their complexes. As it was theoretically predicted and proven in practice the most efficient application of TROSY is achieved in combination with the selective or uniform deuteration of biomolecules, in particular proteins. In this way biomolecular particles with the molecular sizes exceeding 100 kDa have now become amenable for high resolution NMR studies. Due to its simplicity and flexibility the TROSY principle has also been integrated into many other NMR applications not necessary targeting extremely large biomolecules. An impressive repertoire of TROSY-based NMR experiments has been developed for the detection and characterization of weak interactions such as scalar couplings across hydrogen bonds and residual dipolar couplings. The use of TROSY in Nuclear Overhauser spectroscopy might significantly facilitate the spectral analysis by suppressing the strong diagonal peaks benefiting studies of a wide range of biomolecules. Still many other TROSY applications are yet to be developed such as side chain resonances assignment and NOESY experiments for collecting structural constraints in larger proteins. In combination with advanced biochemical methods these developments will propel us into a new era of NMR structure determination.

9. REFERENCES

Andersson, P., Annila, A., and Otting, G., 1998a, *J. Magn. Reson.* **133**, 364-367.
Andersson, P., Gsell, B., Wipf, B., Senn, H. and Otting, G., 1998b, *J. Biomol. NMR* . **11**, 279-288.
Bax, A., 1994, *Curr. Opin. Struct. Biol.* **4**, 738-744.
Blicharski, J. S., 1967, *Phys. Lett.* **24**, 608-610.
Blicharski, J. S., 1972, *Acta Physica Polonica* **41**, 223-236.
Bodenhausen, G., and Ruben, D. J., 1980, *Chem. Phys. Lett.* **69**, 185-189.

Boisbouvier, J., Brutscher, B., Simorre, J. P., and Marion, D., 1999a, *J. Biomol. NMR.* **14**, 241- 252.

Boisbouvier, J., Gans, P., Blackledge, M., Brutscher, B.. and Marion, D., 1999b, *J. Am. Chem. Soc.* **121**, 7700-7701.

Brutscher, B., Boisbouvier, J., Pardi, A., Marion, D., and Simorre, J. P., 1998, *J. Am. Chem. Soc.* **120**, 11845-11851.

Cavanagh, J., Fairbrother, W. J., Palmer, A. G., and Skelton, N. J., 1996, *Protein NMR Spectroscopy: Principles and Practice*, Academic Press, New York.

Clore, G. M., and Gronenborn, A. M., 1998, *Trends Biotechnol.* **16**, 22-34.

Clubb, R. T., Ferguson, S. B., Walsh, C. T., and Wagner, G., 1994, *Biochemistry* **33**, 2761-2772.

Cordier, F., Dingley, A. J. and Grzesiek, S., 1999, *J. Biomol. NMR* **13**, 175-180.

Cornilescu, G., Hu, J. S., and Bax, A., 1999a, *J. Am. Chem. Soc.* **121**, 2949-2950.

Cornilescu, G., Ramirez, B. E., Frank, M. K., Clore, G. M., Gronenborn, A. M., and Bax, A., 1999b, *J. Am. Chem. Soc.* **121**, 6275-6279.

Cotton, G. J., Ayers, B., Xu, R., and Muir, T. W., 1999, *J. Am. Chem. Soc.* **121**, 1100-1101.

Czisch, M., and Boelens, R., 1998, *J. Magn. Reson.* **134**, 158-160.

Dingley, A. J., and Grzesiek, S., 1998, *J. Am. Chem. Soc.* **120**, 8293-8297.

Dingley, A. J., Masse, J. E., Peterson, R. D., Barfield, M., Feigon, J., and Grzesiek, S., 1999, *J. Am. Chem. Soc.* **121**, 6019-6027.

Eletsky, A., Kienhofer, A., and Pervushin, K., 2001, *J. Biomol. NMR* **20**, 177-180.

Ernst, R. R., Bodenhausen, G., and Wokaun, A., 1987, *The Principles of Nuclear Magnetic Resonance in One and Two Dimensions*, Clarendon, Oxford.

Farmer, B. T., and Venters, R. A., 1995, *J. Am. Chem. Soc.* **117**, 4187-4188.

Farmer, B. T., and Venters, R. A., 1996, *J. Biomol. NMR* **7**, 59-71.

Fiala, R., Czernek, J., and Sklenar, V., 2000, *J. Biomol. NMR* **16**, 291-302.

Gardner, K. H., and Kay, L. E., 1998, *Annu. Rev. Biophys. Biomolec. Struct.* **27**, 357-406.

Gardner, K. H., Konrat, R., Rosen, M. K., and Kay, L. E., 1996, *J. Biomol. NMR* **8**, 351-356.

Gardner, K. H., Rosen, M. K., and Kay, L. E., 1997, *Biochemistry* **36**, 1389-1401.

Gardner, K. H., Zhang, X. C., Gehring, K., and Kay, L. E., 1998, *J. Am. Chem. Soc.* **120**, 11738- 11748.

Gerald, R., Bernhard, T., Haeberlen, U., Rendell, J., and Opella, S., 1993, *J. Am. Chem. Soc.* **115**, 777-782.

Goldman, M., 1984, *J. Magn. Reson.* **60**, 437-452.

Goto, N. K., Gardner, K. H., Mueller, G. A., Willis, R. C., and Kay, L. E., 1999, *J. Biomol. NMR* **13**, 369-374.

Griffey, R. H., and Redfield, A. G., 1987, *Quart. Rev. Biophys.* **19**, 51–82.

Grzesiek, S., and Bax, A., 1992, *J. Magn. Reson.* **96**, 432-440.

Grzesiek, S., and Bax, A., 1993, *J. Am. Chem. Soc.* **115**, 12593-12594.

Grzesiek, S., and Bax, A., 1999, *J. Biomol. NMR* **14**, 181-184.

Grzesiek, S., Kuboniwa, H., Hinck, A. P., and Bax, A., 1995a, *J. Am. Chem. Soc.* **117**, 5312-5315.

Grzesiek, S., Wingfield, P., Stahl, S., Kaufman, J. D., and Bax, A., 1995b, *J. Am. Chem. Soc.* **117**, 9594-9595.

Gueron, M., Leroy, J. L., and Griffey, J., 1983, *J. Am. Chem. Soc.* **105**, 7262-7266.

Hennig, M., D'Arcy, A., Hampele, I. C., Page, M. G., Oefner, C., and Dale, G. E., 1998, *Nat. Struct. Biol.* **5**, 357-362.

Ikura, M., Kay, L. E., and Bax, A., 1991, *J. Biomol. NMR* **1**, 299-304.

Kaplan, J. I., and Fraenkel, G., 1980, *NMR of Chemically Exchanging Systems*, Academic Press.

Kay, L. E., and Gardner, K. H., 1997, *Curr. Opin. Struct. Biol.* **7**, 722-731.
Kay, L. E., Keifer, P., and Saarinen, T., 1992, *J. Am. Chem. Soc.* **114**, 10663-10665.
Kelly, M. J. S., Krieger, C., Ball, L. J., Yu, Y. H., Richter, G., Schmieder, P., Bacher, A., and Oschkinat, H., 1999, *J. Biomol. NMR* **14**, 79-83.
Kloiber, K., and Konrat, R., 2000, *J. Biomol. NMR* **18**, 33-42.
Konrat, R., Yang, D. W., and Kay, L. E., 1999, *J. Biomol. NMR* **15**, 309-313.
Kumar, P., and Kumar, A., 1995, *J. Magn. Reson. Ser. A* **115**, 155-164.
Kumar, P., and Kumar, A., 1996, *J. Magn. Reson. Ser. A* **119**, 29-37.
LeMaster, D. M., 1994, *Prog. Nucl. Magn. Reson. Spectrosc.* **26**, 371-419.
Lerche, M. H., Meissner, A., Poulsen, F. M., and Sørensen, O. W., 1999, *J. Magn. Reson.* **140**, 259-263.
Lohr, F., Pfeiffer, S., Lin, Y. J., Hartleib, J., Klimmek, O., and Ruterjans, H., 2000, *J. Biomol. NMR* **18**, 337-346.
Loria, J. P., Rance, M., and Palmer, A., G., 1999a, *J. Magn. Reson.* **141**, 180-184.
Loria, J. P., Rance, M., and Palmer, A. G., 1999b, *J. Biomol. NMR* **15**, 151-155.
Madhu, P. K., Grandori, R., Mandal, P. K., Hohenthanner, K. and Müller, N. 2001, *J Biomol NMR.* **20**, 31-37
Majumdar, A., Kettani, A., and Skripkin, E., 1999a, *J. Biomol. NMR* **14**, 67-70.
Majumdar, A., Kettani, A., Skripkin, E. and Patel, D. J. 1999b, *J. Biomol. NMR* **15**, 207-211.
Marion, D., Ikura, M., Tschudin, R., and Bax, A., 1989, *J. Magn. Reson.* **85**, 393-399.
Mayne, C. L., and Smith, S. A., Eds., 1996, Encyclopedia of NMR: Relaxation Processes in Coupled Spin Systems. Vol. 6. Encyclopedia of NMR. Edited by Grant, D. M., and Harris, R. K., New York: Wiley.
McCallum, S. A., Hitchens, T. K., and Rule, G. S., 1999, *J. Mol. Biol.* **285**, 2119-2132.
McCoy, M. A., and Mueller, L., 1992, *J. Am. Chem. Soc.* **114**, 2108-2112.
Meissner, A., Schulte-Herbruggen, T., Briand, J., and Sørensen, O. W., 1998, *Mol. Phys.* **95**, 1137-1142.
Meissner, A., and Sørensen, O. W., 1999a, *J. Magn. Reson.* **139**, 447-450.
Meissner, A., and Sørensen, O. W., 1999b, *J. Magn. Reson.* **140**, 499-503.
Meissner, A., and Sørensen, O. W., 2000, *J. Magn. Reson.* **142**, 195-198.
Muir, T. W., Sondhi, D., and Cole, P. A., 1998, *Proc. Natl. Acad. Sci. U S A* **95**, 6705-6710.
Otomo, T., Ito, N., Kyogoku, Y., and Yamazaki, T., 1999a, *Biochemistry* **38**, 16040-16044.
Otomo, T., Teruya, K., Uegaki, K., Yamazaki, T., and Kyogoku, Y., 1999b, *J. Biomol. NMR* **14**, 105-114.
Ottiger, M., Zerbe, O., Guntert, P., and Wüthrich, K., 1997, *J. Mol. Biol.* **272**, 64-81.
Pellecchia, M., Sebbel, P., Hermanns, U., Wüthrich, K., and Glockshuber, R., 1999, *Nat. Struct. Biol.* **6**, 336-339.
Permi, P., Heikkinen, S., Kilpelainen, I., and Annila, A., 1999, *J. Magn. Reson.* **140**, 32-40.
Pervushin, K., 2001, *J. Biomol. NMR* **20**, 275-285.
Pervushin, K., Braun, D., and Wüthrich, K., 2000a, *J. Biomol. NMR* **17**, 195-202.
Pervushin, K., Fernandez, C., Riek, R., Ono, A., Kainosho, M., and Wüthrich, K., 2000b, *J. Biomol. NMR* **16**, 39-46.
Pervushin, K., Ono, A., Fernandez, C., Szyperski, T., Kainosho, M., and Wüthrich, K., 1998a, *Proc. Natl. Acad. Sci. U. S. A.* **95**, 14147-14151.
Pervushin, K., Riek, R., Wider, G., and Wüthrich, K., 1997, *Proc. Natl. Acad. Sci. U. S. A.* **94**, 12366-12371.
Pervushin, K., Riek, R., Wider, G., and Wüthrich, K., 1998b, *J. Am. Chem. Soc.* **120**, 6394-6400.
Pervushin, K., Wider, G., Riek, R., and Wüthrich, K., 1999, *Proc. Natl. Acad. Sci. U. S. A.* **96**, 9607-9612.

Pervushin, K., Wider, G., and Wüthrich, K., 1998c, *J. Biomol. NMR* **12**, 345-348.
Pervushin, K. V., 2000, *Q. Rev. Biophys.* **33**, 161-197.
Piotto, M., Saudek, V., and Sklenar, V., 1992, *J. Biomol. NMR* **2**, 661-665.
Platzer, N., Buisson, J. P., and Demerseman, P., 1992, *J. Heterocycl. Chem.* **29**, 1149-1153.
Ramamoorthy, A., Wu, C. H., and Opella, S. J., 1997, *J. Am. Chem. Soc.* **119**, 10479-10486.
Rance, M., 1988, *J. Am. Chem. Soc.* **110**, 1973-1974.
Rance, M., Loria, J. P., and Palmer, A. G., 1999, *J. Magn. Reson.* **136**, 92-101.
Riek, R., Pervushin, K., Fernandez, C., Kainnosho, M., and Wüthrich, K., 2001, *J. Am. Chem. Soc.* **123**, 658-64.
Riek, R., Wider, G., Pervushin, K., and Wüthrich, K., 1999, *Proc. Natl. Acad. Sci. U.S.A.* **96**, 4918-4923.
Roberts, J. E., Harbison, G. S., Munowitz, M. G., Herzfeld, J., and Griffin, R. G., 1987, *J. Am. Chem. Soc.* **109**, 4163-4169.
Rosen, M. K., Gardner, K. H., Willis, R. C., Parris, W. E., Pawson, T., and Kay, L. E., 1996, *J. Mol. Biol.* **263**, 627-636.
Salzmann, M., Pervushin, K., Wider, G., Senn, H., and Wüthrich, K., 1998, *Proc. Natl. Acad. Sci. U.S.A.* **95**, 13585-13590.
Salzmann, M., Pervushin, K., Wider, G., Senn, H., and Wüthrich, K., 1999a, *J. Biomol. NMR* **14**, 85-88.
Salzmann, M., Pervushin, K., Wider, G., Senn, H., and Wüthrich, K., 2000, *J. Am. Chem. Soc.* **122**, 7543-7548.
Salzmann, M., Wider, G., Pervushin, K., Senn, H., and Wüthrich, K., 1999b, *J. Am. Chem. Soc.* **121**, 844-848.
Salzmann, M., Wider, G., Pervushin, K., and Wüthrich, K., 1999c, *J. Biomol. NMR* **15**, 181-184.
Sandstrom, J., 1982, *Dynamic NMR Spectroscopy*, Academic Press.
Scheurer, C., and Bruschweiler, R., 1999, *J. Am. Chem. Soc.* **121**, 8661-8662.
Schulte-Herbruggen, T., Briand, J., Meissner, A., and Sørensen, O. W., 1999, *J. Magn. Reson.* **139**, 443-446.
Shaka, A. J., Keeler, J., and Freeman, R., 1983a, *J. Magn. Reson.* **53**, 313-340.
Shaka, A. J., Keeler, J., Frenkiel, T., and Freeman, R., 1983b, *J. Magn. Reson.* **52**, 335-338.
Shan, X., Gardner, K. H., Muhandiram, D. R., Rao, N. S., Arrowsmith, C. H., and Kay, L. E., 1996, *J. Am. Chem. Soc.* **118**, 6570-6579.
Shimizy, H., 1964, *J. Phys. Chem.* **40**, 3357-3364.
Sørensen, M. D., Meissner, A., and Sørensen, O. W., 1997, *J. Biomol. NMR* **10**, 181-186.
Sørensen, O. W., Eich, G. W., Levitt, M. H., Bodenhausen, G., and Ernst, R. R., 1983, *Prog. Nucl. Magn. Reson. Spectrosc.* **16**, 163-192.
Takahashi, H., Nakanishi, T., Kami, K., Arata, Y., and Shimada, I., 2000, *Nat. Struct. Biol.* **7**, 220-223.
Talluri, S., and Wagner, G., 1996, *J. Magn. Reson. Ser. B* **112**, 200-205.
Tessari, M., Mulder, F. A. A., Boelens, R., and Vuister, G. W., 1997a, *J. Magn. Reson.* **127**, 128-133.
Tessari, M., Vis, H., Boelens, R., Kaptein, R., and Vuister, G. W., 1997b, *J. Am. Chem. Soc.* **119**, 8985-8990.
Tessari, M., and Vuister, G. W., 2000, *J. Biomol. NMR* **16**, 171-174.
Tjandra, N., and Bax, A., 1997a, *Science* **278**, 1111-1114.
Tjandra, N., and Bax, A., 1997b, *J. Am. Chem. Soc.* **119**, 8076-8082.
Veeman, W. S., 1984, *Prog. NMR Spectrosc.* **16**, 193-235.
Vold, R. R., and Vold, R. L., 1975, *J. Chem. Phys.* **64**, 320-332.
Vold, R. R., and Vold, R. L., 1978, *Prog. NMR Spectrosc.* **12**, 79-133.

Vuister, G. W., and Bax, A., 1992, *J. Magn. Reson.* **98**, 428-435.

Wagner, G., 1993a, *Curr. Opin. Struct. Biol.* **3**, 748-754.

Wagner, G., 1993b, *J. Biomol. NMR* **3**, 375-385.

Wang, Y. X., Jacob, J., Cordier, F., Wingfield, P., Stahl, S. J., Lee-Huang, S., Torchia, D., Weigelt, J., 1998, *J. Am. Chem. Soc.* **120**, 10778-10779.

Wider, G., 1998, *Prog. Nucl. Magn. Reson. Spectrosc.* **32**, 193-275.

Wider, G., and Wüthrich, K., 1999, *Curr. Opin. Struct. Biol.* **9**, 594-601.

Wohnert, J., Dingley, A. J., Stoldt, M., Gorlach, M., Grzesiek, S., and Brown, L. R., 1999, *Nucleic Acids Res.* **27**, 3104-3110.

Wu, C. H., Ramamoorthy, A., Gierasch, L. M., and Opella, S. J. 1995, *J. Am. Chem. Soc.* **117**, 6148-6149.

Wüthrich, K., 1986, *NMR of Proteins and Nucleic Acids*, Wiley, New York.

Wüthrich, K., 1998, *Nat. Struct. Biol.* **5**, 492-495.

Wüthrich, K., Spitzfaden, C., Memmert, K., Widmer, H., and Wider, G., 1991, *FEBS Lett.* **285**,   237-247.

Xu, R., Ayers, B., Cowburn, D., and Muir, T. W., 1999, **96**, 388-93.

Yamazaki, T., Lee, W., Arrowsmith, C. H., Muhandiram, D. R., and Kay, L. E., 1994a. *J. Am. Chem. Soc.* **116**, 11655-11666.

Yamazaki, T., Lee, W., Revington, M., Mattiello, D. L., Dahlquist, F. W., Arrowsmith, C. H., and Kay, L. E., 1994b, *J. Am. Chem. Soc.* **116**, 6464-6465.

Yamazaki, T., Otomo, T., Oda, N., Kyogoku, Y., Uegaki, K., Ito, N., Ishino, Y., and Nakamura, H., 1998, *J. Am. Chem. Soc.* **120**, 5591-5592.

Yang, D. W., and Kay, L. E., 1999a, *J. Biomol. NMR* **13**, 3-10.

Yang, D. W., and Kay, L. E., 1999b, *J. Biomol. NMR* **14**, 273-276.

Yang, D. W., and Kay, L. E., 1999c, *J. Am. Chem. Soc.* **121**, 2571-2575.

Yang, D. W., Venters, R. A., Mueller, G. A., Choy, W. Y., and Kay, L. E., 1999, *J. Biomol. NMR* **14**, 333-343.

Zhu, G., Kong, X. M., and Sze, K. H., 1999, *J. Biomol. NMR* **13**, 77-81.

Chapter 2

Segmental Isotopic Labeling: Prospects for a New Tool to Study the Structure-function Relationships in Multi-domain Proteins

Jennifer J. Ottesen*, Ulrich K. Blaschke*, David Cowburn‡, and Tom W. Muir*

Laboratories of Synthetic Protein Chemistry and Physical Biochemistry‡, The Rockefeller University, 1230 York Avenue, New York, NY 10021, USA and ‡New York Structural Biology Center, c/o Box 163, 1230 York Avenue, New York, NY 10021.*

Abstract: Segmental isotopic labeling allows NMR active nuclei to be introduced selectively into a specific region(s) of full-length protein sequences. In this perspective we review briefly the technologies available for the segmental isotopic labeling of proteins and suggest how these strategies might be combined with current NMR techniques to study aspects of protein function that are often difficult to address using conventional structural analysis.

1. INTRODUCTION

Many proteins contain multiple domains that act as quasi-independent folding units connected by long linker regions, much like beads strung on a necklace (Pawson and Nash 2000). The joining of different domains can serve many functions. Two domains may interact with different sites in the same target, thereby increasing the avidity of an interaction. Conversely, different domains may interact with distinct partners, as observed for adaptor proteins in cellular signaling pathways, thereby localizing one binding partner to another. Finally, modular domains can engage in complex intramolecular interactions that control the structure and function of their host protein, including allosteric modulation of enzymatic activity. The importance of intramolecular domain interactions in regulating signaling proteins was initially under-appreciated, but now this situation is quickly changing as we learn more about the detailed structure and biochemistry of large multi-domain proteins (Campbell and Downing 1998). Indeed, there are a growing number of proteins where inter-domain contacts are thought to regulate structure and/or function; for two recent examples elucidated using NMR techniques (see Kim *et al.*, 2000; Shekhtman *et al.*, 2001). Although intramolecular interactions are likely to be a common feature in multi-domain proteins, in many cases these interactions will be difficult to characterize at the molecular level. Both the mobility of the linker regions connecting the domains and the weak affinity of many inter-domain interactions make some proteins difficult to crystallize or otherwise complicates structure elucidation. Although knowing the structures of the individual domains provides important information, for example in describing ligand-binding surfaces, it is often insufficient to understand the mechanisms that regulate the functions of the protein as a whole. Specifically, there is a pressing need to develop approaches that provide a way of extracting detailed structural and functional information on a specific domain within the context of a large multi-domain protein and that allow the interplay between distinct structural elements within the parent protein to be examined. Such information would provide a more detailed view of how the functional properties of large multi-domain proteins are regulated through ligand binding and/or post-translational modifications such as phosphorylation.

NMR solution studies are uniquely suited for analyses of proteins with regions of flexible structure. However, proteins containing multiple functional or regulatory domains tend to be too large for facile structure determination by NMR. Until recently, several issues prevented thorough NMR analysis of proteins over ~30 kDa in size, a limit that multi-domain proteins often exceed (Wider and Wüthrich 1999). The rate of transverse relaxation of ^{1}H, ^{13}C, and ^{15}N nuclei increases in relation to molecular weight and with the strength of the magnetic field used for structural analysis, since chemical shift anisotropy (CSA) relaxation increases with the square of the field strength. This increased relaxation rate is reflected in severe line broadening, reducing both resolution and sensitivity in spectra of larger proteins. One approach to this problem is the incorporation of high levels of deuterium into a protein, which eliminates proton-related relaxation pathways (Gardner and Kay 1998). Deuteration thus improves the line widths, with the unfortunate concomitant loss of protons available for NOE-based distance restraints. The more recently introduced TROSY (transverse relaxation optimized spectroscopy, for molecular weights >30 kDa (Pervushin *et al.*, 1997)) and CRINEPT (Riek *et al.*, 1999) (cross-correlated relaxation-enhanced polarization transfer, for molecular weights >150 kDa) pulse sequences can be incorporated into most standard NMR experiments to optimize transverse relaxation pathways, resulting in narrowed line widths in the NMR spectrum for ^{15}N.

Although these new NMR experiments reduce the line width problem associated with large proteins, they do not address the issue of spectral complexity. The number of signals in an NMR spectrum is proportional to the size of the protein. Uniform and selective, e.g. of methyl groups (Gardner and Kay 1998), heteronuclear labeling schemes are common in NMR studies, permitting dispersion in multiple dimensions, reviewed in (Sattler *et al.*, 1999). As the density of the signals increases, spectral analysis becomes increasingly more difficult. This issue is specifically addressed by segmental isotopic labeling. Segmental isotopic labeling is a new method that allows the specific labeling of a discrete region within a full-length protein with ^{2}H, ^{13}C, and/or ^{15}N isotopes (Yamazaki *et al.*, 1998; Xu *et al.*, 1999). These isotopes can be used as a probe to examine only the labeled part of the protein using suitable heteronuclear NMR experiments, essentially a "spectroscopic filter". Since segmental isotopic labeling allows the signals from only the labeled portion of the protein to be separated out and displayed, it reduces the spectral complexity significantly. In principle, segmental isotopic labeling can be used for the *de novo* NMR structure determination of discrete regions of very large proteins. The approach is also ideally suited to study binding interactions and inter-domain structural and functional modulation in the context of large, multi-domain proteins where the structures of the isolated domains may already be known. The TROSY NMR spectra of a large protein with one labeled domain can be reduced to the equivalent of that one domain. Since all of the domain-domain interactions that occur in the context of the full-length protein will still be reflected in the spectra, this allows for the facile identification of such interactions. Similarly, a segmental labeled protein could be used to rapidly screen for potential ligands *in trans* by established techniques currently used to identify ligands for smaller proteins (Shuker *et al.*, 1996).

Several NMR methods have been developed recently to exploit differences in isotopic labeling between subunits for the study of protein-protein and protein-ligand interactions in macromolecular complexes. The availability of segmental isotopic labeling strategies now allows many of these NMR methods to be applied to the analysis of intra-molecular contacts as well. Moreover, the technique permits analysis of the dynamics and interactions of the linker regions between domains that are not always visible in crystallographic structures. In this perspective we review briefly the technologies available for the segmental isotopic labeling of proteins and suggest how these strategies might be combined with current NMR techniques to study aspects of protein function that are often difficult to address using conventional structural techniques.

2.   PROTEIN LIGATION TECHNIQUES

In order to label a single domain or section within a protein, it is necessary to express the protein in segments, and then stitch together the individual pieces to generate the full-length protein. In segmental labeling schemes, an individual protein domain is expressed in media containing the desired isotopic precursors containing ^{2}H, ^{15}N, or ^{13}C, while the rest of the protein sequence is expressed in unlabeled (or otherwise differently labeled) media. Ligation of the independently labeled sequences will give rise to the single labeled domain in the context of the full-length protein. Two separate semi-synthetic methods, *expressed protein ligation* (EPL) and *trans-splicing*, can be exploited for segmental labeling. These methods have been discussed in depth in several recent reviews (Cotton and Muir 1999; Noren *et al.*, 2000; Muir 2001; Cowburn and Muir 2001). Both of these methods take advantage of elements of the natural protein splicing process in which a protein fragment (termed an intein) self-cleaves from the surrounding sequence (the N- and C-terminal exteins), resulting in the two extein sequences being joined together by a normal peptide bond.

2.1   Expressed Protein Ligation

Expressed protein ligation is an extension of the well-established native chemical ligation approach (Dawson *et al.*, 1994), originally developed for the total chemical synthesis of proteins from unprotected synthetic peptide fragments. Native chemical ligation is based on the chemoselective reaction between a peptide containing an N-terminal cysteine (α-Cys) and a peptide containing a C-terminal thioester moiety resulting in the generation of a normal peptide bond at the ligation site. EPL is a semi-synthetic version of native chemical ligation in which one or both of the building blocks is a recombinant polypeptide (Figure 1A). Simple manipulations allow the necessary reactive groups for ligation to be introduced into expressed protein fragments (Cotton and Muir 1999; Noren *et al.*, 2000; Muir 2001). A single ligation step permits assembly of a full-length protein from two pieces, allowing terminal segments of the protein sequence to be labeled with isotopes (*exo-segmental labeling*). As an example of this, EPL was used to prepare NMR sample quantities of a fragment of the tyrosine kinase, c-Abl

(Xu *et al.*, 1999), in which a single Src homology 2 (SH2) domain was labeled with ^{15}N. Importantly, the ligation procedures employed allowed the two folded fragments of the protein to be chemically ligated in high yield. Comparison between the ^{1}H{^{15}N} HSQC NMR spectra of the fully labeled and segmental labeled proteins revealed clearly the improvement in spectral resolution that segmental labeling provides (Xu *et al.*, 1999).

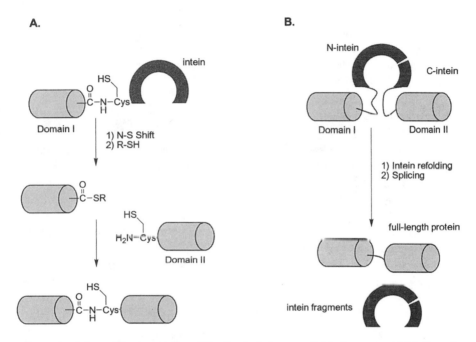

Figure 1. Schematic representation of ligation techniques available for segmental labeling of a terminal protein segment (Exo-Segmental Labeling). **A.** *Expressed Protein Ligation.* The α-thioester of Domain I is generated with an intein-fusion strategy. Domain II, containing an N-terminal Cys residue, is added and a native ligation reaction produces the full-length protein. **B.** *Trans-Splicing.* The two domains are expressed as fusion proteins with the N- or C-terminal portions of a split-intein. Intein reconstitution allows the protein splicing process to proceed, resulting in the two domains being joined together.

Although, the current discussion will emphasize the segmental labeling of discrete domains of a protein, we would like to emphasize that the ligation chemistry in no way precludes splitting a protein within a domain. Indeed, native chemical ligation and EPL have been used to prepare numerous protein domains from constituent fragments (Dawson and Kent 2000). Nonetheless, the choice of ligation site within a domain is, in general, more critical than between domains; for a full discussion of the design criterion see (Camarero and Muir 1999). It should also be noted that EPL does not distinguish between synthetic peptide sequences and expressed protein fragments, given the presence of the necessary α-thioester and α-Cys groups. Thus, the ability to incorporate phosphorylated, glycosylated, or other unnatural amino acids through solid phase peptide synthesis may be combined with segmental labeling techniques for the study of proteins that are not easily accessible by other methods.

2.2   Trans-splicing

Trans-splicing is based on the observation that protein splicing can be triggered by reconstituting inactive N- and C-terminal halves of an intein, reviewed in (Paulus 2000). This provides a means of joining any two protein fragments *in vitro;* the N-terminal fragment would be expressed as a fusion protein with the N-terminal half of the intein, and the C-terminal fragment of the intein would be expressed fused to the C-terminal portion of the protein. Reconstitution of these fusion proteins results in generation of the desired full-length protein (Figure 1B). In the first example of segmental isotopic labeling using trans-splicing, Yamazaki and coworkers used an artificially split PI-*PfuI* intein for the segmental labeling of the C-terminal domain of the *E. coli* RNA polymerase α subunit with ^{15}N (Yamazaki *et al.*, 1998). Most trans-splicing systems involve artificially split inteins, all of which require a refolding step to initiate protein splicing (Paulus 2000). The recent discovery of a naturally occurring split intein, *Ssp* DnaE, will circumvent the need for the reconstitution step. However, as with some of the artificial systems there may be certain sequence requirements at the ligation junction in order for efficient splicing (Evans *et al.*, 2000).

In many cases it may be desirable to label internal regions (*endo-segmental labeling*) of a protein with NMR probe nuclei. Fortunately, both trans-splicing and EPL can be used to generate proteins in which a central domain is isotopically labeled. Yamazaki and co-workers were again the first to report a trans-splicing approach for endo-segmental labeling (Otomo *et al.*, 1999). The 370-residue maltose binding protein, labeled with ^{15}N at residues 161-295, was synthesized using a clever tandem trans-splicing technique (Figure 2B). This approach utilized two orthogonal split inteins,

PI-*pfuI* and PI-*pfuII*, from *Pyrococcus furiousus*. The locations of the split sites within the inteins were carefully selected to maintain the fidelity of intein reconstitution, allowing the two splicing reactions to occur simultaneously. Indeed, an elegant feature of this approach is the self-assembly of the four intein fragments which not only allows the one-pot synthesis of the product, but significantly reduces the concentration dependence of this nominally third order reaction.

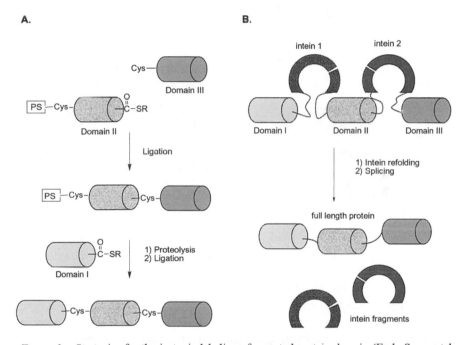

Figure 2. Strategies for the isotopic labeling of a central protein domain (Endo-Segmental Labeling). **A.** *Sequential Expressed Protein Ligation.* The central domain (II) contains both a α-thioester and a cryptic N-terminal Cys protected with a proteolytically removable pro-sequence. This allows the directed attachment of the flanking domains in two step ligation approach. **B.** *Tandem Trans-Splicing.* The three protein domains are expressed as fusions with two orthogonal sets of intein fragments. The central domain (II) is fused to both the C-terminus of intein 1 and the N-terminus of intein 2. Intein refolding and protein splicing at both intein locations leads to the generation of the full-length protein in a one-pot process.

A.

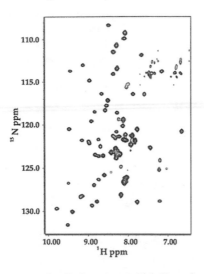

B.

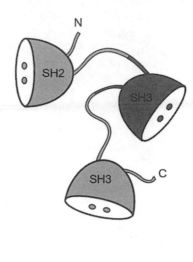

Figure 3. Endo-segmental labeling of a multi-domain protein. **A.** $^{1}\text{H}\{^{15}\text{N}\}$ TROSY-HSQC NMR spectrum of the 304-residue protein c-Crk-II recorded at 600 MHz. **B.** Domain architecture and labeling strategy for c-Crk-II. The central SH3 domain and regions of the surrounding linker (residues 125-207, shown in darker hatching) were selectively labeled with ^{15}N using a sequential EPL approach.

The ability to perform sequential EPL steps to couple multiple domains was demonstrated recently by the assembly of the adaptor protein c-Crk-II from three unlabeled fragments (Blaschke *et al.*, 2000). c-Crk-II is a 304-residue protein containing three separate Src homology domains connected by long linker regions, with a SH2-SH3-SH3 domain architecture. The modular nature of the protein and several unanswered questions about the nature of the interactions between the protein domains and linker regions made c-Crk-II an ideal test case for endo-segmental labeling by EPL. Two ligation sites were introduced by making the conservative changes $\text{Ser}^{125} \rightarrow \text{Cys}$ and $\text{Ser}^{208} \rightarrow \text{Cys}$ in the flexible linker regions surrounding the central SH3 domain. The central fragment was expressed fused to an N-terminal factor Xa signal sequence in order to protect the α-Cys group and to a C-terminal intein sequence that could be cleaved to generate the α-thioester moiety (Figure 2A). Thus, the initial ligation step connected the central domain to the C-terminal domain. A proteolytic step cleaved off the N-terminal protecting group from this intermediate product, generating the requisite α-Cys moiety for the final ligation step. We have recently used this sequential EPL approach to generate multi-milligram amounts of endo-labeled c-Crk-II (Ottesen *et al.*, unpublished). Figure 3 shows the $^1H\{^{15}N\}$ TROSY-HSQC spectrum of the full-length c-Crk-II with ^{15}N labeling of residues 125-207, which includes the central SH3 domain and portions of the surrounding linker regions.

3.   NMR TECHNIQUES

Once a labeled protein has been produced, several NMR techniques are available to study domain interactions and orientations. This review will focus on methods that benefit from the use of segmental isotopic labeling of a single domain. However, other methods are available, such as the use of residual dipolar coupling constants to determine inter-domain orientations, that might require two or more domains to be labeled in a protein (Fischer *et al.*, 1999). NMR approaches that benefit from segmental isotopic labeling fall into two broad categories; those in which segmental labeling is used to simplify spectral analysis by highlighting the signals of a domain of interest (e.g. chemical shift perturbation methods), and those in which segmental labeling is an integral part of the methodology (e.g. saturation transfer methods).

3.1   Chemical Shift Perturbation and Amide Exchange Methods

The local environment of an NMR-active nucleus in a protein can change significantly when that protein interacts with a binding partner, since residues at the binding interface undergo a transition from a surface position to a buried position. Two of the most common methods for determining a contact surface, chemical shift perturbation and amide exchange coefficients, monitor how these changes in the local environment affect spectral properties. Chemical shift perturbation is a simple, indirect method of identifying the residues at a protein-protein interface (Pellecchia *et al.*, 1999). The chemical shift of an NMR-detectable nucleus is extremely sensitive to the surrounding environment. Chemical shifts may be compared between two structural states of a protein, and significant modifications in chemical shift between the two states may be mapped against a known protein structure in order to determine the binding interface (Figure 4). A protein segment or domain may have significantly different properties in the context of the full-length (i.e. parent) protein. These may be subtle effects related to functional interactions, or may be more dramatic changes due to the stabilization of a specific tertiary structure. Thus, chemical shift mapping is probably only generally effective for small changes on binding; large refolding effects preclude such simple analysis.

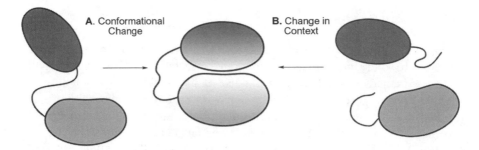

Figure 4. Chemical shift perturbation and hydrogen/deuterium exchange. These phenomena are sensitive to environmental changes in a molecule such as those that occur at a binding interface (indicated by the shading gradient). By simplifying spectral analysis, segmental isotopic labeling allows discrete regions of a protein to be studied in different states. Examples include: (**A**) A ligand-binding event or post-translational modification that induces a conformational change; (**B**) The effect of context, i.e. the isolated domain *versus* the domain in the parent protein.

Hydrogen-deuterium exchange rates may also be measured for a domain in different contexts. The rate of exchange of amide protons with the bulk solvent is dependent on protection from solvent and specific hydrogen bond interactions (Wüthrich and Wagner 1979) that are properties of protein structure, and should change significantly between a bound and free domain surface. Thus, significant changes of this exchange rate can be mapped to a region of the domain surface.

Both chemical shift perturbation and hydrogen/deuterium exchange methods involve a comparison between the spectra of two structural states of a protein that are thought to involve a change in domain interactions. The two states could be simply the domain in solution and the domain in the context of a full-length protein if binding occurs in the basal protein state. If a structural change were expected due to ligand binding, the two states might be the ligand-free and ligand-bound forms of the protein. Similarly, if phosphorylation leads to a functional change, the phosphorylated protein could be constructed and the spectra compared to those of the unphosphorylated protein. EPL has proven a valuable tool for generating homogenously phosphorylated proteins through the ligation of synthetic phosphopeptides into protein sequences (Muir *et al.*, 1998; Huse *et al.*, 2000). This technique can be combined with segmental labeling via EPL to generate proteins with specific phosphorylation patterns. Equally, EPL may be utilized to incorporate other moieties, such as carbohydrates, that may have effects on protein structure but are difficult to introduce through molecular biology techniques.

While environment-dependent chemical shift perturbation and amide exchange rate mapping are arguably the simplest methods available for determining a binding surface, they are indirect methods of determining an interface and therefore inaccuracies can be introduced into the analysis (Takahashi *et al.*, 2000). These methods assume that changes in the local environment of an NMR sensitive nucleus occur only at the interaction surface between ligand and receptor. However, protein domains do not act exclusively as rigid interacting bodies. A binding event can trigger compensatory conformational changes within surface loops that propagate beyond the immediate environment of the interface. Thus, residues within the protein interior or surrounding the protein interface would reflect a change in the environment despite not being a part of the contact surface.

3.2   Paramagnetic Spectral Broadening

Methods that provide direct confirmation of the relationship between the two domains would alleviate the problems associated with relying on the indirect effects of environmental changes to detect associations. Examination of the spectral broadening effect of paramagnetic groups is an example of one such approach. It has long been known that a paramagnetic group, such as a nitroxide spin label, has a distance-dependent effect on the line width of NMR resonances at distances of tens of angstroms from the spin label (Gillespie and Shortle 1997a). Recently, Gillespie and Shortle used nitroxide spin labels to model the denatured state of staphylococcal nuclease (Gillespie and Shortle 1997b). In another recent example, paramagnetic spectral broadening was exploited to generate distance restraints that were used to determine the structure of the 25 kDa protein eIF4E (Battiste and Wagner 2000). In this example, site-directed spin labeling was used to introduce nitroxide probes at five different locations in the protein. Cross peak heights within the $^1H\{^{15}N\}$ HSQC spectra were quantified in the oxidized (paramagnetic) and reduced (non-paramagnetic) states of the protein. The ratio of these peak heights were then used to sort the broadened resonances into distance spheres of <14 Å, 14-23 Å, and > 23 Å from the sites of spin label incorporation. This technique may be particularly useful for determining distance restraints in large proteins labeled with high levels of deuterium, in which short-range quantitative distance restraints from NOE data are not available.

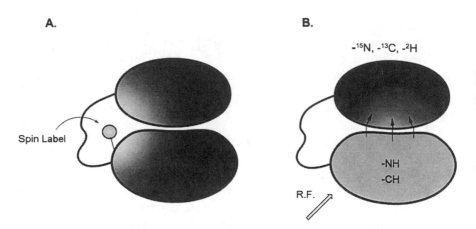

Figure 5. **A**. *Paramagnetic Line Broadening.* Insertion of a spin label into the protein leads to a distance-dependent broadening effect on NMR signals (indicated by the hatching gradient). **B**. *Saturation transfer.* Irradiation of aliphatic protons causes saturation of the nonlabeled domain (lower), and of the amide protons in the labeled domain located at the interface (indicated by the hatching gradient).

Paramagnetic spectral broadening, in combination with segmental labeling, should also be useful for the determination of domain interfaces (Figure 5A). If a spin label is incorporated in a non-labeled domain of a multi-domain protein, any paramagnetic broadening of the resonances in the labeled domain indicates that the global protein fold brings that domain within range of the spin label. Since paramagnetic line broadening effects occur over a large distance, there is some latitude in placement of the spin label. Additionally, the distances that can be extracted from the broadening effects can provide specific information on the orientation and relative position of the labeled domain relative to the spin label. A series of sequences incorporating the spin label at different locations in the protein can be used to triangulate the orientation of domains with known structures and to begin to determine the structure of local linker regions. Accordingly, it should be straightforward to incorporate spin labeled amino acids directly into a protein sequence in a site-specific manner using EPL. Several probes such as, 2,2,6,6-tetramethylpiperidine-1-oxyl-4-amino-4-carboxylic acid (TOAC) (Marchetto et al., 1993; Tonolio et al., 1998), are available in forms that can be readily introduced into synthetic peptides and therefore into large proteins via EPL methods.

3.3   Saturation Transfer Experiments

While techniques such as chemical shift perturbation and paramagnetic line broadening exploit segmental labeling primarily to simplify spectral analysis, distinct isotopic labeling strategies are required for other methodologies that could be used to identify domain interfaces. Recently, a new method was introduced that relies on saturation transfer across the interface between a non-labeled (^1H) protein and a uniformly ^2H and ^{15}N labeled binding partner to identify the residues at the interface of a protein complex (Figure 5B) (Takahashi et al., 2000). In this approach, the non-labeled protein is irradiated selectively at the frequency of the aliphatic proton. Spin-diffusion rapidly causes the saturation of the amide protons within this subunit. The uniformly deuterium-labeled protein contains no aliphatic protons and does not experience direct saturation. However, due to the close spatial proximity of amino acids at the interface between the two proteins, cross-relaxation and saturation occur across this contact site. Thus, the cross peaks attributable to the amide protons at the protein interface will display a reduced peak volume. Assuming sufficient levels of deuteration in the labeled sample and sufficiently low proton density in the solvent (10% H_2O/90% D_2O) to minimize non-specific saturation transfer, cross relaxation is limited to the interface between the two proteins. Since the protein labeled with ^{15}N is also fully deuterated throughout the side chains, a $^1H\{^{15}N\}$ HSQC experiment will display only the exchangeable N-linked protons in the labeled protein. Thus, the intensity of these protons can be monitored, and those protons that display a reduced intensity upon irradiation (i.e., saturation) can be mapped onto a known structure to determine the interaction surface.

The saturation transfer principle has been utilized to determine the interface between two proteins in a 64 kDa complex (Takahashi et al., 2000). A similar method was exploited to study protein-nucleic acid complexes, taking advantage of selective saturation at nucleic acid-specific resonances to eliminate the need for deuteration (Ramos et al., 2000). Additionally, the term saturation transfer difference (STD) NMR has been coined for analysis of saturation transfer from a large protein to a small ligand (Mayer and Meyer 1999). The common theme between these different applications is the ability to distinguish between two members of a complex for the specific saturation of one member of the bound complex, either by frequency selection or through elimination of protons within one binding partner. Given a suitable segmental labeling scheme, saturation transfer should also be applicable to the detection of interactions within a multi-domain protein. The individual domain of interest would be labeled with ^{2}H and ^{15}N and ligated into the full-length protein. Saturation of proton resonances in the surrounding, unlabeled protein sequence would then be reflected in the residues of the labeled domain that make contacts to any other region of the protein.

3.4   Direct NOE Detection

The study of oligomeric protein complexes is complicated by the difficulty of distinguishing between intra- and inter-molecular interactions. Asymmetry has been introduced into these systems through mixtures of differentially isotopically labeled samples (e.g., Walters et al., 1997). A wide variety of heteronuclear filtered and edited experiments have been developed to directly observe contacts between different nuclei, and thus distinguish interactions between labeled and unlabeled proteins (Otting and Wüthrich 1990; Breeze 2000). These techniques are immediately applicable to segmental labeled protein samples, which can be considered in this case as differently labeled proteins connected covalently.

While the large number of available heteronuclear filtered and edited experiments precludes a full discussion here, it is worthwhile to note that the segmental labeling scheme proposed for cross-saturation experiments is suitable for very simple ^{15}N-edited NOESY experiments specific for inter-domain interactions. In this labeling scheme, one domain is uniformly labeled with ^{15}N and ^{2}H while the remainder of the sequence maintains natural isotopic abundance (^{1}H and ^{14}N). Due to solvent exchange, the only NMR-active hydrogen atoms in the labeled domain will be the amide protons. The remainder of the protein contains side chain protons, but the amide protons are bound to ^{14}N nuclei. Thus, if a ^{15}N-edited NOESY spectrum is acquired, the only cross peaks observed outside of the amide-amide region will be those between the amide protons of the labeled domain, and the side chain resonances of the nonlabeled domain. These observed NOE interactions would be due solely to local structure around the ligation site and the domain-domain interactions caused by the global fold of the protein. Since NOE interactions provide information about both interacting nuclei, groups on both sides of the interface are identified in one experiment. However, to gain the full value of this experiment, a full assignment of the protein side chain resonances would be required. In the absence of such an assignment, only the amide protons involved in the NOE would be identified.

4.   CONCLUSIONS

The primary advantage of segmental labeling is that it allows NMR techniques that were developed for individual proteins or macromolecular complexes to be applied to domains within a large protein. While we have discussed several types of studies that could benefit from the ability to perform segmental labeling, new NMR techniques are constantly under development. Many of these techniques could be easily modified for use with a segmentally labeled protein to provide insights into the structural properties of multi-domain proteins. Additionally, the benefits of segmental labeling are not limited to studies of protein interfaces. For example, ^{15}N spin relaxation parameters could be measured on a labeled domain to examine backbone dynamics in the context of a very large protein, to determine both local flexibility, and interdomain orientation (Fushman and Cowburn 1998; Fushman and Cowburn 2002). While segmental labeling does require a significant investment in time and materials to produce sufficient quantities of labeled proteins for NMR studies, it can provide vital information for understanding key interactions within multi-domain proteins that is not available through any other method.

4.1   Acknowledgments

We thank Mike Goger and members of the Muir and Cowburn groups for help in the preparation of this manuscript. The references cited here are representative of contemporary work in these areas. This is a prospective article rather than a review, and no attempt at historical completeness has been made. This work was supported by the National Institutes of Health (GM55843, T.W.M.; GM47021, D.C.).

5. REFERENCES

Battiste, J. L., and Wagner, G., 2000, *Biochemistry* **39**:5355-5365.
Blaschke, U. K., Cotton, G. J., and Muir, T. W., 2000, *Tetrahedron* **56**:9461- 9470.
Breeze, A. L., 2000, *Prog. Nucl. Mag. Res. Sp.* **36**:323-372.
Camarero, J. A., and Muir, T. W., 1999, *Current Protocols in Protein Science* **18**:1-21.
Campbell, I. D., and Downing, A. K., 1998, *Nat. Struct. Biol.* **5**:496-499.
Cotton, G. J., and Muir, T. W., 1999, *Chem. Biol.* **6**:R247-256.
Cowburn, D., and Muir, T. W., 2001, *Meth. Enzymol.* **339**:41-54.
Dawson, P. E., and Kent, S. B. H., 2000, *Annu. Rev. Biochem.* **69**:923-960.
Dawson, P. E., Muir, T. W., Clark-Lewis, I., and Kent, S. B. H., 1994, *Science* **266**:776-779.
Evans, T. C., Jr., Martin, D., Kolly, R., Panne, D., Sun, L., Ghosh, I., Chen, L., Benner, J., Liu, X. Q., and Xu, M. Q., 2000, *J. Biol. Chem.* **275**:9091-9094.
Fischer, M. W. F., Losonczi, J. A., Weaver, L. J., and Prestegard, J. H., 1999, *Biochemistry* **38**:9013-9022.
Fushman, D., and Cowburn, D., 1998, in *Structure, Motion, Interaction and Expression of Biological Macromolecules*: Adenine Press
Fushman, D., and Cowburn, D., 2002, Chapter 3, this volume.
Gardner, K. H., and Kay, L. E., 1998, *Ann. Rev. Biophys. Biomol. Str.* **27**:357-406.
Gillespie, J. R., and Shortle, D., 1997a, *J. Mol. Biol.* **268**:158-169.
Gillespie, J. R., and Shortle, D., 1997b, *J. Mol. Biol.* **268**:170-184.
Huse, M., Holford, M. N., Kuriyan, J., and Muir, T. W., 2000, *J. Am. Chem. Soc.* **122**:8337-8338.
Kim, A. S., Kakalis, L. T., Abdul-Manan, N., Liu, G. A., and Rosen, M. K., 2000, *Nature* **404**:151-158.
Marchetto, R., Schreier, S., and Nakaie, C. R., 1993, *J. Am. Chem. Soc.* **115**:11042-11043.
Mayer, M., and Meyer, B., 1999, *Angew. Chem. Int. Ed.* **38**:1784-1788.
Muir, T. W., 2001, *Synlett* **6**:733-740.
Muir, T. W., Sondhi, D., and Cole, P. A., 1998, *Proc. Natl. Acad. Sci. USA* **95**:6705-6710.
Noren, C. J., Wang, J. M., and Perler, F. B., 2000, *Angew. Chem. Int. Ed.* **39**:450-466.
Otomo, T., Ito, N., Kyogoku, Y., and Yamazaki, T., 1999, *Biochemistry* **38**:16040-16044.
Otting, G., and Wüthrich, K., 1990, *Quart. Rev. Biophys.* **23**:39-96.
Paulus, H., 2000, *Annu. Rev. Biochem.* **69**:447-496.
Pawson, T., and Nash, P., 2000, *Genes Dev.* **14**:1027-1047.
Pellecchia, M., Sebbel, P., Hermanns, U., Wüthrich, K., and Glockshuber, R., 1999, *Nat. Struct. Biol.* **6**:336-339.
Pervushin, K., Riek, R., Wider, G., and Wüthrich, K., 1997, *Proc. Natl. Acad. Sci. USA* **94**:12366-12371.
Ramos, A., Kelly, G., Hollingworth, D., Pastore, A., and Frenkiel, T. A., 2000, *J. Am. Chem. Soc.* **122**:11311-11314.
Riek, R., Wider, G., Pervushin, K., and Wüthrich, K., 1999, *Proc. Natl. Acad. Sci. USA* **96**:4918-4923.
Sattler, M., Schleucher, J., and Griesinger, C., 1999, *Prog. Nucl. Mag. Res. Sp.* **34**:93-158.
Shekhtman, A., Ghose, R., Wang, A. C., Cole, P. A., and Cowburn, D., 2001, *J. Mol. Biol.* **314**: 129-138.
Shuker, S. B., Hajduk, P. J., Meadows, R. P., and Fesik, S. W., 1996, Science **274**:1531-1534.
Takahashi, H., Nakanishi, T., Kami, J., Arata, Y., and Shimada, I., 2000, *Nat. Struct. Biol.* **7**:220-223.
Toniolo, C., Crisma, M., and Formaggio, F., 1998, *Biopolymers* **47**:153-158.
Walters, K. J., Matsuo, H., and Wagner, G., 1997, *J. Am. Chem. Soc.* **119**:5958-5959.
Wider, G., and Wüthrich, K., 1999, *Curr. Opin. Struct. Biol.* **9**:594-601.
Wüthrich, K., and Wagner, G., 1979, *J. Mol. Biol.* **130**:1-18.

Xu, R., Ayers, B., Cowburn, D., and Muir, T. W., 1999, *Proc. Natl. Acad. Sci. USA* **96**:388-393.
Yamazaki, T., Otomo, T., Oda, N., Kyogoku, Y., Uegaki, K., and Ito, N., 1998, *J. Am. Chem. Soc.* **120**:5591-5592.

Chapter 3

Characterization of Inter-Domain Orientations in Solution Using the NMR Relaxation Approach

David Fushman[†], and David Cowburn[*]
[†]Dept. of Chemistry and Biochemistry, University of Maryland, College Park, MD 20742, USA ; [*]New York Structural Biology Center, and The Rockefeller University, Box 163, 1230 York Avenue, New York, NY 10021-6399, USA

Abstract   The relative orientation of domains in a multidomain protein can be derived by measuring the orientational tensors of individual subdomains, and aligning them. Using orientational tensors derived from ^{15}N relaxation data, the overall structure and changes on ligand binding can be calculated.

1.   INTRODUCTION

Knowledge of the detailed three-dimensional structure of any given macromolecule is critical for optimizing and/or regulating the use of that macromolecule, be it a protein that is malfunctioning in a metabolic pathway, or a synthetic polymer used in microchip technology. Currently, there are two major strategies for determining the detailed three-dimensional structure of a macromolecule: X-ray crystallography and nuclear magnetic resonance. X-ray crystallographic analysis requires the time-consuming process of preparing high quality crystals, whereas classical NMR three-dimensional analysis is limited to macromolecules that are less than 35 kilo Daltons (Yu, 1999).

Furthermore, such methods of high-resolution structure determination are generally applicable to macromolecules formed by tight contacts between the individual, well-structured components of the macromolecule. These methods have more limited applicability in those cases where there are

53

weaker interactions between the components; examples include the relatively transient associations formed in complexes involved in signal transduction, or in transcriptional control. Packing forces comparable to the interdomain interactions might bias crystal structures of such complexes, while the precision and accuracy of the conventional NMR structural approaches are necessarily limited by the restricted number of nuclear Overhauser effect (NOE) contacts and by interdomain flexibility rendering the available NOE information uninterpretable.

Recently proposed NMR approaches (Tolman et al., 1995; Bruschweiler et al., 1995; Broadhurst et al., 1995; Tjandra et al., 1997a; Tjandra and Bax 1997) are potentially capable of improving both the accuracy and precision of structure determination in solution and might prove to be the method of choice in those cases when the number of available short-range NOE contacts is limited. These methods are based on `long-range` structural information in the form of inter-nuclear vector constraints with respect to an overall, molecular reference frame. These constraints may arise from correlation with the anisotropic hydrodynamic properties of the molecule (Bruschweiler et al., 1995; Broadhurst et al., 1995; Tjandra et al., 1997a), or from weak alignment of molecules in solution caused by either their interaction with the magnetic field (Tolman et al., 1995) or by the liquid crystalline characteristics of the medium (Tjandra and Bax 1997). The NMR relaxation approach (Bruschweiler et al., 1995; Broadhurst et al., 1995; Tjandra et al., 1997a), which takes advantage of the anisotropic character of the overall rotation, is most generally applicable to a wide range of macromolecules in their native milieu. The magnetic alignment method (Tolman et al., 1995; Tjandra et al., 1997b) requires macromolecules to possess a sufficiently high anisotropy of the magnetic susceptibility, and is not, therefore, widely applicable. The approaches based on weak alignment of macromolecules in liquid crystalline medium may be restricted by possible interactions between the molecule under investigation and the medium. For a list of intractable target proteins by this method using lipid bicelles see footnote 8 in (Clore et al., 1998), although more recent alignment methods may alleviate this issue (Clore et al., 1998; Hansen et al., 1998; Koenig et al., 1999; Sass et al., 1999).

Naturally occurring polymers such as nucleic acids and proteins are macromolecules that have distinct three-dimensional structures. Indeed, the ability of any given protein to carry out its physiological role, regardless of whether it functions as a structural element, a binding partner, and/or a biochemical catalyst, requires that the protein assume a specific conformation. This conformation is dependent on the three-dimensional folding of the protein into specific domains and the orientation of these domains to each other, as well to the corresponding domains of other proteins.

The binding of a ligand to a protein (e.g., a substrate to an enzyme), generally results in a local alteration of the three-dimensional structure of the protein. In addition, the binding of the ligand to one site of a protein can also alter the structure of other regions of the polypeptide (see e.g., Kempner, 1993). Indeed the relative orientation and motions of domains within many proteins are key to the control of multivalent recognition, or the assembly of protein-based cellular machines. Therefore, it is not surprising that there has been a long and continuous effort to determine the structures of nucleic acids and proteins, not only in their resting state, but also in their more dynamic state in their native environment.

In recent years, it has become apparent that there is a large but finite number of protein structural domains that are shared throughout nature. These domains are used by the proteins to carry out their biological roles. One such pair of domains is the Src homology domains SH2 and SH3. Eukaryotic cellular signal-transduction pathways that are initiated by trans-membrane receptors with associated tyrosine kinases rely on these two small protein domains for mediating many of the protein-protein interactions that are necessary for transmission of the signal (Cantley et al., 1991; Schlessinger and Ullrich, 1992; Pawson and Schlessinger, 1993). These domains were first discovered in cytoplasmic (non-receptor) protein tyrosine kinases such as the src oncogene product, thus leading to the term 'src homology domains' (Sadowski et al., 1986).

The unique importance of these domains became clear with the discovery of the crk oncogene product, which consists of little more than an SH2 and an SH3 domain fused to the viral gag protein, but is capable of transforming cells (Mayer et al., 1988). SH2 and SH3 domains have been identified in molecules with distinct functions that act downstream from the receptors for, among others, epidermal growth actor (EGF), platelet-derived growth factor (PDGF), insulin and interferon, and the T-cell receptor (Koch et al., 1991).

An important aspect of the role of protein domains such as the SH2 and SH3 domains is their ability to recognize particular amino acid sequences in their target proteins: SH2 domains bind tightly to phosphorylated tyrosine residues (Anderson et al., 1990; Matsuda et al., 1990; Moran et al., 1990; Mayer et al., 1991; Songyang et al., 1993) whereas SH3 domains bind to proline rich segments forming a short helical turn in the complexes (Kuriyan and Cowburn, 1997). The modular nature of these domains is made clear by the fact that they occur in different positions in the polypeptide chains of the intact proteins of which they are a part, and that the binding functions can often be reproduced by isolated domains. Although SH2 and SH3 domains frequently occur close together in sequence, some proteins have only one or the other domain, and some have more than one version of either domain. Proteins that contain more than one of these domains do not always maintain a strict spacing or particular order between the domains.

Even for the SH2 and SH3 domains for which the individual structural properties and ligand specificities are fairly well understood, the structural organization and interactions between them in the multidomain complexes are complex, and difficult to elucidate. These interactions are likely to be of significance, in particular, in view of the frequency of protein constructs containing adjacent SH2 and SH3 domains. Examples of structural studies of the multiple SH3/SH2 domain constructs are the Abelson protein tyrosine kinase SH(32) or Abl SH(32); Lck SH(32); Grb2 SH(323); Hck SH(321); Src and Src SH(32) (reviewed in Kuriyan and Cowburn, 1997; Sicheri and Kuriyan, 1997). Structural approaches to these complexes are complicated by the limited contacts and energies of the interdomain interactions. While the crystal structures of the src-family SH(321) kinase systems have shed significant insight into unexpected kinases/SH3 interactions, and demonstrated the allosteric nature of kinase inhibition by intramolecular phosphorylation, these structures of the down-regulated, inactive forms of the enzymes do not provide a detailed understanding of the mechanism of regulation, or the roles of domains in substrate recognition (Sicheri and Kuriyan, 1997; Mayer et al., 1995). This issue of interdomain flexibility in solution is a general one for large multidomain proteins (Campbell and Downing, 1998).

Therefore, there is a need for determining the structural organization and interactions of components of macromolecules including monitoring enzymatic reactions, DNA-protein interactions, ligand binding, and protein folding. Furthermore, there is a need to exploit such determinations in order to be able to design more potent drugs, pharmaceutical therapies and diagnostic agents. In addition, there is a need to further elucidate the complex structural characteristics of synthetic chemical polymers in solution.

2.   THEORY

Here we describe a method for determining the relative orientation of the individual components of a macromolecule in solution with respect to the global molecular coordinate frame of the macromolecule. This method relies on the basic property of a molecule, its shape. An ideal sphere tumbles isotropically in solution, so that all orientations within a molecule are equivalent. A non-spherical shape of a protein molecule will result in the anisotropy of its overall rotational diffusion in solution, which, in turn, will lead to the orientation dependence (Fushman *et al.*, 1999b) of the processes of nuclear spin relaxation caused by the orientation-dependent interactions, e.g., dipolar interactions or chemical shift anisotropy. The sensitivity of spin relaxation rates to internuclear vector orientations within a molecule constitutes the physical basis of the approach used here. As a result, the measured spin relaxation parameters contain significant structural information that could be retrieved in the form of orientational constraints and used for structure determination.

From the physical point of view, this orientational dependence is caused by differences in the apparent tumbling rates sensed by various internuclear vectors in an anisotropically tumbling molecule. Assume we have a molecule with the principal components of the overall rotational diffusion tensor D_x, D_y, and D_z (x, y, and z denote the principal axes of the diffusion tensor), and let $D_x < D_y < D_z$. For an internuclear vector parallel to the z-axis, its reorientations will be caused by molecular rotations around the x- and y- axes, but not z-axis, so that the apparent rotational diffusion rate for the vector will be determined by $D(z) = (D_x+D_y)/2$ (assuming small degree of anisotropy). On the contrary, reorientations of a vector perpendicular to the z-axis (e.g. along the x-axis) will be determined by $D(x) = (D_z+D_y)/2 > D(z)$ and hence will proceed faster than for the previous internuclear vector. These differences in the tumbling rates will lead to differences in the spin relaxation rates for the corresponding pairs of nuclei; the effect will increase with the anisotropy of the molecule. Theoretical expressions describing this phenomenon and relating spin relaxation rates to the overall rotational diffusion tensor can be found in (Woessner, 1962).

This phenomenon has been used to accurately analyze protein dynamics in the case of overall rotational anisotropy (e.g. Tjandra *et al.*, 1995; Fushman *et al.*, 1998; Copie *et al.*, 1998; Pfeiffer *et al.*, 2001). The structural information in the form of internuclear vector orientations can also be used in the structure refinement procedures, where the relaxation-derived orientational constraints for individual bond vectors could complement the 'short-range' NOE information derived using 'conventional' (twentieth century) NMR approaches (Tjandra *et al.*, 1997a).

The approach described here (the main idea is outlined in Fig.1) is somewhat different in that it is based on

(a) a representation of a macromolecule as a set of two or more relatively rigid components (segments);

(b) the determination of the orientation of the molecular frame for the selected components of the macromolecule in solution, by determining a set of rotational diffusion axes for the selected components, and

(c) the determination of the inter-component orientation in the macromolecule based on the orientation of the selected components with respect to the global molecular coordinate frame, and aligning the set of rotational diffusion axes of the components. This will result in the determination of the relative orientation of the selected components with respect to the global molecular coordinate frame of the macromolecule.

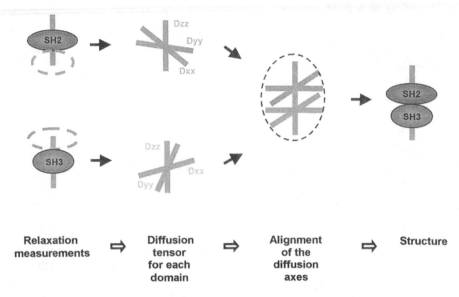

Figure 1. Schematic representation of the idea of orientational alignment of segments in a molecule (in this case, individual domains in a multidomain system) based on the determined orientation of the principal axes of the overall rotational diffusion tensor of the molecule.

2.1   Underlying Chemical Physics: Basic Equations

Several approaches to determination of the overall rotational diffusion tensor from ^{15}N relaxation data were suggested in the literature (Bruschweiler *et al.*, 1995; Tjandra *et al.*, 1995; Lee *et al.* 1997; Copie *et al.*, 1998; Blackledge *et al.*, 1998; Fushman *et al.*, 1999b). The approach applied here uses the orientational dependence of the ratio of the spectral densities (Fushman *et al.*, 1999b).

$$\rho \equiv \frac{3J(\omega_N)}{4J(0)} = \frac{3/4}{1+(\omega_N\tau_x)^2}\left\{1+\frac{(\omega_N\tau_x)^2}{(\omega_N\tau_x)^2+\left(1+\frac{1}{6}\varepsilon\right)^2}\times\right.$$

$$\left.\frac{\varepsilon\sin^2\theta}{3+2\varepsilon+\left[1+\frac{1}{3}\varepsilon(2-3\sin^2\theta)\right]^2}\left[4+3\varepsilon+\frac{2}{9}\varepsilon^2-\varepsilon\sin^2\theta\left(1+\frac{4+\frac{11}{3}\varepsilon+\frac{19}{18}\varepsilon^2+\frac{5}{54}\varepsilon^3}{(\omega_N\tau_x)^2+\left(1+\frac{2}{3}\varepsilon\right)^2}\right)\right]\right\}$$

(1)

Here $\varepsilon \equiv D_{\parallel}/D_{\perp}-1$, $\tau_x^{-1} \equiv 6D_{\perp}$, θ is the angle between an NH bond and the unique axis (associated with $D_{\parallel}$) of the rotational diffusion tensor, assumed axially symmetric here, and ω_N is the ^{15}N resonance frequency. Given the orientation $\{\Theta, \Phi\}$ of the principal axes of the diffusion tensor, the angle θ can be determined as

$$\theta = \cos^{-1}(\lambda_{ix}\cos\Phi\sin\Theta + \lambda_{iy}\sin\Phi\sin\Theta + \lambda_{iz}\cos\Theta) \qquad (2)$$

where $\{\lambda_{ix}, \lambda_{iy}, \lambda_{iz}\}$ are the coordinates (in the protein coordinate frame) of a unit vector in the direction of the NH bond. These basic equations assume that the diffusion tensor is axially symmetric. In the general case of rotational anisotropy (presence of the rhombic component) the corresponding equations are more complex (see e.g. Woessner, 1962)), but the general idea that structural information is encoded in the relaxation rates still holds.

The orientation-dependent quantity ρ in Eq.1 can be directly derived from a set of standard relaxation measurements, R_1, R_2, NOE, as

$$\rho = \left(\frac{2R_2'}{R_1'}-1\right)^{-1}, \qquad (3)$$

assuming conformational exchange contributions to R_2 are negligible. Here R_2' and R_1' are the transverse and longitudinal relaxation rates modified to subtract the contributions from high-frequency motions (see e.g. Fushman et al., 1999a, Fushman et al., 1999b):

$$R'_1 = R_1 - 7\left(\frac{0.921}{0.87}\right)^{-2} P_{HF} = 3(c^2 + d^2)J(\omega_N) \tag{4a}$$

$$R'_2 = R_2 - \frac{13}{2}\left(\frac{0.955}{0.87}\right)^{-2} P_{HF} = (c^2 + d^2)\left[2J(0) + \frac{3}{2}J(\omega_N)\right] \tag{4b}$$

where $\quad P_{HF} = d^2 J(0.87\omega_H) = -\frac{\gamma_N}{\gamma_H}(1 - NOE)R_1/5.$ (4c)

These equations were obtained from the standard expressions (Abragam, 1961), under the assumption that the spectral density function scales as $J(\omega) \propto \omega^{-2}$ at $\omega \approx \omega_H$ (Ishima and Nagayama, 1995; Farrow et al., 1995). Here d and c represent contributions from ^1H-^{15}N dipolar interaction and ^{15}N chemical shift anisotropy (CSA), respectively, and γ_N and γ_H are gyromagnetic ratios for ^{15}N and ^1H.

Equation (1) is exact in the absence of local motion. It provides a good approximation for the protein core residues, characterized by restricted mobility in the protein backbone (in this case both $J(0)$ and $J(\omega_N)$ scale as the squared order parameter (Lipari and Szabo, 1982b; Lipari and Szabo, 1982a), so the contribution from local motion to ρ in Eq.1 is canceled out), and is more general than those used previously (Lee et al., 1997; Copie et al., 1998), which are valid only for small anisotropies ($\varepsilon < 1$) and for $\omega_N \tau_x \gg 1$. Figure 2 illustrates how well Eq.1 describes the expected values for R_2'/R_1' in practically the whole range of possible degrees of anisotropy. This then extends the approach to accommodate both small, fast tumbling molecules (e.g., $\omega_N \tau_x \sim 1.7$ for the free Abl SH3 domain) and molecules with significant rotational anisotropy.

It is worth mentioning that the R_2'/R_1' ratio is insensitive, to a first order approximation, to modulation of residue-specific ^{15}N chemical shift anisotropy tensor and/or dipolar interaction, as the $(d^2 + c^2)$ term is canceled out. The noncollinearity of the two tensors will require corrections to Eqs. (4a,b) for high degrees of rotational anisotropy ($D_{\parallel}/D_{\perp} > 1.5$), as described in detail in (Fushman and Cowburn, 1999).

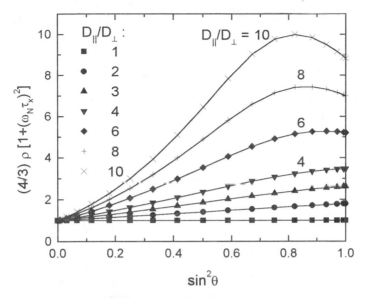

Figure 2.   Demonstration of the accuracy of Eq.1 in a wide range of rotational anisotropies, $D_\parallel/D_\perp$, from 1 to 10. Symbols correspond to synthetic experimental data generated assuming model-free (Lipari and Szabo, 1982a; Lipari and Szabo, 1982b) local motions with $S^2=0.87$, $\tau_{loc} =20$ ps (typical parameters for restricted local backbone dynamics in protein core), $\tau_c = 5$ ns, and ^1H resonance frequency of 600 MHz, and various degrees of anisotropy as indicated. The solid lines correspond to Eq.1.

2.2   Tensor Representation of Segments

As obvious from Eqs.(1) and (3), the R_2/R_1 ratio contains vital structural information in the form of the angles of orientation of bond vectors with respect to the overall molecular coordinate frame. This can be used to derive structural information in the form of bond orientation within a protein that then can be incorporated in the process of structure determination (e.g. Tjandra *et al.*, 1997a) for proteins with unknown structure. Another aspect of the approach -- and this is the focus of this paper -- is grouping bond vectors on a bigger scale, based on relatively rigid segments with known structure (e.g. domains in a multidomain system or helices in a multi-helix bundle), in order to determine the orientation of the rotational diffusion axes for each segment. These axes can then be aligned, since the segments are parts of the same molecular system -- this process will provide the proper orientation of the segments.

The approach described here is based on representation of macromolecules as an assembly of relatively rigid, well-structured components (segments). Given the structure of these components, their orientation within the molecule can be derived from relaxation measurements. In a protein, these segments could range from aromatic rings and peptide planes to elements of the secondary structure (α-helices, β-sheets). Other examples of the rigid elements could be nucleotide bases, sugar rings, and other rigid cyclic structures. On a larger scale, one could consider individual domains in a multidomain protein or individual proteins in a protein/protein complex or a protein/DNA complex. In these latter examples, relatively rigid, well-structured core of the individual moieties can be used as a structural basis for the characterization of the relative orientation of the domains. Which components could be considered 'rigid' depends on the time scale of the intra- and inter-segmental processes with respect to the characteristic time frame of the method used. Two time scales are involved in the method discussed here:

(1) the time scale of the overall rotational diffusion and (2) the time frame of spin relaxation processes. Accurate application of the method requires that the orientation of a given component be fixed within a molecule on both time scales. In reality, inter-segmental motions are always present, which results in a time-averaged diffusion tensor. The limitations of this approximation will be discussed in the later sections.

On the time scale of the overall rotational diffusion, relevant here, (several nanoseconds and slower) all these elements can be considered rigid and therefore are amenable for the approach discussed here. In this paper we, however, focus on the multidomain proteins, and therefore will consider individual domains as relatively rigid segments. A similar approach might be useful for determining relative orientation of helices in a helix bundle, where each segment (helix) can be characterized by a vector along the helix axis (e.g., McDonnell *et al.*, 1999).

The main idea of the approach is illustrated in Fig.1. In all these cases mentioned above, the individual segments, considered relatively rigid, can be characterized, from analysis of relaxation data, by a rotational diffusion tensor, which includes both the principal values and the orientation of the principal axes of the tensor with respect to the molecular frame of the segment. Because all these individual segments participate in the overall rotational diffusion of the molecule (complex), all of them should be sensing the same hydrodynamic characteristics of the overall tumbling. In other words, these individual diffusion tensors are the same tensor seen by different segments in the complex. This concept then allows one to align the individual diffusion axes frames derived for each segment. Since these frames are rigidly attached to the molecular frames for the individual components, this transformation will automatically orient molecular frames of the individual segment and thus provide their orientation with respect to one another in a complex. It is worth mentioning that this discussion here does not consider possible motions that take place in the molecule: internal dynamics within each segment and inter-segmental motions. While the effect of local dynamics is eliminated, to first order, in the approach described here, the inter-segmental motions can influence the results; see the discussion below.

Two assumptions are made in the approach to structure determination of multi-domain systems discussed here: (1) that the relative flexibility of the individual domains (segments) is significantly reduced (compared to the case of independent beads on a string), so that the concept of the common overall rotational diffusion tensor is justified; and (2) that the structure of the individual domains is the same as in their free state. While the first assumption is critical for the approach discussed here, the second one is merely a useful simplification, allowing the use of preexisting knowledge of structures of individual components to determine their relative orientation.

2.2.1   How can we obtain the structural information?

The following two sections describe the necessary steps and procedures involved in the determination of the inter-segmental orientation.

2.2.1.1   How do we derive the diffusion tensor characteristics?

Equation (1) allows the determination of the principal values of the diffusion tensor if the orientation of its principal axes frame is known. The latter information -- and this is the one we are actually interested in for solving the domain orientation problem -- is included there in implicit form, via Eq. (2). The problem is that neither the principal axes nor principal values are known *a priori*. The approach used here is then to determine all components of the tensor from the minimization of the target function,

$$\chi^2 = \sum_i \left[\frac{\rho_i^{exp} - \rho_i^{calc}}{\sigma_{\rho i}} \right]^2 \tag{5}$$

Here ρ^{exp} is the experimentally determined value of $(2R_2'/R_1'-1)^{-1}$, derived from the experimental data according to Eqs.(4a-c), and ρ^{calc} is the predicted value of this parameter, calculated using Eqs.1, 2 for a given set of the principal axes and principal values of the diffusion tensor; σ_ρ denotes the experimental errors in ρ^{exp}; and the sum runs over all available internuclear vectors.

The minimization of the target function Eq. (5) for an axially symmetric tensor requires a search in a four-dimensional space of the principal values $\{D_\parallel, D_\perp\}$ and the Euler angles $\{\Phi, \Theta\}$. Because of the periodic boundary conditions for $\{\Phi, \Theta\}$, it is convenient to perform this optimization as two algorithmically different (however, not independent) search procedures. In our approach, we derive the optimal values of $D_\parallel$ and $D_\perp$ using the simplex algorithm (Press *et al.*, 1992); for each test set of $\{D_\parallel, D_\perp\}$, the optimal values of the Euler angles are obtained by a 1°-step grid search in the $\{\Phi, \Theta\}$ space, as illustrated in Fig.3.

In the most general case of a completely anisotropic diffusion tensor, six parameters have to be determined for the rotational diffusion tensor: three principal values and three Euler angles. This determination requires an optimization search in a six-dimensional space, which is a significantly more CPU-demanding procedure than that for an axially symmetric tensor. Possible efficient approaches to this problem suggested recently include a simulated annealing procedure (Hus *et al.*, 2000) and a two-step procedure (Ghose *et al.*, 2001). The latter is essentially an extension of the method presented here. To increase the efficiency of the search procedure, a hybrid, two-step approach was designed in (Ghose *et al.*, 2001). In the first step, a singular value decomposition approach (Press *et al.*, 1992) is applied to obtain a preliminary estimate of the diffusion tensor. Being accurate for small degrees of anisotropy, this method, however, yields a reasonably good estimate for the orientation angles even for a diffusion tensor that cannot be considered weakly anisotropic. These preliminary results are then used as input for the second step, to accurately determine the complete diffusion tensor based on the same optimization algorithm as described here, extended for a six-dimensional search. The preliminary knowledge of the Euler angles obtained on the first step allows one to significantly reduce the size of the search region in the angular space, thus significantly reducing the necessary CPU time.

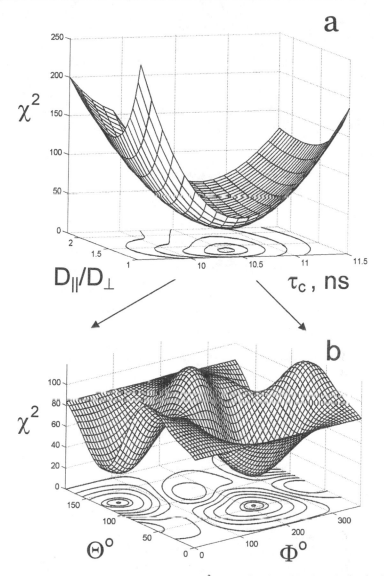

Figure 3. Landscape of the target function χ^2, Eq.5, for the SH3 domain in the bound SH(32) in the coordinates (**a**) τ_c and $D_\parallel/D_\perp$ and (**b**) Θ and Φ. The target function in panel **b** corresponds to the minimum of χ^2 in panel **a** at τ_c=10.5 ns and $D_\parallel/D_\perp$ = 1.3. This figure also illustrates the separation of the minimization procedure, in the principal values and angular space.

2.2.1.2   How do we align the segments?

Assume we have done the preliminary work, and determined the orientation angles $\{\Theta_i, \Phi_i\}$ (three Euler angles $\{\alpha_i, \beta_i, \gamma_i\}$ in the general case) for the principal axes of the rotational diffusion tensor for each segment. How do we reorient the segments? The set of Euler angles determines a rotational matrix, $R(PDB \rightarrow D_i)$ that relates the PDB coordinate frame with the rotational diffusion axes 'seen' by each segment. This means, for example, that the principal axes (column vectors) $\{\xi_i, \eta_i, \zeta_i\}$ of the overall diffusion tensor 'seen' by segment I can be determined as

$$\vec{\xi}_i = R(\alpha_i, \beta_i, \gamma_i)\begin{bmatrix}1\\0\\0\end{bmatrix}; \quad \vec{\eta}_i = R(\alpha_i, \beta_i, \gamma_i)\begin{bmatrix}0\\1\\0\end{bmatrix}; \quad \vec{\zeta}_i = R(\alpha_i, \beta_i, \gamma_i)\begin{bmatrix}0\\0\\1\end{bmatrix} \quad (6)$$

Since all these axes seen by different segments are the same axes, these should be aligned. This can be done by applying an inverse rotation matrix to atom coordinates for each segment:

$$\begin{bmatrix}x_k{}'\\y_k{}'\\z_k{}'\end{bmatrix} = R(PDB \rightarrow D_i)^{-1}\begin{bmatrix}x_k^i\\y_k^i\\z_k^i\end{bmatrix}; \qquad k=1..N_i \qquad (7)$$

This procedure will reorient all segments in a molecule with respect to the overall rotational diffusion coordinate frame.

Alternatively, it might prove more useful to fix one of the segments (e.g. segment i), and reorient all other segments with respect to this one. The corresponding transformation for atoms belonging to segment j can then be described by the matrix

$$R(j \rightarrow i) = R(PDB \rightarrow D_i) * R(PDB \rightarrow D_j)^{-1} \qquad (8)$$

These procedures described here account for reorientation of individual segments (domains) in a molecule. In addition, the necessary transformation includes translation, which can be performed based on interatomic distances derived from NOE constraints. This use of rigid frames with limited NOE constraints has been used for large-scale structure determinations of complexes (Clore, 2000).

2.2.2   Solving the Orientation-degeneracy Problem

Because the rotational diffusion tensor is not sensitive to the directionality of the axis, the solutions in terms of the Euler angles are usually degenerate: the pairs of angles $\{\Theta, \Phi\}$ and $\{180°- \Theta, \Phi +180°\}$ are equivalent. This two-fold degeneracy for each segment provides a problem of increased degeneracy when it comes to relative orientation of multiple domains. Thus, for a two-domain system (Figure 4), the apparent degeneracy is 4. However, by fixing orientation of one of the domains (e.g. SH3), it is easy to show that the actual degeneracy is 2. This degeneracy can be further reduced for the SH(32) system considered here by applying the chemical connectivity requirement, i.e. that the C-terminus of SH3 and the N-terminus of SH2 be close in space.

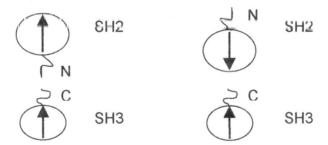

Figure 4.   Solving the orientation degeneracy problem. This is a schematic illustration of the degeneracy problem in the determination of the relative orientation of individual domains from alignment of the principal axes of the rotational diffusion tensor. Since the derived rotational diffusion axis does not have directionality, both 'up' and 'down' orientations are possible for the SH2 domain relative to SH3. The backbone connectivity requirement, however, helps solve this problem: only the orientation shown on the left is consistent with the fact that the C-terminus of SH3 is connected to the N-terminus of the SH2 domain.

3.   PRACTICAL EXAMPLE

3.1   Determination of the Relative Orientation of SH2 and SH3 Domains in the Abl SH(32) Construct:

The approach described above has been applied to a dual domain system from Abelson kinase. We studied Abl SH(32) construct, alone and in complex with a consolidated ligand (Cowburn et al., 1995), comprising the individual ligands for SH3 and SH2 domains (3BP-2 and 2BP-1, respectively) connected via a $(Gly)_6$ linker. We also studied the individual molecules of SH2 and SH3 domains free in solution, as a control. Details of the experimental procedures and sample conditions are specified in (Fushman et al., 1999b).

For those residues with restricted local dynamics, both the overall hydrodynamic characteristics of the molecule and the local inter-nuclear vector orientation can be directly derived from the observed values of the R_2'/R_1' ratio, using Eq.1. Figure 5 presents the ratio, R_2'/R_1', of the experimentally determined ^{15}N relaxation rates for individual amides in the proteins analyzed here. The observed relatively uniform distribution of R_2'/R_1' values along the backbone in the free domains is consistent with the relatively low rotational anisotropy of these molecules, expected from their 3D structures. On the contrary, one can see well-pronounced variations in the R_2'/R_1' values versus residue number for the SH(32) construct, both free and bound, indicating a significant degree of rotational anisotropy in these molecules. Variations in the individual values of R_2'/R_1' along the protein backbone principally reflect different orientation of the individual NH bonds with respect to the rotational diffusion frame. It is worth noting that the observed patterns of R_2'/R_1' distribution along the backbone differ significantly between free and bound states of Abl SH(32), suggest differences in the orientation of the rotational diffusion tensor, hence different interdomain orientation, in these two cases.

The apparent correlation time, τ_{app}, can be obtained from R_2'/R_1' for each amide group without any reference to protein structure (Fushman et al., 1994):

$$\tau_{app} = (2\omega_N)^{-1}\sqrt{6R_2'/R_1'-7}. \qquad (9)$$

In the case of isotropic rotational diffusion ($D_\parallel = D_\perp$), τ_{app} coincides with the overall (isotropic) correlation time τ_c. The observed average levels of apparent isotropic τ_{app} for the free SH3, SH2, SH(32) and for the SH(32)/ligand complex increase linearly with the molecular weight of these proteins (see Figure 5B).

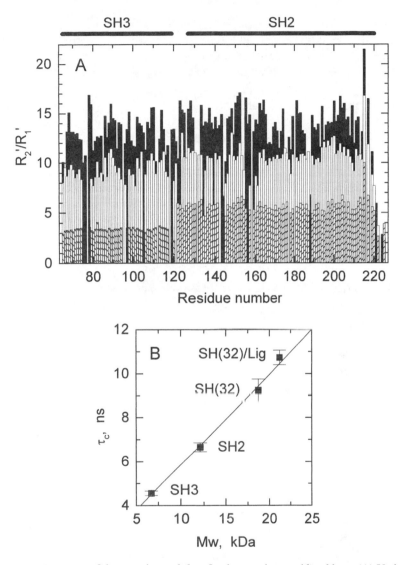

Figure 5. Summary of the experimental data for the proteins considered here. (A) Variation in the experimentally determined backbone ^{15}N R_2'/R_1' ratio versus protein sequence. Vertical bars represent the data for the unligated (white), and bound (black) SH(32), and for the free SH3 and SH2 domains (hatching). Horizontal bars on the top indicate location of the individual domains in the Abl SH(32) dual domain sequence. (B) Observed molecular weight (MW) dependence of the overall rotational correlation time (see Fushman et al., 1999) for the free SH3 and SH2 domains and for the SH(32) construct, free and bound. The observed linear dependence with MW indicates that two domains in SH(32) are not tumbling independently of each other (Hansen et al., 1998). The solid line corresponds to a molecular weight dependence $\tau_c = 1.76 \, (\pm 0.20) + 0.41 (\pm 0.02) \, Mw$. The observed slope is comparable to 0.33, calculated from the Stokes-Einstein equation for protein density, 0.73 cm^3/g and hydration, 0.34 g/g H$_2$O.

Higher values of the apparent rotational correlation time for the individual domains in the dual domain construct compared to the free domains in solution, suggest the presence of restricted rotational diffusion of the domains imposed by the linker. Differences in the average levels of τ_{app} in the SH3 and SH2- parts of the free dual domain construct, although small, indicate some degree of interdomain flexibility in SH(32). No significant difference was observed between the average τ_{app} values for the two domains in the SH(32)/ligand complex, consistent with restriction in the interdomain flexibility expected upon binding of the consolidated ligand. This relation between the overall tumbling times for individual domains in a dual domain protein is different from that reported in (34), where large differences between the correlation times of the individual domains in a two-domain protein were observed, consistent with those of independent beads on a string. The rotational correlation times of the SH2 and SH3 domains in Abl SH(32) observed here suggest presence of certain orientational constraints between the two domains. This, therefore, justifies the proposed method of characterization of interdomain orientation by applying the concept of the common overall rotational diffusion tensor for all domains in a molecule (Figure 1).

Figure 6 (overleaf).   Contour maps of the optimized target function, χ^2, in angular coordinates Φ and Θ, for the (**a, b**) SH3 and (**c, d**) SH2 domains, in the unligated and bound SH(32). The minima are indicated by letters A and B; the shaded areas around the minima correspond to 68.3% confidence levels for the joint distribution of Φ and Θ. Numbers near the contour lines indicate the values of χ^2. Two sets of $\{\Phi, \Theta\}$-angles, indicated on the map as A and B, are consistent with the data, because of the axial degeneracy of the analysis (see Fig.4). These two sets are equivalent, related by the symmetry transformations: $\Phi_A = \Phi_B$ +180°, $\Theta_B = 180° - \Theta_A$, and correspond to the opposite orientations of the $D_{\parallel}$-axis of the rotational diffusion tensor. In the case of dual domain construct, the selection of proper pairs from the $\{\Phi, \Theta\}$ angles determined for the individual domains can be done based on restrictions imposed by their chemical linkage. The selected pairings, as shown in Figure 8, panels **a** and **b**, are A_{SH3}--B_{SH2} and B_{SH3}--A_{SH2}.

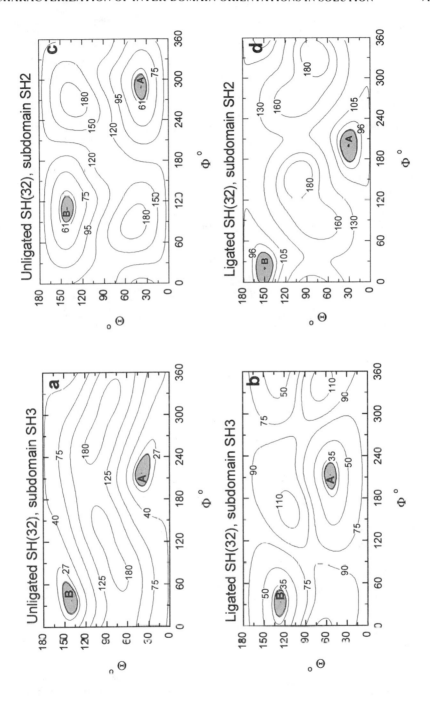

To quantify these observations, we determined the average orientation of the overall rotational diffusion tensor for each of the two domains, using the measured R_2'/R_1' ratio. The approach, illustrated in Figure 1, assumes that the backbone tertiary structure of the individual domains is substantially preserved in the dual domain. The amide ^{1}H and ^{15}N chemical shifts in the individual domains and in SH(32) are nearly identical (Gosser et al., 1995). For example, only 6% of the amide groups in the SH2 and 6% in the SH3 domain exhibit a total chemical shift difference between the free state and the dual domain construct of more than 30 Hz at 600 MHz. A similar picture is observed for the bound SH(32): only 10% of amide signals are shifted by more than 30 Hz in the SH(32)/Ligand complex compared to the free SH2 bound to the ligand. Changes in orientation of amides between free and bound SH3 in crystal state are very small (Musacchio et al., 1994); most changes are located in the RT-Src loop. Similarly, the backbone structure in the core of the Abl SH2 domain remains practically unaltered upon its binding to 2BP-1 (Francart and Cowburn, unpublished). This is also consistent with the comparison of the crystal structures of the other SH2 domains (from Src (Waksman et al., 1993) and from SYP tyrosine phosphatase (Lee et al., 1994)) in the free form and bound to various phosphotyrosine-containing peptides, indicating only minor changes in the structure of the protein core. No significant changes in the result were observed when those residues in both domains involved in ligand binding were excluded from the analysis for the bound SH(32). While clearly an approximation, the assumption that the axes can be determined from the orientations of the core of the Abl SH2 and SH3 domains is then entirely reasonable (see Figs. 6-8). Full structure determination, as separate domains, from the NOE data for SH(32) is in progress. The treatment used here derives a time-averaged set of anisotropic axes, and the relative orientational dependence derived is not dependent on inter-domain scalar effects, such as NOE's. This analysis then fundamentally extends that of (Tjandra et al., 1997a) which assumed a time-independent, fixed relationship between sub domains.

4.   POSSIBLE LIMITATIONS OF THE APPROACH

A variety of factors could affect the precision and accuracy of the structural approach described here. These factors can be separated in two groups: (1) factors related to the experimental approach and data analysis, and (2) factors depending on a particular molecular system under consideration. The first group includes the obvious problems related to the experimental measurements (sensitivity, temperature and sample stability, artifacts related to off-resonance effects etc), data analysis (data processing errors, spectral overlap) and possible limitations of the underlying theory. The latter include inaccuracy in the current treatment of the CSA and dipolar tensors, in particular, in the presence of significant rotational anisotropy (Fushman and Cowburn, 1999), conformational exchange contributions, influence of other, not directly bound nuclei etc. While these problems could be solved, at least in principle, by improving the experimental and analytical approaches, the problems related to the second group are intrinsic to a molecular system under consideration. These include molecular dynamics (both intra- and interdomain motions), and limited sampling of the orientational space by a given set of internuclear vectors. These problems, briefly discussed below, are not specific for the method discussed here, and are common for any approach (e.g. residual dipolar couplings) based on orientational properties of proteins or other molecules.

4.1 Influence of Protein Dynamics.

With regard to internal backbone dynamics, the approach described here is independent, as pointed out above, of local motions typically present in the well-defined elements of protein structure. The interdomain (intersegmental) dynamics, on the other hand, could influence both the principal values and the orientation of the principal axes of the overall diffusion tensor seen by each domain (segment). In fact, the observed rotational anisotropy $(D_\parallel/D_\perp)$ of the dual domain (1.24 for SH3 and 1.16 for SH2 in the unligated SH(32)) is somewhat lower than expected from rigid-body models (Koenig, 1975) (e.g., compared to $D_\parallel/D_\perp=1.87$ for a prolate ellipsoid of revolution with the axial ratio of 2:1), indicating that presumably the observed values are averaged by interdomain dynamics on the NMR time scale (10 ns to 100 ms). This is consistent with the different degrees of rotational anisotropy for the individual domains; the $D_\parallel/D_\perp$ values for SH2 are somewhat smaller than for the SH3 part. Even in the SH(32)/ligand complex, where the average characteristics of rotational diffusion for the two domains are similar (average τ_{app} of 10.61 ns for SH3 and 10.89 ns for SH2), their apparent anisotropies are somewhat different (1.30 for SH3 and 1.20 for SH2), possibly due to flexibility of the ligand itself or caused by interconversion between bivalent and monovalent bound forms. Similar interpretations involving interdomain mobility were reported earlier (Bruschweiler et al., 1995; Hansen et al., 1994; Barbato et al., 1992). In this regard, the present results demonstrate that the relative orientation of SH2 and SH3 domains in Abl SH(32) is not fixed, and can be changed by the protein interaction with a consolidated ligand. Ligand binding studies (Xu et *al.*, 1999) suggest flexibility of the dual domain SH(32) construct to accommodate several relative orientations of the binding sites.

Figure 7 (overleaf). Differences in orientations of the rotational diffusion tensor axis (shown by rods) determined from NMR relaxation data for (A) SH3 and (B) SH2 domains, free in solution and in the context of the SH(32) dual domain system, free and bound.

Figure 8 (overleaf). Relative orientation of the SH3 and SH2 domains in the Abl SH(32) construct (a) unligated and (b) dual ligated in solution and (c) unligated in the crystal structure (Nam *et al.*, 1996). The structure of protein backbone is represented by ribbons colored green for the SH2 and blue for the SH3; the N- and C-termini are indicated. For each domain, rods represent experimental orientations of the unique diffusion axis (corresponding to the parallel component $D_{\parallel}$ of the rotational diffusion tensor) in the case of the dual domain construct, unligated (gold) and ligated (pink). Relative orientation of the individual domains in the dual domain construct in solution was obtained by aligning the corresponding diffusion axes, as shown in panels **a** and **b**. The angle of rotation of each domain around its diffusion axis cannot be determined from these data, because of the assumed axial symmetry of the model. In the ligated SH(32), the orientation shown was chosen to ensure proximal positioning of the ligand binding sites of the two domains (residues implicated in binding are colored red). This orientation is consistent with that derived from a rhombohedral fitting of the data (Ghose *et al.*, 2001). Also shown in **b**, for comparison, are orientations of the diffusion axes of the two domains as they are in the unligated SH(32). The observed change in orientation of the overall rotation axis upon ligation is 15° for the SH3, and 47° for the SH2 parts of the dual domain construct. Panel **c** depicts the orientation of the individual domains in the crystal structure (Nam *et al.*, 1996) of unligated SH(32); the rods represent the derived rotational diffusion axes for the individual domains obtained by fitting relaxation data in solution to this structure. The data indicate differences in the relative orientations of the domains between crystal and solution studies, likely to be caused by packing forces in the crystal.

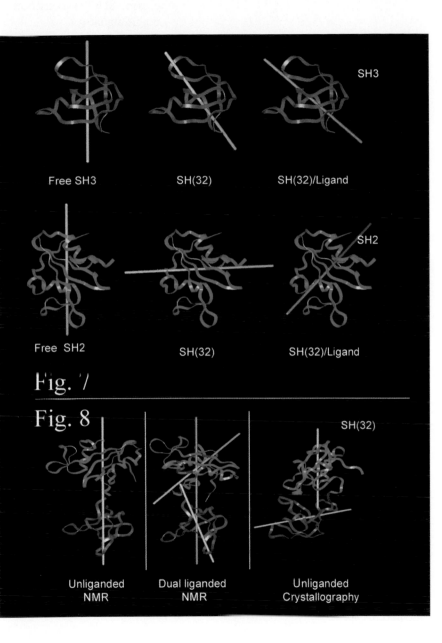

Free SH3   SH(32)   SH(32)/Ligand

SH3

Free SH2   SH(32)   SH(32)/Ligand

SH2

Fig. 7

Fig. 8

Unliganded   Dual liganded   Unliganded
NMR   NMR   Crystallography

SH(32)

4.2 Limited Orientational Sampling

The problem of limited sampling is caused by the fact that only a limited set of internuclear vectors is available for any given molecular structure. Depending on a particular three-dimensional structure of a molecule, these vectors might not sample all orientations uniformly. A particular example is the α-helix, where the backbone NH-vectors are oriented almost parallel to the helix axis. While providing a good sampling for the principal axis of a rank-2 tensor (both rotational diffusion or alignment tensors are of this kind) in the direction of the helix, this oriented set of vectors will, however, render inaccurate the determination of a principal axis of the tensor that is orthogonal to the helix axis. A general mathematical formalism for treating the incomplete sampling problem was suggested recently (Fushman *et al.*, 2000). This approach also allows the assessment of the accuracy of the tensor determination from a given set of vectors. The analysis of various sets of internuclear vectors in the known protein structures revealed significant differences in their sampling properties, related to the role these vectors and the corresponding atoms play in the protein. For example, the distribution of the NH vectors could be far from uniform, reflecting the intrinsic feature of a folded protein, where amide hydrogens play essential roles in the hydrogen-bonding networks of the protein fold.

5. CONCLUSIONS

The approach presented here is based on the hydrodynamic characterization of proteins by ^{15}N relaxation. This provides methods for determining the relative orientation of the individual components of a macromolecule with respect to the global molecular coordinate frame of the macromolecule. Such determinations both complement and extend the currently known methods of calculating the three dimensional structures of macromolecules. Therefore, the methods discussed here can be performed alone or in conjunction with other approaches including conventional methods of structure determination that use local constraints (interatomic distances, torsion angles, and/or hydrogen bonds) and other new methods based on vector orientations derived from residual dipolar coupling (Tjandra and Bax, 1997). In addition, the methods of structural analysis provided here can be performed in conjunction with synthetic structural refinements of the individual components of the macromolecule in order to design and construct modified macromolecules that have improved and/or more desirable properties.

6.   REFERENCES

Abragam, A. 1961. in *The Principles of Nuclear Magnetism*. Oxford: Clarendon Press
Anderson, D., Koch, C.A., Grey, L., Ellis, C., Moran, M.F., and Pawson, T., 1990, *Science* **250**:979-982.
Barbato, G., Ikura, M., Kay, L.E., Pastor, R.W., and Bax, A., 1992, *Biochemistry* **31**:5269-5278.
Blackledge, M., Cordier, F., Dosset, P., and Marion, D., 1998, *J.Am.Chem.Soc.* **120**:4538-4539.
Broadhurst, R.W., Hardman, C.H., Thomas, J.O., and Laue, E.D., 1995, *Biochemistry* **34**:16608-16617.
Bruschweiler, R., Liao, X., and Wright, P.E., 1995, *Science* **268**:886-889.
Campbell, I.D., and Downing, A.K., 1998, *Nat. Struct. Biol.* **5**:496-499.
Cantley, L.C., Auger, K.R., Carpenter, C., Duckworth, B., Graziani, A., Kapeller, R., and Solthoff, S., 1991, *Cell* **64**:281-282.
Clore, G.M., 2000, *Proc. Natl. Acad. Sci. USA* **97**:9021-9025.
Clore, G.M., Starich, M.R., and Gronenborn, A.M., 1998, *J. Am. Chem. Soc* **120**:10571-10572.
Copie, V., Tomita, Y., Akiyama, S.K., Aota, S., Yamada, K.M., Venable, R.M., Pastor, R.W., Krueger, S., and Torchia, D.A., 1998, *J. Mol. Biol.* **277**:663-682.
Cowburn, D., Zheng, J., Xu, Q., and Barany, G., 1995, *J. Biol. Chem.* **270**:26738-26741.
Farrow, N., Zhang, O., Szabo, A., Torchia, D., and Kay, L., 1995, *J. Biomol. NMR* **6**:153-162.
Fushman, D., and Cowburn, D., 1999, *J. Biomol. NMR* **13**:139-147.
Fushman, D., Ghose, R., and Cowburn, D., 2000, *J. Am. Chem. Soc.* **122**:10640-10649.
Fushman, D., Najmabadi-Haske, T., Cahill, S., Zheng, J., LeVine, H., 3rd, and Cowburn, D., 1998, *J. Biol. Chem.* **273**:2835-2843.
Fushman, D., Tjandra, N., and Cowburn, D., 1999a, *J. Am. Chem. Soc.* **121**:8577-8582.
Fushman, D., Weisemann, R., Thuring, H., and Ruterjans, H., 1994, *J. Biomol. NMR* **4**:61-78.
Fushman, D., Xu, R., and Cowburn, D., 1999b, *Biochemistry* **38**:10225-10230.
Ghose, R., Fushman, D., and Cowburn, D., 2001, *J. Magn. Reson.* **149**:204-217.
Gosser, Y.Q., Zheng, J., Overduin, M., Mayer, B.J., and Cowburn, D., 1995, *Structure* **3**:1075-1086.
Hansen, A.P., Petros, A.M., Meadows, P.P., and Fesik, S.W., 1994, *Biochemistry* **33**:15418-15424.
Hansen, M.R., Rance, M., and A., P., 1998, *J. Am. Chem. Soc.* **120**:11210-11211.
Hus, J.C., Marion, D., and Blackledge, M., 2000, *J. Mol. Biol.* **298**:927-936.
Ishima, R., and Nagayama, K., 1995, *Biochemistry* **34**:3162-3171.
Kempner, E.S., 1993, *FEBS Lett.* **326**:4-10.
Koch, C.A., Anderson, D., Moran, M.F., Ellis, C., and Pawson, T., 1991, *Science* **252**:668-674.
Koenig, B.W., Hu, J.-S., Ottiger, M., Bose, S., Hendler, R.W., and Bax, A., 1999, *J. Am. Chem. Soc.* **121**:1385-1386.
Koenig, S., 1975, *Biopolymers* **14**:2421-2423.
Kuriyan, J., and Cowburn, D., 1997, *Ann. Rev. Biophys. Biomol. Str.* **26**:259-288.
Lee, C.-H., Kominos, D., Jacques, S., Margolis, B., Schlessinger, J., Shoelson, S.E., and Kuriyan, J., 1994, *Structure* **2**:423-438.
Lee, L.K., Rance, M., Chazin, W.J., and Palmer, A.G., III., 1997, *J. Biomol. NMR* **9**:287-298.
Lipari, G., and Szabo, A., 1982a, *J.Amer. Chem.Soc.* **104**:4546-4559.
Lipari, G., and Szabo, A., 1982b, *J. Amer. Chem.Soc.* **104**:4559-4570.
Matsuda, M., Mayer, B.J., Fukui, Y., and Hanafusa, H., 1990, *Science* **248**:1537-1539.

Mayer, B., Jackson, P., and Baltimore, D., 1991, *Proc.Natl.Acad.Sci.USA* **88**:627-631.

Mayer, B.J., Hamaguchi, M., and Hanafusa, H., 1988, *Nature* **332**:272-275.

Mayer, B.J., Hirai, H., and Sakai, R., 1995, *Cur. Biol.* **5**:296-305.

McDonnell, J.M., Fushman, D., Milliman, C.L., Korsmeyer, S.J., and Cowburn, D., 1999, *Cell* **96**:625-634.

Moran, M.F., Koch, C.A., Anderson, D., Ellis, C., England, L., Martin, G.S., and Pawson, T., 1990, *Proc. Natl. Acad. Sci. USA* **87**:8622-8626.

Musacchio, A., Saraste, M., and Wilmanns, M., 1994, *Nat. Struct. Biol.* **1**:546-551.

Nam, H.J., Haser, W.G., Roberts, T.M., and Frederick, C.A., 1996, *Structure* **4**:1105-1114.

Pawson, T., and Schlessinger, J., 1993, *Curr. Biol.* **3**:434-442.

Pfeiffer, S., Fushman, D., and Cowburn, D., 2001, *J. Am. Chem. Soc.*

Press, W.H., Teukolsky, S.A., Vetterling, W.T., and Flannery, B.P. 1992. in *Numerical Recipes in C*, NY: Cambridge University Press

Sadowski, I., JC, S., and T, P., 1986, *Mol. Cell. Biol.* **6**:4396-4408.

Sass, J., Cordier, F., Hoffmann, A., Rogowski, M., Cousin, A., Omichinski, J.G., Löwen, H., and Grzesiek, S., 1999, *J. Am. Chem. Soc.* **121**:2047-2055.

Schlessinger, J., and Ullrich, A., 1992, *Neuron* **9**:383-391.

Sicheri, F., and Kuriyan, J., 1997, *Curr. Opin. Str. Biol.* **7**:777-785.

Songyang, Z., Shoelson, S.E., Chaudhuri, M., Gish, G., Pawson, T., Haser, W.G., King, F., Roberts, T., Ratnofsky, S., Lechleider, R.J., Neel, B.G., Birge, R.B., Fajardo, J.E., Chou, M.M., Hanafusa, H., Schaffhausen, B., and Cantley, L.C., 1993, *Cell* **72**:767-778.

Tjandra, N., and Bax, A., 1997, *Science* **278**:1111-1114.

Tjandra, N., Feller, S.E., Pastor, R.W., and Bax, A., 1995, *J. Am. Chem. Soc.* **117**:12562-12566.

Tjandra, N., Garrett, D.S., Gronenborn, A.M., Bax, A., and Clore, G.M., 1997a, *Nat. Struct. Biol.* **4**:443-449.

Tjandra, N., Omichinski, J.G., Gronenborn, A.M., Clore, G.M., and Bax, A., 1997b, *Nat. Struct. Biol.* **4**:732

Tolman, J.R., Flanagan, J.M., Kennedy, M.A., and Prestegard, J.H., 1995, *Proc. Natl. Acad. Sci. USA* **92**:9279-9283.

Waksman, G., Shoelson, S.E., Pant, N., Cowburn, D., and Kuriyan, J., 1993, *Cell* **72**:779-790.

Woessner, D., 1962, *J. Chem. Phys* **37**:647-654.

Xu, Q., Zheng, J., Xu, R., Barany, G., and Cowburn, D., 1999, *Biochemistry* **38**:3491-3497.

Yu, H., 1999, *Proc. Natl. Acad. Sci. USA* **96**:332-334.

Chapter 4

Global Fold Determination of Large Proteins using Site-Directed Spin Labeling

John L. Battiste, John D. Gross and Gerhard Wagner
Department of Biological Chemistry and Molecular Pharmacology, Harvard Medical School, Boston, MA 02115.

Abstract: Protein structures up to 25 kDa may be obtained using distance and torsion angle restraints provided by the nuclear Overhauser effect (NOE) and J couplings respectively. Alternatively, distances may be obtained between a paramagnetic center and a nucleus through the electron-nuclear dipolar interaction. A paramagnetic center may be introduced into an intrinsically diamagnetic protein by covalently attaching a nitroxide spin label to cysteine. This technique, referred to as site-directed spin labeling (SDSL), has been employed extensively in electron-paramagnetic resonance. We review applications of SDSL to obtain global folds of proteins and macromolecular complexes within the frame work of high-resolution NMR. Issues pertaining to protein engineering, interpretation of paramagnetic broadening and combination of paramagnetic broadening with limited NOE data will be discussed. Paramagnetic broadening effects (PBEs) in conjunction with limited NOE data and/or residual dipolar couplings and TROSY triple resonance methodology promise to extend the size limit of proteins amenable to structural studies by NMR.

1.   INTRODUCTION

The core structural information used for determination of macromolecular structures by NMR consists of distance and torsion angle restraints obtained from NOE and J-coupling data, respectively. Success of this NOE based methodology is impaired with increasing molecular weight.

This is due to increased peak overlap and the fact that shorter transverse relaxation times limit sensitivity and resolution. Deuteration of aliphatic carbons is used routinely for increasing both sensitivity and resolution though the number of measurable NOEs is reduced considerably. The latter problem is partially alleviated by reintroducing ^1H spins into non-exchangeable CH_3 sites of highly perdeuterated proteins (Gardner and Kay, 1997a; Gardner and Kay, 1997b; Goto et al., 1999; Metzler et al., 1996; Rosen et al., 1996; Smith et al., 1996) . Despite these important advances only a handful of proteins over 30 kDa have been solved using the above approach though backbone resonances of many proteins within this molecular weight range, or beyond, have been assigned (Ferentz and Wagner, 2000).

In parallel there has been a major emphasis on discovering new methodologies-apart from the NOE-to obtain additional types of NMR data that could be used as modeling restraints.. Much attention has been focused on obtaining residual dipolar couplings (RDCs) in partially aligned samples (Tjandra and Bax, 1997; Tjandra et al., 1997; Tolman et al., 1995) for obtaining bond vector orientations relative to a molecule fixed ordering frame. RDCs have been tremendously valuable in refining structures and providing long range orientation information that NOE data often cannot provide. More recent work has focused on obtaining relative orientation of protein domains in multidomain proteins or protein complexes using RDCs and rigid body docking where NOE data are sparse (Clore, 2000; Fischer et al., 1999; Mueller et al., 2000; Skrynnikov et al., 2000) . In addition, fits of measured RDCs of an unknown protein relative to protein fragments from structural databases have allowed global fold determination in the absence of NOE data (Delaglio et al., 2000; Meiler et al., 2000). Approaches for obtaining backbone protein structure using exclusively RDC data have also been described (Hus et al., 2001). Such approaches for obtaining global folds in the absence of NOE data are attractive as only backbone resonance assignments and RDCs are required. The advent of TROSY (Pervushin et al., 1997; Salzmann et al., 1998; Salzmann et al., 1999; Yang and Kay, 1999) has extended the molecular weight range for which backbone assignments can be completed. Any methods that support global fold determination based on only backbone information are then particularly useful for large proteins or macromolecular complexes.

In light of this, it would be useful to have an alternative method to the NOE for obtaining *distance* restraints which is sensitive, robust and applicable to larger molecular weight systems where only backbone resonance assignments are obtainable. Similar to the NOE, paramagnetic broadening due to the electron-nuclear dipolar coupling has a r^{-6} distance dependence on the proximity of the nuclear spin to an unpaired electron. For inherently paramagnetic metallo-proteins, these broadening effects, in addition to information obtained from other paramagnetic phenomena such as pseudo-contact shifts, electron-nuclear cross correlation effects and weak alignment, have been incorporated with a high degree of success into molecular modeling calculations of small to medium sized proteins (Bertini *et al.*,1996; Bertini *et al.*, 1997; Hus *et al.*, 2000). Progress on the structure determination of paramagnetic metallo-proteins has been reviewed in detail elsewhere (Bertini *et al.*, 1999), and will not be covered here. While the results with metallo-proteins have been impressive, they are not generally applicable for a majority of proteins. An alternative method for producing a paramagnetic protein is to attach a nitroxide spin label, a technique that has enjoyed a rich history in EPR spectroscopy. In fact, the potential for obtaining distance restraints from NMR of nitroxide spin labeled proteins was demonstrated over a decade ago (Kosen *et al.*, 1986; Schmidt and Kuntz, 1984). The effort required to obtain nitroxide labeled samples, and the small number of distance restraints per spin label, made this methodology impractical compared to the now standard NOE/J-coupling techniques. As mentioned previously, TROSY triple resonance techniques (Salzmann *et al.*, 1998; Salzmann *et al.*, 1999; Yang and Kay, 1999) in combination with perdeuteration of isotopically labeled protein samples permits backbone resonance assignment of very large molecular weight proteins though in many cases difficulty obtaining sufficient NOEs for structural calculations is encountered In this situation, only the secondary structure can be reliably calculated, and an opportunity arises for reinvestigation of the utility of spin-label distance restraints for determining at least the global fold of the protein. Particularly in light of the recent emphasis on structural genomics, the ability to determine global folds has significant importance for comparative proteomics studies.

In the past several years, a few groups have demonstrated proof of principle that nitroxide spin labels can be used to determine the global folds of small to medium proteins in the absence of extensive NOE data (Battiste and Wagner, 2000; Gaponenko *et al.*, 2000). This review will focus on aspects of the structure determination process using nitroxide spin labels from sample preparation to modeling considerations, as well as discuss issues necessary for extending this technique to very large molecular weight systems.

2.   SAMPLE PREPARATION

2.1   Design of Mutation Sites

The core requirement for utilizing site-directed spin labels in quantitative NMR analysis is to have selective modification at a unique site in the protein. The most straightforward way to accomplish the modification is to have single-cysteine mutations of the protein, where a spin label can be attached with very selective chemistry via a disulfide bond (see section 2.2). For a protein to be amenable to spin-label studies, it must not have any critical cysteines that stabilize the fold of the protein. In general, this does not severely limit the range of proteins that can be studied, as has been demonstrated by the large number of proteins that have been characterized by site-directed spin labeling and EPR spectroscopy (similarly for many fluorescently labeled proteins). The most common occurrence of structurally important cysteines are those involved in metal ligation. In this case substitution with paramagnetic metal ions (if the native metal is not paramagnetic) would be a more logical choice for quantitative analysis of paramagnetic broadening effects (as well as contact shifts). Cysteines are also structurally important in the disulfide bonds of many extracellular proteins. The presence of disulfide bonds is problematic for the spin-label technique, although it may be possible to develop conditions on a case by case basis for attachment of the spin label without rearrangement of the core disulfides. Another common occurrence of conserved cysteines are those important for catalysis in an active site. Substitution with serine will destroy the catalytic activity of the protein, but it may retain a very similar structure. In this case, the use of catalytic activity as a check of mutations having little effect on structure will be lost, but other means may be used to evaluate structural conservation (see section 3.1). Lastly, non-conserved cysteines can be typically substituted by alanine (buried) or serine (exposed) with very little effect on the structure or thermodynamic stability of the protein. Even in cases where some cysteines cannot be mutated out, the protein can still be utilized for spin label studies as long as the important cysteines are buried in the hydrophobic core of the protein or a solvent occluded active site where they would not be significantly reactive compared to the desired site on the surface. Further purification and/or mass spectrometry would be required in these cases to ensure modification of the protein at a unique site.

Once it has been demonstrated that wild-type cysteines can be removed with minimal perturbation of the wild-type structure, sites on the protein need to be chosen for introduction of the nitroxide. The published results for using spin-labeled proteins to determine global folds were done on model systems of known three-dimensional structure, which greatly simplified choosing suitable places to introduce the spin label (Battiste and Wagner, 2000; Gaponenko *et al.*, 2000). The following text outlines a protocol for introducing spin-labels in the absence of any prior structural information. The most logical progression is to assign the backbone of a perdeuterated $^{13}C/^{15}N$-labeled protein, with all necessary disulfides mutated, in order to determine the secondary structure from analysis of the backbone chemical shifts (CSI or TALOS) and/or NOEs from 3D or 4D ^{15}N-NOESY-HSQC experiments. From the secondary structure and hydrophobicity patterns, sites on the surface of the protein that do not cause large structural perturbation upon cysteine mutation should be chosen with a very high degree of success. Conservation patterns among related proteins (if known) may also be helpful, since non-conserved hydrophilic positions are ideal choices, as they are most likely to be on the surface of the protein with the side-chain extended into solvent. Hydrophilic loops are a good first choice; however, surfaces of helices may also be good sites, since it has been shown that the nitroxide side chain packs very well in this secondary structure element and does not negatively effect helix stability (Bolin *et al.*, 1998) Ampiphatic patterns of the helices may be used to deduce surface exposed residues. The number of spin labeled samples should probably be at least 1 per 30-40 amino acids in the protein sequence (i.e. $\sim$ 5 - 7 spin-labeled samples for a 200 amino acid protein).

The choice of mutation sites for bimolecular complexes is guided by an empirical approach based on chemical shift perturbations induced upon complex formation. Residues with substantial chemical shift changes in complex relative to the free protein are likely to be at or very near the complex interface. Similarly, results of alanine scanning mutagenesis and subsequent characterization of complex formation may reveal key surface residues. These approaches may be carried out in the absence of structural information though a structure of at least one component of the complex can facilitate interpretation. In the latter case, identification of conserved surface exposed hydrophobic residues in orthologous proteins is useful. Ultimately, mutation sites are chosen so as to be within broadening range of the complex binding partner while keeping structural and binding perturbations at a minimum. A specific example of this approach (eIF4E/eIF4G-4EBD) is discussed in Section 5.

2.2   Covalent Modification of Protein

For EPR or NMR studies, spin labels need to be introduced at a specific site in the protein. Historically, spin labels were attached via alkylation of basic side chains, such as cysteine or lysine via halogenated reactants. The specificity of these reactions was usually low and further purification was necessary with careful analysis to determine the reaction site. The field was greatly advanced with the development of methanethiosulfonate reagents (Figure 1) that specifically react with cysteine residues to form a disulfide bond between the protein and spin label (Berliner *et al.*, 1982).

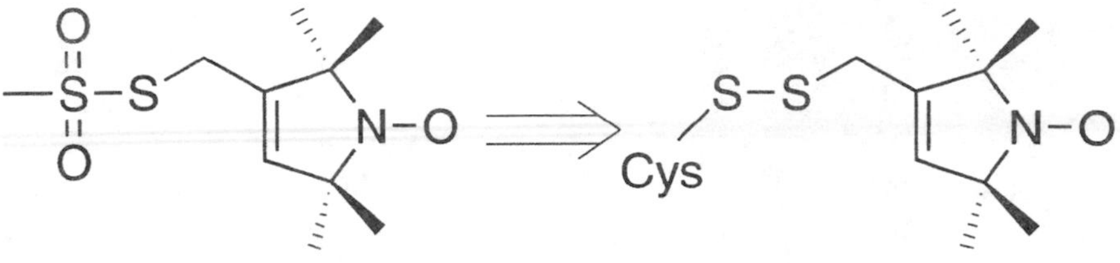

Figure 1. Chemical structure of the spin label reagent (1-oxyl-2,2,5,5-tetramethyl-Δ^3-pyrroline-3-methyl) methanethiosulfonate (MTSL), and the reaction to form a disulfide bond with cysteine.

This reagent reacts nearly quantitatively with solvent exposed cysteines in <12 hours at room temperature. There are comparatively few cysteines in proteins, and it is relatively easy to engineer single cysteines into a protein to specifically label it with minimal subsequent purification necessary. The systematic approach of introducing spin labels to multiple positions across a protein for analysis of structure and dynamics ("site-directed spin labeling") was popularized by Hubbell and co-workers (Hubbell and Altenbach, 1994; Hubbell *et al.*, 1996). In order to use spin labels for NMR studies it is necessary to have >95-99% of the protein sample containing the spin label. Therefore, it is necessary to remove unlabeled protein using either an organo-mercury column that binds free thiols with high affinity or some other chromatographic technique.

For protein complexes it is recommended that the spin label reaction be carried out prior to complex formation, if possible. In this case, ill-chosen placement of cysteine, and therefore MTSL, would be obvious from chemical shift perturbations in the reduced spectrum of the complex. If the reaction is carried out on the intact complex, there is the risk that an SH group buried within or near the interface would be obstructed. In this case formation of the MTSL-protein adduct would be inhibited and one might incorrectly interpret the absence of broadening. Fortunately, the success of the spin label reaction can be confirmed by mass spectrometry or EPR spectroscopy. Moreover, in the latter scenario useful information about the binding surface is obtained if failure of the spin-label reaction is confirmed.

2.3   Intermolecular Relaxation Effects/Aggregation

For monomeric systems, the desired result is to have relaxation of nuclear spins caused only by intramolecular interactions with the nixtroxide; however intermolecular relaxation effects are possible if the protein concentration is too high or aggregation occurs. Intermolecular broadening affects were observed for the protein eIF4E on a hydrophobic surface encapsulated by a CHAPS micelle (Battiste and Wagner, 2000). The hydrophobic spin label probably non-specifically aggregated on this surface. Except at high protein concentrations, the effect was localized to a single side chain tryptophan NH resonance. Intermolecular effects from aggregation will be heavily time averaged so that only very short distances will have observable broadening. Therefore, it will produce an isolated broadening effect, most likely on a surface side-chain NH. It should be noted that the spin label reaction is carried out followed by extensive dialysis and that CHAPS was added after concentrating the sample for NMR; therefore, manifestation of intermolecular effects is possible in other systems. Suspected intermolecular broadening effects can be easily tested by varying the protein concentration and monitoring either NMR or EPR signals. If variance as a function of concentration is observed, broadening can be extrapolated to zero protein concentration. However, it is probably more prudent not to use these distances as modeling restraints. In general, it is probably preferable to not use the high concentrations typically employed for sequence assignment and/or NOE analysis (at least 1 mM). On high field instruments, suitable ^{15}N-HSQC spectra can be obtained on samples in the 100-400 μM range where aggregation will be minimized.

3.   NMR ANALYSIS

3.1   Assignment/Structural Perturbation

Chemical shift assignments of the NH protons in each spin labeled sample are necessary for both obtaining distances and to evaluate the extent of structural perturbation that might be caused by the mutation to cysteine and/or addition of the nitroxide side chain. Rigorous assignments using ^{13}C/^{15}N-labeled protein are only essential for the wild-type protein sample, as well as the protein with all natural cysteines removed that is the precursor for introduction of cysteines for nitroxide attachment. Each individual spin label sample can be assigned mostly by comparison to the wild-type ^{15}N-HSQC spectra. 3D TOCSY-HSQC or NOESY-HSQC experiments can be used to verify assignments, although these will be less useful with larger proteins, especially if perdeuterated. The ease of assignment by this method is in itself a check of the lack of structural perturbation caused by the spin label. The only large changes that should be present will be within 2-3 amino acids of the site of attachment, as might be expected for primary sequence effects. It is not necessary to spend effort assigning these peaks close in primary sequence, since quantitation of their broadening would not provide any useful structural restraints. Minor effects may be observed for positions that are relatively close in tertiary space to the spin label side chain. It is difficult to quantitate how much chemical shift perturbation is permissible to assume a lack of "significant" perturbation. Certainly, the majority of the backbone amides should have virtually identical chemical shifts, and individual judgement will have to be used in situations where this is not the case. When the ultimate goal is to obtain a "global fold" for the protein, a moderate amount of structural alteration for loops where the nitroxide is introduced would not have an impact on the success of the technique anyway. In ambiguous situations, molecular modeling may be a better way to ascertain whether the structure is altered by the spin label, since gross errors should result in structures that are not able to satisfy all restraints. Numerous site-directed mutagenesis studies of proteins have shown that single point mutations rarely change the overall global configuration or lead to complete denaturation (Matthews, 1995; Matthews, 1993). It should not go unmentioned that of course functional assays for the protein of interest would provide clear evidence that the global fold of the protein is unchanged by mutation. However, the inverse is not necessarily true, since it is possible to disrupt the function of a protein (i.e. catalytic activity) without significantly changing its structure.

3.2   Quantitation of Paramagnetic Relaxation Enhancements

The theoretical description of relaxation of nuclear spins by paramagnetic electrons is described by the Solomon and Bloembergen equations (Solomon, 1955). A complete description of the relevant terms, simplifications and assumptions for use of these equations in analysis of spin-labeled biomolecules has been presented in previous reviews (Kosen, 1989; Krugh, 1976). The final equations for nuclear transverse and longitudinal relaxation caused by paramagnetic spins ($R2^p$ and $R1^p$) are

$$R2^p = \frac{K}{r^6}(4\tau_c + \frac{3\tau_c}{1+\omega_h^2\tau_c^2}) \text{ and } R1^p = \frac{2K}{r^6}(\frac{3\tau_c}{1+\omega_h^2\tau_c^2}) \tag{1}$$

where r is the distance between the electron and nuclear spins, τ_c is the correlation time for the electron-nuclear interaction, ω_h is the Larmor frequency of the nuclear spin (proton), K is a set of constants equal to 1.23 x 10^{-32} cm^6-sec^{-2} (Kosen, 1989). Rigid-body isotropic rotational diffusion has been assumed. The correlation time is a function of both the isotropic correlation time (τ_r) and the electronic longitudinal relaxation time (τ_s).

$$\frac{1}{\tau_c} = \frac{1}{\tau_s} + \frac{1}{\tau_r} \qquad (2)$$

Nitroxides usually have τ_s values of greater than 10^{-7} s, such that $\tau_c \approx \tau_r$ for small to medium sized proteins where $\tau_r \sim 10^{-9}$-10^{-8} s.

It is impractical to directly determine the relaxation parameters R2^p and R1^p from a NMR experiment on a single sample, since the total relaxation is a sum of the intrinsic diamagnetic nuclear relaxation and paramagnetic induced relaxation. The traditional approach is to acquire spectra on the paramagnetic or "oxidized" spin-label sample as well as after reduction of the sample to a diamagnetic form. Assuming that the diamagnetic relaxation rates are the same in the two states, the paramagnetic relaxation terms can calculated as a difference between the two spectra. There are several experimental methods to determine relaxation parameters, which are a detailed in the following sections.

3.2.1   ΔT1/T2 Relaxation Rates

The most straight forward approach is to directly measure T1 or T2 relaxation rates from inversion recovery or spin-echo 1D pulse sequences. However, this has the disadvantage for biomolecules of not being able to resolve the majority of the resonances in the sample. For global fold determination of larger ^{15}N-labeled proteins, ^{15}N-HSQC based ^{1}H-relaxation experiments can be performed for the backbone and side chain amides in the protein. This approach has the obvious advantage of increased resolution and application to large proteins. The first report to use ^{15}N-labeled proteins with spin-labels was on the unfolded state of staphylococcal nuclease (Gillespie and Shortle, 1997). The ^{15}N-label was essential to resolve amide protons that are poorly dispersed in ^{1}H for denatured proteins. By measuring both T2 and T1 values, correlation times can be calculated for each amide from equation 1 and used to calculate distances. Paramagnetic broadening of amide protons far removed in primary sequence from the site of spin-label attachment identified that there was residual long range structure present in the denatured state. Molecular modeling using these restraints showed that the "denatured" protein might be in a partially collapsed state rather than a complete random coil. For experiments on ^{15}N-labeled barnase Rosevear and co-workers used only T1 measurements for determining distances (Gaponenko et al., 2000). Due to slower relaxation times, 2D-measurements of T1 are much more sensitive than T2, which may be an important issue for very large proteins where T2 experiments will be insensitive and difficult to quantitate even with perdeuteration of the carbon-bound protons. Correlation times for the individual electron-nuclear interactions can still be calculated even if only T1 is measured by running the experiments at different field strengths (Gaponenko et al., 2000; Weber, Mullen and Mildvan, 1991). However, diamagnetic T1 relaxation in macromolecules is generally multiexponential which complicates precise determination of the paramagnetic contribution (Solomon, 1955). In contrast, this is not as serious an issue for T2.

3.2.2   Intensity Ratios

Prior to utilizing isotopic labeling for measurement of T2 and T1 relaxation times directly, indirect methods for extraction of relaxation parameters from 1D difference spectra (Campbell et al., 1973; Schmidt and Kuntz, 1984) or homonuclear two-dimensional experiments (Girvin and Fillingame, 1994; Kosen, 1989; Kosen et al., 1986) were developed. The basic premise is that for Lorentzian line shapes the difference in T2 relaxation can be easily related to the relative intensities or peak heights. It is also possible to iteratively apply exponential broadening to the reduced protein spectrum until the best fit with the oxidized spectrum is obtained.

Indirect methods of determining relaxation parameters based on peak heights can also be applied to ^{15}N-HSQC experiments. The advantage over direct measurement of relaxation parameters is significantly less spectrometer time and data analysis per sample. Higher throughput will be important for larger proteins where more spin-labels per protein will be required to determine the global fold. The first demonstration of using peak heights in HSQC spectra to determine T2^p relaxation was by Gillespie and Shortle (Gillespie and Shortle, 1997). In this study they also directly determined T2^p from 2D relaxation experiments and found that the two methods correlated well. Extraction of data from a heteronuclear experiment has some simplifications from a 2D homonuclear experiment, where the reduction in intensity can be from either proton in the cross peak (therefore only an average distance to two protons is obtained). However, because of the relatively long INEPT delays where ^{1}H magnetization is in the transverse plane, the reduction in cross peak intensity for the oxidized sample has to be modeled considering relaxation during both the acquisition period and

INEPT delays.   Due to ^{15}N's 10-fold lower gyromagnetic ratio, ^{15}N $T2^p$ relaxation can be considered negligible compared to ^{1}H given that the atoms will always be at a similar distance from the spin label.   In addition, ^{1}H $T1^p$ relaxation is considered negligible compared to $T2^p$.   Under these conditions, $R2^p$ ($1/T2^p$) is extracted from the intensity ratio as:

$$\frac{I_{ox}}{I_{red}} = \frac{R2\exp(-R2^p\Delta)}{R2 + R2^p} \tag{3}$$

where $R2$ is the diamagnetic proton transverse relaxation rate and Δ is the total INEPT delay ($\Delta \approx 1/J_{NH}$). With this value of $R2^p$, the electron-nuclear distance is obtained through equation 1.

Figure 2 shows idealized $R2^p$ enhancements expected from observed intensity ratios in HSQC spectra for various initial linewidths and correlation times.

Battiste and Wagner (2000) applied this methodology to five spin-labeled samples of eIF4E, adding two dimensional peak fitting to deconvolute partially overlapping peaks.   An example of the NMR spectra for the oxidized and reduce spectra are shown in Figure 3.   In this case, the high resolution structure was known, and there was a good correlation between expected and observed distances.

Figure 2 (overleaf).   Curves of equations used for conversion of intensity ratios into distances for modeling calculations. (A) R2 spin enhancement ($R2^p$) due to spin label as a function of intensity ratio in ^{15}N-HSQC with total INEPT delays of 9 ms. The $R2^p$ for a particular intensity ratio is also dependent upon the intrinsic linewidth of the amide in the absence of the paramagnetic spin (reduced sample). Curves are shown for two different intrinsic half-height linewidths (Δv (Hz) = $\pi/R2$) of 20 Hz (☐) and 40 Hz (○). (B) Conversion of $R2^p$ into distances. The distance for a particular $R2^p$ is also dependent upon the correlation time for the interaction. Curves are shown for two different correlation times of 16 ns (☐) and 30 ns (○). 16 ns is the global correlation time for eIF4E in 25 mM CHAPS that is used for this study. An intensity ratio cutoff of 0.85 (for definitive presence of a broadening effect) is indicated by a dashed line for the 16 ns curve. Value of cutoff is arbitrary and may vary for different systems. (C) The relationship between intensity ratio in HSQC and calculated distance for two correlation times (both assuming an intrinsic R2 of 20 Hz).

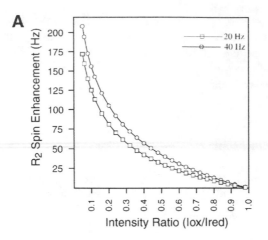

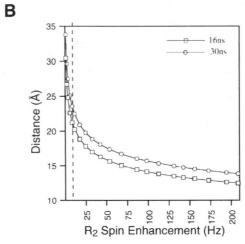

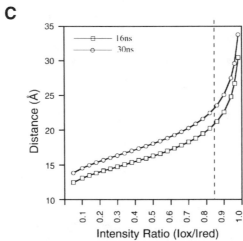

4.   RESTRAINTS AND MOLECULAR MODELING

4.1   Conversion of Relaxation Enhancements into Distances

Once relaxation rates have been determined, the only unknowns in equation 1 are the distance and the correlation time for the interaction. As was mentioned above, the correlation times can be determined from either measurement of both T2 and T1, or by measuring one relaxation rate at multiple fields. The other alternative is to assume that the global isotropic correlation time is a good approximation for the individual electron-nuclear interactions. Gaponenko report electron-nuclear correlation times varying over a tight range from 0.6 to 1.8 ns (Gaponenko *et al.*, 2000). Gillespie and Shortle (1997) reported a wider range of correlation times from 1 to 15 ns, but mostly attributed this to compounding of the error in the measured relaxation parameters by comparing results obtained using either $T2^P$ or $T1^P$ at 500 and 600 MHz. In principle, measuring the individual correlation times may improve the accuracy of the calculated distances; however it is not clear whether it is worth the effort, since a factor of 2 error in τ_c only results in ~10% error in the distance. Other factors such as minor structural perturbations and r^{-6} averaging from flexible spin label side chains probably warrant against "tight" or precise bounds during modeling calculations anyway (see section 4.1). A good analogy may be the approach

---

Figure 3 (overleaf).   Regions of ^{15}N-HSQC spectra of spin-labeled eIF4E at 750 MHz. (A-D) Spin-label at residue 120, and (E-H) spin-label at residue 200. Panels (A,B,E,F) are oxidized or paramagnetic samples with a free radical on the nitroxide group. Panels (C,D,G,H) are reduced or diamagnetic samples taken after reduction of the spin label with ascorbic acid. Resonance assignments are given in the reduced spectra. The suffixes ε and δ1/δ2 indicate side chain HN groups of tyrptophan and asparagine residues, respectively. Numbers in oxidized spectra are values for intensity ratios (I_{ox}/I_{red}) that are less than 0.85. Boxes indicate undetectable crosspeaks with an upper limit for the intensity ratio estimated from the noise in the oxidized spectrum.

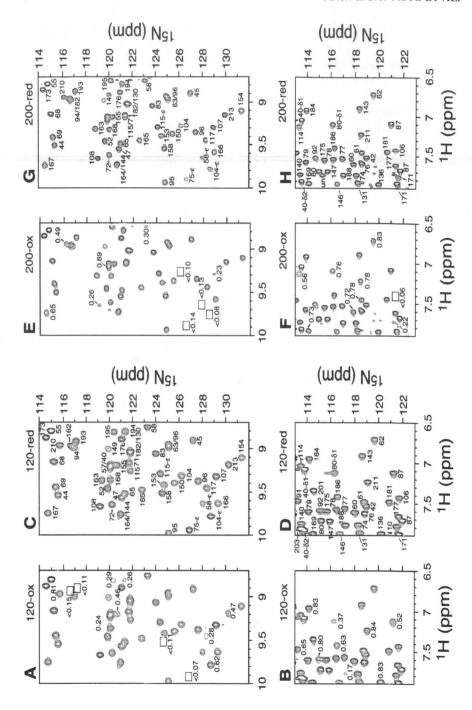

adopted for using NOE restraints in modeling calculations, where qualitative distances are favored over precise ones because of many complicating factors. A large number of qualitative restraints still does a good job in computationally converging the structures into a narrow range of possibilities. There is no reason to think that the same approach should not hold true with spin-label restraints given that a large number of restrains are obtained from many spin label samples. Battiste and Wagner found that tightening the bounds on paramagnetic broadening effects (PBEs) from ± 4,0 Å to 3.0 Å decreased the rmsd's of calculations with limited NOE data, yet increased the rmsd's in reference calculations including all NOEs from the high resolution structure (also resulted in more and larger PBE violations). Thus, in highly underdetermined systems, care must be taken not to bias the calculations with overly precise bounds, since rmsd's and number of violations may not be good indicators of structural quality.

In order to increase the number of qualitative restraints for modeling calculations, some groups have used "negative" restraints for protons that did not have detectable broadening within experimental error (Battiste and Wagner, 2000; Girvin and Fillingame, 1994). Unlike NOEs, there are not any obvious mechanisms that would lead to a proton not having observable broadening if its average distance were close enough to the spin label. For these protons a lower limit on the distance can be calculated and the proton can be modeled to be any distance greater than the lower limit. While these are not "high-value" restraints, since they are easily satisfied, the majority of protons in a protein fall into this class and a very large number of restraints can be obtained (~2/3 of amide protons for the 25 kDa protein eIF4E; a much higher percentage would be expected for larger proteins). Modeling calculations with and without these restraints showed that they contributed significantly to lowering rmsd values (by about 0.5-1.0 Å) (Battiste and Wagner, unpublished results), without having any negative impact on the accuracy or precision of the reference calculations with the full NOE data set.

4.2   NOE Data for Larger Proteins

The intent of the recent work on using nitroxide spin labels for structure determination is to ultimately obtain global folds for large proteins where NOE data are difficult to obtain. However, it should be possible to obtain high quality data and many useful NH-NH NOE restraints from 3D NOESY-HSQC and 4D HSQC-NOESY-HSQC experiments on ^{15}N-labeled proteins up to very high molecular weights by using perdeuteration (Grzesiek et al., 1995) and/or transverse relaxation enhanced (TROSY) type experiments (Pervushin et al., 1997; Pervushin et al., 1999). Some work has gone into obtaining global folds using only this limited NH-NH NOE data (Venters *et al.*, 1995); however, it cannot reliably determine global folds, particularly for proteins with high α-helical content. Recent methods have selectively protonated methyl or aromatic groups in otherwise highly deuterated proteins to obtain additional NOE data for hydrophobic protons in the core of the protein (Gardner *et al.*, 1997; Metzler *et al.*, 1996; Rosen *et al.*, 1996; Smith *et al.*, 1996). This technique shows great promise for determining global folds of large proteins; however, it is not clear how high in molecular weight it will be possible to obtain sufficient restraints. Regardless, addition of spin label distances to these limited NOE data sets should be complimentary and increase the precision and accuracy of the calculated structures. The long range affect of the spin label can define many structural features better than short range NOEs. Some hypothetical data has been generated to show that the combination of NH-Methyl/Methyl-Methyl NOEs and spin label PBEs improve the quality of structures calculated (Battiste and Wagner, 2000).

4.3   Dihedral Angle Restraints

Direct measurement of backbone dihedral angles in large proteins is difficult; however, indirect methods can qualitatively restrict angles to certain regions. Analysis of ^{13}C backbone chemical shifts (CSI) (Wishart and Sykes, 1994) and patterns of NH-NH NOEs can accurately deduce secondary structure elements, which can be restrained as $\pm 50\text{-}80°$ around average values for helices and beta-sheets. Furthermore, the program TALOS can be used to more precisely determine ϕ/ψ angles independent of secondary structure by comparing chemical shift of tri-peptide segments to a database (Cornilescu *et al.*, 1999). Both of these methods simply utilize backbone chemical shifts that will obtainable for any large protein that is sequentially assigned.

5.   EXAMPLES OF GLOBAL FOLD DETERMINATION

5.1   F1-ATPase domain

The first demonstration of using nitroxide spin labels for modeling calculations of a protein was the subunit c from F_1F_0 ATP synthase (Girvin and Fillingame, 1994). Subunit c is a small protein of ~80 amino acids that does not form a tight globular fold. As a result, there was a paucity of NOEs to determine higher resolution structures, and a single spin label at Asp61 was used to provide additional modeling restraints. Thus, while the aim of this study was not global folds of large proteins, it shares a common problem of lacking sufficient NOE data. Distances were obtained from analysis of 2D DQF-COSY spectra, where an iterative fitting method was used to determine the paramagnetic broadening contribution. The spin-label distances with NOEs showed that the subunit was forming an extended helical hairpin structure with extensive helical-helical contacts. The structures were calculated with 89 intrahelical NOEs, 15 interhelical NOEs, 13 interhelical PBEs, 12 intrahelical PBEs, and 29 "negative" PBEs to unbroadened protons. Rms deviations of ~1.0 Å for backbone atoms in the ordered helical segments were observed. It was not determined what were the rms deviations of calculations in the absence of either PBE or NOE data (if convergence were possible).

5.2   Barnase

Barnase is a 110 amino acid protein that was used as a model system for examining whether nitroxide spin labels could determine its global fold in the absence of NOE data (Gaponenko et al., 2000). The spin label was attached at two positions (15 and 102) and distances were determined from direct measurement of $T1^p$ relaxation times. Using this method distances up to a remarkable 35 Å were measured. An average of ~70 restraints out of a possible 100 were obtained per spin label. With the addition of secondary structure restraints the global fold was accurately determined with an rmsd of 2.2 Å (2.9 Å to crystal structure). Calculations with the secondary structure restraints alone were not able to determine the global fold with a rmsd of ~12 Å. Their results were promising in that they only used two spin label probes to obtain the global fold; however, a threading algorithm was necessary to eliminate mirror image structures. The authors attributed this to the low number of spin label reference points, yet similar problems were observed for the protein eIF4E with 5 spin labels (Battiste and Wagner, unpublished data) , and it may be a general problem of the technique. In this case, a small number of spin label restraint violations (mainly to side chain NHs) produced enough energetic differences for distinction; however, it may not always be as clear cut as desired. If ambiguities arise during initial structure calculations, it should be possible to design a single additional mutation to distinguish between the two structures .

5.3   eIF4E

eIF4E is a 213 amino acid protein that has been characterized by NMR in a CHAPS micelle of MW ~45-50 kDa (Matsuo et al., 1997). Thus it provides a good model system for larger proteins to which this global fold technique will hopefully be applied. Five spin labels were made and analyzed by an indirect intensity-ratio technique (Battiste and Wagner, 2000; Gillespie and Shortle, 1997). Distances up to 23 Å were measured. A total of 174 PBEs, 341 "negative" PBEs, 403 NH-NH NOEs (as expected for a large perdeuterated protein), and loose ϕ/ψ angle restraints for secondary structure elements from CSI analysis were able to correctly determine the global fold of eIF4E with an rmsd of 2.3 Å (3.2 Å relative to a reference NMR structure). The NH-NH NOEs and angle restraints were not able to determine the global fold properly (rmsd 8.3 Å, Figure 4).

A hypothetical modeling calculation with simulated NOE data for a highly deuterated methyl-protonated protein and PBE restraints produced reasonable quality structures that faithfully produced many long range structural features of the nucleotide-binding cleft.

5.4   U1A/PIE mRNA Complex

The structure of the 38 kDa U1A dimer bound to the PIE mRNA element provides another example of utilizing nitroxide spin labels for modeling restraints where NOEs fall short (Varani et al., 2000). The majority of the system was determined by NOE analysis; however, a critical protein-protein contact in the dimer was not defined by the NOE data. Qualitative distances determined from spin-labeling of this extended helix showed that they were forming a helix-helix interaction at the apex of the mRNA bulge. It is often noted that extended structures or large clefts between domains are ill-defined by short range NOE data. The long range effect of the nitroxide spin label is one technique to compliment NOE data for these difficult questions (as well as orientational data from dipolar couplings), and should find many applications.

Figure 4 (overleaf) Superpositions of backbone atoms for families of structures calculated with (A) full NOE data set, (B) reduced NOE data set (i.e. from perdeuterated protein), and (C) reduced NOE data set plus PBEs. The family of 20 low energy structures for each are shown in thin black lines, while the average structure of the "reference" calculation with the full NOE data set is shown by a thick black line in all parts.

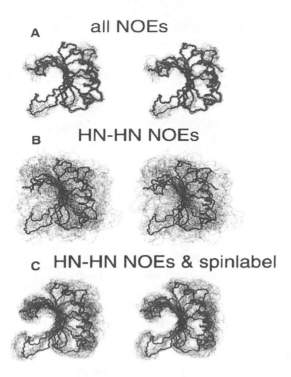

A   all NOEs

B   HN-HN NOEs

c   HN-HN NOEs & spinlabel

6.   FUTURE DIRECTIONS

An obvious extension of the SDSL method reviewed here is to furnish distance restraints across a bimolecular complex interface.   Unambiguous intermolecular NOEs may be obtained through half-filter NOESY experiments though the sensitivity of this approach is often insufficient for the application to large proteins (Ferentz and Wagner, 2000).   Isotope labeling schemes may be employed to obtain unambiguous intermolecular NOEs without recourse to the half-filter (Takahashi *et al*., 2000; Walters, Matsuo and Wagner, 1997) experiments. An alternative method would be to measure PBEs across a complex interface. Combination of RDCs, intermolecular PBEs and rigid body docking may allow determination of complex structures in the absence of intermolecular NOE data.   PBEs together with RDCs and/or Me-NH, Me-Me and NH-NH NOE data may facilitate global fold determination in cases where a single approach is insufficient. One promising application might be for protein-nucleic acid complexes, where many features of nucleic acid structure are often underdetermined by NMR data. Similar line broadening analysis as was described for many proteins above was applied to a paramagnetic platinum adduct of DNA, and the long range distances helped define a characteristic bend to the DNA structure (Dunham *et al.*, 1998).

In our laboratory, we have observed paramagnetic broadening across the interface of a large (effective MW ~ 50 kDa) protein complex between the cap binding protein eIF4E and a domain of the adapter protein eIF4G (eIF4G-4EBD) suspended in a CHAPS micelle; the results so far are promising. As mentioned previously , five cysteine mutants of eIF4E were constructed based on the principles outlined in Section 2.1. Positioning of mutants on the dorsal surface of eIF4E was also guided by titrations with the 4E Binding Protein-1 (4EPB-1) (Fletcher *et al.*, 1998; Matsuo *et al.*, 1997), a regulatory protein that acts as a molecular mimic of eIF4G (Marcotrigiano *et al.*, 1999) in a conserved region of residues YxxxxΦL. Such studies have coarsely defined the 4EBP , and by extension, the eIF4G-4EBD binding surface on eIF4E. In addition, mutations of residues found on the dorsal surface of eIF4E abrogate binding to both 4EBPs and eIF4G and these data are consistent with the NMR titration data (Ptushkina *et al.*, 1998; Pyronnet *et al.*, 1999). Accordingly, mutation sites, such as R132C, were chosen to be near the complex interface. Depicted in Figure 5 are sections of the ^{15}N-HSQC of ^{15}N labeled eIF4G-4EBD in complex with the R132C MTSL adduct of eIF4E at natural abundance.

Additional residues, found in the interface of murine eIF4E in complex with a 17 amino acid fragment from eIF4GII are also broadened (Marcotrigiano *et al.*, 1999). It is noteworthy that R132 is near W75, a key interfacial residue defined by mutational studies and NMR titrations. In sum, intermolecular PBEs are observed in R132C, R120C and S169C mutants of eIF4E. The total of 46 PBEs and 144 "negative" intermolecular constraints have been obtained. Further structural characterization of this interesting system using Me/Me and Me/NH intramolecular NOEs is underway.

While using site-directed spin labeling for molecular modeling calculations is still in its infancy, it is exciting to think of its potential applications to high-throughput structural genomics projects. Relatively little spectrometer time and subsequent analysis is necessary per sample once the resonance assignments have been obtained for the unmodified protein. Also, a lower protein concentration is required than for NOE analysis, allowing a wider range of proteins to be analyzed. As with other aspects of structural genomics, the bottleneck may be in sample preparation, which for spin labeling requires additional cloning and molecular biology. If combined with RDCs, however, the number of spin label modifications required to determine the fold may be dramatically reduced. This would allow all structural information to be extracted from a few 2D correlation experiments while observations of PBEs in CRINEPT TROSY (Riek *et al.*, 1999) type experiments would allow studies of particle sizes in excess of 100 kDa.

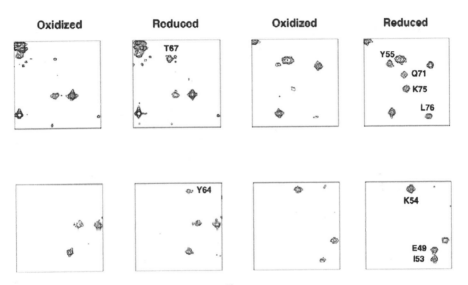

Figure 5. Selected spectral regions from ^{15}N HSQC spectrum of R132C MTSL labeled eIF4E (at natural abundance) in complex with ^{15}N labeled eIF4G-4EBD at 750 MHz. A conserved motif YxxxxΦL, corresponding to residues YGPTFLL of yeast eIF4G-4EBD and numbered 64-70, have been shown to be in the complex interface for murine eIF4E/eIF4GII peptide complex.

6.1   Acknowledgments

Support for this work was provided by NIH grants GM 47467, CA 68262 and RR 00995. J.D.G acknowledges support from an American Cancer Society postdoctoral fellowship.

7.   REFERENCES

Battiste, J. L. and Wagner, G., (2000), *Biochemistry* **39**: 5355-5365.
Berliner, L. J., Grunwald, J., Hankovszky, H. O., and Hideg, K., (1982), *Anal. Biochem.* **119**: 450-455.
Bertini, I., Couture, M. M. J., Donaire, A., Eltis, L. D., Felli, I. C., Luchinat, C., Piccioli, M., and Rosato, A., (1996), *Eur. J. Biochem.* **241**: 440-452.
Bertini, I., Donaire, A., Luchinat, C., and Rosato, A., (1997), *Proteins* **29**: 348-358.
Bertini, I., Rosato, A., and Turano, P., (1999), *Pure Appl. Chem.* **71**: 1717-1725.
Bolin, K. A., Hanson, P., Wright, S. J., *and* Millhauser, G. L., (1998), *J. Magn. Reson.* **131**: 248-253.
Campbell, I. D., Dobson, C. M., Williams, R. J. P., and Xavier, A. V., (1973), *J. Magn. Reson.* **11**: 172.
Clore, G. M., (2000), *Proc. Nat.l. Acad. Sci. U.S.A.* **97**: 9021-9025.
Cornilescu, G., Delaglio, F., *and* Bax, A., (1999), *J. Biomol. NMR* **13**: 289-302.
Delaglio, F., Kontaxis, G., and Bax, A., (2000), *J. Am. Chem. Soc.* **122**: 2142-2143.
Dunham, S. U., Dunham, S. U., Turner, C. J., and Lippard, S. J., (1998), *J. Amer. Chem. Soc.,* **120**: 5395-5406.
Ferentz, A. E., and Wagner, G., (2000), *Quart. Rev. Biophys.* **33**: 29-65.
Fischer, M. W. F., Losonczi, J. A., Weaver, J. L., and Prestegard, J. H., (1999), *Biochemistry* **38**: 9013-9022.
Fletcher, C. M., McGuire, A. M., Gingras, A. C., Li, H. J., Matsuo, H., Sonenberg, N., and Wagner, G., (1998), *Biochemistry* **37**: 9-15.
Gaponenko, V., Howarth, J. W., Columbus, L., Gasmi-Seabrook, G., Yuan, J., Hubbell, W. L., and Rosevear, P. R., (2000), *Protein Science* **9**: 302-309.
Gardner, K. H., and Kay, L. E., (1997a), *J. Am. Chem. Soc.* **119**: 7599-7600.
Gardner, K. H., and Kay, L. E., (1997b), *Curr. Opin. in Structural Biology* **7**: 722-731.
Gardner, K. H., Rosen, M. K. and Kay, L. E., (1997), *Biochemistry* **36**: 1389-1401.
Gillespie, J. R., and Shortle, D., (1997), *J. Mol. Biol.* **268**: 158-169.
Girvin, M. E., and Fillingame, R. H., (1994), *Biochemistry* **33**: 665-674.
Goto, N. K., Gardner, K. H., Mueller, G. A., Willis, R. C., and Kay, L. E., (1999), *J. Biomol. NMR* **13**: 369-374.
Grzesiek, S., Wingfield, P., Stahl, S., Kaufman, J. D., and Bax, A., (1995), *J. Amer. Chem., Soc.* **117**: 9594-9595.
Hubbell, W. L., and Altenbach, C., (1994), *Curr. Opin. Struct. Biol.* **4**: 566-573.
Hubbell, W. L., Mchaourab, H. S., Altenbach, C., and Lietzow, M. A., (1996), *Structure* **4**: 779-783.
Hus, J., Marion, D., and Blackledge, M., (2000), *J. Mol. Biol.* **298**: 927-36.
Hus, J.-C., Marion, D., and Blackledge, M., (2001), *J. Am. Chem. Soc.* **123**: 1541-1542.
Kosen, P. A., (1989), *Meth. Enzymol.* **177**: 86-121.
Kosen, P. A., Scheek, R. M., Naderi, H., Basus, V. J., Manogaran, S., Schmidt, P. G., Oppenheimer, N. J., and Kuntz, I. D., (1986), *Biochemistry* **25**: 2356-2364.
Krugh, T. R., (1976), In *Spin Labeling: Theory and Applications* (ed. L. J. Berliner), pp. 339-372. Academic Press, New York.
Marcotrigiano, J., Gingras, A.-C., Sonenburg, N., and Burley, S. K., (1999), *Mol. Cell* **3**: 707-716.
Matsuo, H., Li, H., McGuire, A. M., Fletcher, C. M., Gingras, A.-C., Sonenberg, N., and Wagner, G., (1997), *Nat. Struct. Biol.* **4**: 717-724.
Matthews, B., (1995), *Adv. Protein Chem.* **46**: 249-278.
Matthews, B. W., (1993), *Ann. Rev. Biochem.* **62**: 139-160.

Meiler, J., Peti, W., and Griesinger, C., (2000), *J. Biomol. NMR* **17**: 283-294.

Metzler, W. J., Witteking, M., Goldfarb, V., Mueller, L., and Farmer, B. T., (1996), *J. Amer. Chem. Soc.* **118**: 6800-6801.

Mueller, G. A., Choy, W. Y., Yang, D., Forman-Kay, J. D., Venters, R. A., and Kay, L. E., (2000), *J. Mol. Biol.* **300**: 197-212.

Pervushin, K., Riek, R., Wider, G., and Wüthrich, K., (1997), *Proc. Nat'l. Acad. Sci. USA* **94**: 12366-71.

Pervushin, K., Wider, G., Riek, R., and Wüthrich, K., (1999), *Proc. Nat'l. Acad. Sci. USA* **96**: 9607-9612.

Ptushkina, M., Haar, T. V. D., Vasilescu, S., Birkenhager, R., and McCarthy, J. E., (1998), *EMBO J* **17**: 4798-4808.

Pyronnet, S., Imataka, H., Gingras, A.-C., Fukunaga, R., Hunter, T., and Sonenberg, N., (1999), *EMBO J.* **18**: 270-279.

Riek, R., Wider, G., Pervushin, K., and Wüthrich, K., (1999), *Proc. Natl. Acad. Sci. USA* **96**: 4918-4923.

Rosen, M. K., Gardner, K. H., Willis, R. C., Parris, W. E., Pawson, T., and Kay, L. E., (1996), *J. Mol. Biol.* **263**: 627-636.

Salzmann, M., Pervushin, K., Wider, G., Senn, H.,and Wüthrich, K., (1998), *Proc. Natl. Acad. Sci. USA* **95**: 13585-13590.

Salzmann, M., Wider, G., Pervushin, K., Senn, H., and Wüthrich, K., (1999), *J. Am. Chem. Soc.* **121**: 844-848.

Schmidt, P. G., and Kuntz, I. D., (1984),*Biochemistry* **23**: 4261-4266.

Skrynnikov, N. R., Goto, N. K., Yang, D., Choy, W.-Y., Tolman, J. R., Mueller, G. A., and Kay, L. E., (2000), *J Mol Biol.* **295**:1265-73.

Smith, B. O., Ito, Y., Raine, A., Teichmann, S., Ben-Tovim, L., Nietlispach, D., Broadhurst, R. W., Terada, T., Kelly, M., Oschkinat, H., Shibata, T., Yokoyama, S., and Laue, E. D., (1996), *J. Biomol. NMR* ., **8**: 360-368.

Solomon, I., (1955), *Phy. Rev.* **99**: 559-565.

Takahashi, H., Nakanishi, T., Kami, K., Arata, Y., and Shimada, I., (2000), *Nature Struct. Biol.* **7**: 220-223.

Tjandra, N., and Bax, A., (1997), *Science* **278**: 1111-1114.

Tjandra, N., Omichinski, J. G., Gronenborn, A. M., Clore, G. M., and Bax, A., (1997), *Nature Struct. Biol.* **4**: 732-728.

Tolman, J. R., Flanagan, J. M., Kennedy, M. A., and Prestegard, J. H., (1995), *Proc. Nat. Acad. Sci. USA* **92**: 9279-9283.

Varani, L., Gunderson, S. I., Mattaj, I. W., Kay, L. E., Neuhaus, D., and Varani, G., (2000), *Nature Struct. Biol.* **7**: 329-35.

Venters, R. A., Metzler, W. J., SPicer, L. D., Mueller, L., and Farmer, B. T., (1995), *J. Amer. Chem. Soc.* **117**: 9592-9593.

Walters, K., Matsuo, H., and Wagner, G., (1997), *J. Am. Chem. Soc.* **119**: 5958-5959.

Weber, D. J., Mullen, G. P., and Mildvan, A. S., (1991), *Biochemistry* **30**: 7425-7427.

Wishart, D. S., and Sykes, B. D., (1994), *J. Biomol. NMR* **4**: 171-180.

Yang, D., and Kay, L. E., (1999), *J. Am. Chem. Soc.* **121**: 2571-2575.

Chapter 5

Solid State NMR Studies of Uniformly Isotopically Enriched Proteins

Ann McDermott
Columbia University, Department of Chemistry, New York, NY

Abstract   Solid state NMR experiments for structure determination and functional characterization of uniformly ^{13}C, ^{15}N isotopically enriched proteins have undergone major advances recently. Several studies demonstrate that extensive site-specific assignments can be established for domains of 50-100 residues. For these studies, high-field instrumentation and recently developed ^{13}C-^{13}C and ^{13}C-^{15}N correlation experiments were combined to provide intra-residue and inter-residue contacts. Experimental details that influence the line-shape and resolution are discussed in this review. The assigned spectra are immediately useful for assessing secondary structure, ligand binding and conformational dynamics. Promising experiments for obtaining torsional structural constraints and tertiary contacts in uniformly labelled materials have been discussed in the recent literature. The challenge for the upcoming years will be to extend these efforts to biologically interesting targets such as intrinsic membrane proteins.

1.   SSNMR: A CHEMICALLY DETAILED PROBE OF STRUCTURE AND FUNCTION

Many mechanistic biochemical investigations remain the unique domain of nuclear magnetic resonance. NMR has the ability to characterize chemical mechanism, dynamics and structure in an integrated fashion, thus providing unique links between structure and function. NMR provides extraordinarily efficient screens (Hajduk *et al.*, 1999) and detailed characterization (Mulder *et al.*, 2000) for ligand binding to proteins and is the chief method for in-situ studies of chemically activated enzymatic intermediates (McDowell *et al.*, 1996; Studelska *et al.*, 1997) and conformationally distorted ligands on protein receptors (Feng *et al.*, 1996a). NMR remains the method of choice for studying hydrogen bonds, including enzyme active sites (Gu *et al.*, 1999; Wang *et al.*, 1996) and proton pumps (Griffiths *et al.*, 2000). NMR has allowed detailed characterization of partially structured, folded or assembled systems (Dyson and Wright, 1998; Wright and Dyson, 1999) and provides elegant methods to probe the timescales and extent of conformational exchange in proteins (Hill *et al.*, 2000; Lee and Wand, 2001) including enzyme active sites (Ishima *et al.*, 1999; Rozovsky and McDermott, 2001). NMR is arguably our strongest tool for studying chemically detailed mechanisms.

Solid-like or insoluble protein complexes are common in the crowded interior of the cell. It is well known that typical cellular conditions are very concentrated in proteins (>300mg/ml), and the typical protein encounters many neighbors in an extremely crowded and hydrated solid-state-like environment. It is also well known that many enzymes retain their function in the crystal, and in other solid state formulations; microcrystals are even often highly stabilized and enhanced in activity relative to dilute solution. In addition, many proteins lose function if dissolved and separated from their cellular partners (Bray, 1998; Minton, 2000). A protein's insolubility can make it a difficult target for structural and mechanistic studies. Increasingly complex systems are tackled structurally by X-ray crystallography (Ramakrishnan and Moore, 2001), and solution NMR has undergone extraordinary growth with respect to the methods and specialized techniques for studies of large and complex proteins (Gardner and Kay, 1998; Prestegard *et al.*, 2000; Riek *et al.*, 2000; Wider and Wüthrich, 1999) and important applications for membrane proteins are forthcoming (Fernandez *et al.*, 2001; Rastogi and Girvin, 1999). Nonetheless, many important proteins are not studied by X-ray or solution NMR methods because of size or insolubility. For example, few structures of membrane proteins have been solved (Sakai and Tsukihara, 1998; von Heijne, 1999), despite their crucial importance in cell biology and as drug targets, and their high abundance in the genome. The slow progress is mainly because these systems require specialized and solid-like environments; solubilization and crystallization remain exceedingly difficult. Solid-like systems, i.e. proteins for which the whole-body hydrodynamic tumbling is insignificant on the NMR timescale, can be studied by high resolution solid state NMR, formulated as raw membranes or native multienzyme complexes, micelle-protein suspensions, frozen glasses, or hydrated microcrystals.

Until recently, solid state NMR studies have typically involved site-specific isotopic enrichment at strategically selected sites (Schaefer, 1999).

The use of engineered isotopes obviates the need for peak assignments. These experiments often test a specific hypothesis, involving a particular set of atoms of interest in the macromolecule. These methods are useful for focusing on specific ligands or domains in very large structures, for example in the case of taxol conformational studies on tubulin (Li *et al.*, 2000), or the case of chemotaxis receptors (Murphy *et al.*, 2001). The strength in many of these studies is the unique ability to probe a system under highly biologically relevant conditions; the weakness is the inefficient rate at which structural information is gathered, and the lack of information on unlabelled sites. High-resolution structures of macromolecules, solved without prior detailed models, rely on the availability of many structural constraints per residue, which is not practically possible with site-specific labeling protocols. Recently, efforts to study uniformly or extensively isotopically enriched biological solids and assign the spectra have met with an encouraging level of success in several laboratories, and it is clear that many of the pulse sequences originally developed for isolated pairs of spins can give good performance in uniformly labeled materials with minor modifications (Griffin, 1998). The implication of these preliminary reports is that the broad range of samples typically studied with solid state NMR can now be studied inclusively and efficiently. The use of uniform enrichment represents an important change in experimental planning. This review will attempt to offer an overview of the current protocols for studying uniformly labeled samples by solids NMR, from sample preparations, and benchmark linewidths to data reduction and chemical and dynamic characterization, with an eye to the prospects for studying interesting biological targets and detailed mechanisms.

1.1   Linewidth and Resolution

Multidimensional NMR of uniformly labeled solid proteins has been little pursued until a few years ago mainly because of pessimism about sensitivity and resolution. Although the discipline is at an early stage, there is much that is now understood about how to achieve well resolved spectra, and the prospects are better than originally expected. Solid state NMR experiments typically involve detection of "heteronuclei": ^{13}C or ^{15}N spectra are usually recorded. Proton NMR of solids exhibit poor spectral resolution (typically 0.5-2.0 ppm linewidths) due to the complex network of proton couplings. Proton detected SSNMR (Bielecki *et al.*, 1990; Ishii *et al.*, 2001; van Rossum *et al.*, 1997) of biological solids has had some promising discussion in the recent literature but has not yet been extensively utilized in the context of spectral assignment. Because of the limited sensitivity of the detection of low-gyromagnetic ratio nuclei, experiments ideally involve 0.1 to 5 micromoles of isotopically enriched material, which can be made economically by over-expression in *E. coli*. The linewidths of ^{13}C-^{15}N and ^{13}C-^{13}C spectra will be discussed in this section.

The sample's physical formulation is known to have a significant effect upon the inhomogeneous linewidths. Lyophilized materials generally exhibit linewidths in excess of 2 ppm for both ^{13}C and ^{15}N, and the detected linewidth is apparently limited by structural inhomogeneities (Pauli *et al.*, 2000). Microcrystalline proteins often exhibit much narrower linewidths for both nuclei, and the linewidths are limited by spin couplings. Rehydrated or lyoprotected lyophilized materials, frozen solutions and precipitated materials have been compared, and can have linewidths intermediate between these two extremes (Straus *et al.*, 1998).

Spin couplings have an important effect on the linewidths of uniformly enriched materials. While protons are typically decoupled during direct and indirect detection of ^{13}C and ^{15}N, the homonuclear and ^{13}C-^{15}N couplings have not yet been adequately addressed. ^{13}C-^{13}C homonuclear couplings can contribute significantly to the ^{13}C linewidths, and can be the major source of linebroadening in crystalline materials. For example, based on J values measured in solution, (35 Hz $^{13}C\alpha$-$^{13}C\beta$ and 55 Hz $^{13}C\alpha$-^{13}CO coupling) the $C\alpha$ peak is expected to acquire almost 100 Hz in apparent width. The same issue of course arises for carbon spectra in solution NMR of uniformly labeled materials. Here, pulse sequences that remedy this problem involve measurement of ^{13}C shifts during constant time indirect evolution periods where the J coupling evolves and is refocused, but the isotropic chemical shift is refocused to a variable extent as a function of the t1 value, using a moving π pulse (Ernst *et al.*, 1991). The ^{13}C T_2 or dephasing times must be moderately long in comparison with the J couplings to make these constant time experiments practical. Alternatively, selective decoupling pulses can be used. Pulse sequences appropriate or removing these couplings for uniformly labeled solids have been proposed (Straus *et al.*, 1996). The relative merits of constant time vs. selective (WURST) decoupling have been demonstrated for indirect carbon detection in solution NMR of uniformly ^{13}C enriched proteins (Matsuo *et al.*, 1996). Carbon direct detection in solution NMR of uniformly ^{13}C enriched proteins is somewhat unusual, but the merits of these experiments have been discussed recently; the use of signal processing methods to remove expected homonuclear couplings from the directly detected dimension was highlighted (Serber *et al.*, 2001).

Alternatively, the homonuclear couplings can be removed or reduced through selective isotopic labeling. As an alternative to nonspecific labeling protocols, sparse global labeling protocols can have some advantages. Hong and coworkers have suggested a variation of LeMaster's "checkerboard" labeling protocol. These labeling strategies (LeMaster, 1994) involve use of 1,3-^{13}C or 2-^{13}C glycerol (LeMaster, 1996) or 3-^{13}C-pyruvate (Rosen *et al.*,

1996) media to derive even-odd or methyl-group labeling schemes. For such samples a linewidth improvement was observed due to the removal of many of the one-bond homonuclear splittings. The impact is potentially the greatest for directly detected carbon spectra, and most important for the case of the aromatic rings, where small chemical shift differences in neighboring residues give rise to AA' type spectra and severe broadening. Such protocols can represent an important compromise between uniform and site-specific labeling.

Residual proton couplings can have profound impact on the homogeneous component of the linewidth and on the efficiency of magnetization transfer steps. Applied proton decoupling fields typically do not exceed the homonuclear proton couplings by a significant margin (typically 80-120 kHz vs. 10-30 kHz respectively); increasing the power above this level can result in significant contributions to sample heating for aqueous and salt-containing samples. Thus, proton irradiation fails to completely remove heteronuclear couplings. Composite pulse alternatives to simple CW decoupling have been designed; "two point" phase modulation or TPPM (Bennett *et al.*, 1995) has had a significant impact on the quality of biological solids NMR and allows high quality spectra to be collected at moderate (80 kHz) irradiation levels. With these modest decoupling powers linewidths of less than 0.3 ppm can be achieved commonly, for ^{13}C or ^{15}N lines in any functional group in natural abundance amino acids or selectively labeled proteins (i.e. for dilute spins). Is it important to further refine the proton decoupling protocol? Better proton decoupling might result in better spectral resolution if homonuclear couplings are effectively removed from the spectra. Stronger decoupling is certainly important for optimization of transfer efficiencies, particularly for spin pairs with weak couplings and long transfer times, and transfer sequences with relatively strong spin-locks on the detected channel. Alternatives to high power irradiation (for example selective deuteration and more efficient decoupling schemes) are needed, so that protons may be decoupled without excessive sample heating.

1.2   Larger Magnets and Related Hardware

However carefully the sample is prepared, and however carefully the decoupling sequences are selected, the magnetic field strength remains one of the most important variables in the quality of a MAS spectrum of a uniformly labeled protein. Comparisons of SSNMR spectra of proteins at various applied field strengths are not many in the literature. In at least one case, the expected gain in resolution is seen and affords strong enthusiasm about the use of high field magnets for biological solid state NMR experiments. This improvement partly reflects the importance of the homonuclear couplings, which although invariant with respect to changes in applied field, have a lesser impact on sensitivity and resolution at higher applied fields. Regardless of whether the couplings can be removed by pulse sequence enhancements for indirect dimensions, they are unlikely to be removed satisfactorily in the directly detected dimension, thus the benefit of high field for carbon detection spectroscopy is crucial.

A dominant procedural consideration for uniformly labeled samples relates to the densely populated ^{13}C NMR spectrum and the selection of magic angle spinning speed. The key issue at hand is to avoid rotational resonance conditions, i.e. conditions wherein spinning sidebands of one carbon spectrally overlap the center band or isotropic shift of a strongly coupled (directly bonded) neighboring site, or fall within a range dictated by the strength of the coupling (i.e. within a range of 5-10 ppm for directly bonded pairs at high magnetic field strengths). When overlap of this type occurs, not only do the spectra become more congested, but also dipolar interactions are partially restored and broaden the lines significantly. The rotational resonance conditions occur for the backbone ^{13}CO-^{13}Cα carbons at speeds corresponding to $n\omega_r = \Delta\sigma_{iso} = 110$-$140$ ppm, implying that 35-45, 55-70, and 110-140 ppm for ^{13}C are problematic speeds where linebroadening and additional unwanted phenomena can occur. In addition, many ^{13}CO-^{13}Cβ rotational resonance conditions occur in the range $\Delta\sigma_{iso} = 130$-160 ppm; for many aromatic residues there can also be additional problematic speeds near to $\Delta\sigma_{iso} = 70$-110 ppm. Spinning at rates greater than 180 ppm for the ^{13}C spectrum would be fully above all rotational resonance conditions, and the best resolved lineshapes are expected. The fastest speed requires state-of-the art spinning devices and presents restrictions on sample volumes (ca. 10 microliters for high field spectroscopy). Other minor technical difficulties associated with spinning for long acquisition times at these rates include the need for additional sample cooling because of the spinning-associated sample heating, and the need to provide a moisture-tight seal. Alternatively, rotor speeds of 85-100 ppm for the ^{13}C spectrum corresponds to a sample rotational speed ω_r between $n = 1$ and $n = 2$ rotational resonance conditions for the ^{13}CO-^{13}Cα pair ($\Delta\sigma_{iso} > n\omega_r > \Delta\sigma_{iso}/2$), and might have moderate or negligible effects from rotational resonance. The translates into 10–20 kHz for magnetic fields of field 400 - 800 MHz, which is routinely feasible with commercial equipment, and compatible with roughly 40 microliters of sample volume. A lower spinning rate, 45-52 ppm, would be expected to be favorable for most residues (although weak spinning sidebands of the carbonyl and methane groups intrude into the methyl region of the spectrum). In fact, a wide variety of spinning speeds have been used successfully for the spectroscopy of uniformly labeled samples reported to date. The studies of α spectrin (Pauli *et al.*, 2001) involved a speed of 43 ppm ^{15}N^{13}Cα^{13}Cx spectra and 64 ppm for ^{15}N-^{13}CO spectra. Studies of uniformly labeled antaminide involved speeds 200 ppm and 133 ppm. Studies of BPTI (McDermott *et al.*, 2000) and ubiquitin involved a speed of 95-100 ppm. All of these studies resulted in high quality spectra that are compatible with site-specific assignments.

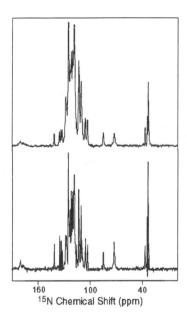

^{15}N Chemical Shift (ppm)

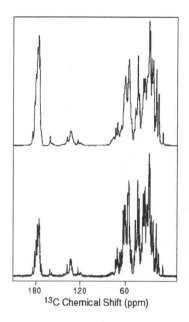

^{13}C Chemical Shift (ppm)

Figure 1. Spectra of uniformly enriched ubiquitin measured at 400 MHz (top) and 800 MHz (bottom) for ^{15}N (left) and ^{13}C (right). Spectral resolution is much better at higher magnetic fields in part because of the presence of many unresolved ^{13}C-^{13}C N splittings.

2.   A STRATEGY AND TOOLKIT FOR ASSIGNMENT

With high quality samples and spectra in hand, the major issue for the field has been to identify an efficient and practical protocol for assigning the spectrum. The dominant approach for resonance assignments of solid state proteins borrows conceptually from solution methods (Bax, 1989), in that the primary tool would be multi-dimensional correlation of the backbone nitrogen and the side-chain Cα and Cβ isotropic shifts. Isotropic shifts are measured with magic angle spinning of unoriented samples; the shift correlation spectra allow for unambiguous identification of many of the sidechains on the basis of isotropic carbon chemical shifts. Linewidths and inherent resolution at high magnetic fields are sufficient to identify individual amino acids in a simple homonuclear two-dimensional experiment, and count the numbers of residue types in some cases by inspection. ^{13}C-^{13}C spectra thereby provide a simple tool for validating samples.

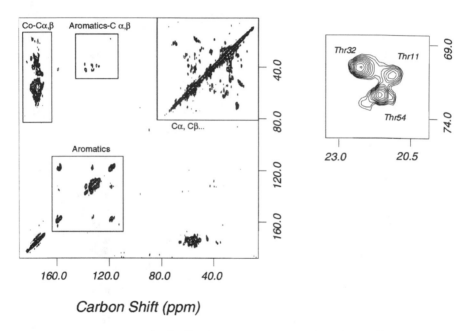

Figure 2.   Two-dimensional ^{13}C-^{13}C correlation spectrum of uniformly ^{15}N-^{13}C enriched BPTI, measured at 800 MHz, as reported previously (McDermott *et al.*, 2000). Sample assessment and sidechain identification are possible with a single two dimensional spectrum.

^{13}Cα and ^{13}Cβ isotropic shifts also provides information on the secondary structure context (Westler *et al.*, 1988; Wishart and Sykes, 1994). The past decade has seen very important developments in the pulse sequences that are available for correlating directly bonded atoms in biological solids. Both spectrally directed transfers and broadbanded transfers have been developed. Transfer efficiencies normally exceed 20% and sometimes exceed 50%. Multidimensional experiments achieve significant dispersion from the amide ^{15}N and ^{13}Cβ measurements in particular, and inter-residue correlations; thus two and three dimensional experiments are sufficient for site-specific backbone and sidechain assignments. These approaches have been partially or fully demonstrated for several proteins recently (McDermott *et al.*, 2000; Pauli *et al.*, 2001); although the methods are far from mature and optimized, the feasibility for favorable proteins is clear at this point.

Homonuclear transfer sequences operating via dipolar interactions during Magic Angle Spinning have been used to identify the spin systems associated with each amino acid sidechain. Many homo-nuclear methods have been developed over the past decade, including DRAMA (Tycko and Dabbagh, 1990), SEDRA (Gullion and Vega, 1992) Rotational Resonance (Colombo *et al.*, 1988; Raleigh *et al.*, 1988), RFDR,(Bennett *et al.*, 1992) Rotational Resonance Tickling (Costa *et al.*, 1997b), C7 (Lee *et al.*, 1995), POST-C7 (Hohwy *et al.*, 1998), and CMR7 (Rienstra *et al.*, 1998), SPC-5 (Hohwy *et al.*, 1999), MELODRAMA (Sun *et al.*, 1995b), DRAWS (Gregory *et al.*, 1995), USEME (Fujiwara *et al.*, 1995), RIL (Baldus *et al.*, 1994), and DREAM (Verel *et al.*, 2001). This lengthy roster in part reflects the complexity of spin systems that are at the cusp of high-field truncation, or in other words have comparable importance of chemical shifts, RF and dipolar parts of the Hamiltonian. The long list also partly reflects the various needs of a protein spectroscopist: sometimes one wants the most broadbanded sequence conceivable, and for other circumstances one wants a high degree of selectivity. The relative merits of many of these methods, especially transfer efficiencies, have been discussed recently (Baldus *et al.*, 1998a). For purposes of assignment of high resolution spectroscopy, we mainly consider sequences consistent with rapid sample spinning (i.e. between 50 and 250 ppm at ^{13}C, as discussed above) and recoupling of well resolved spin pairs. It is worth noting however, that a number of interesting sequences have appeared in the literature to solve a range of protein structure problems formally outside the scope of this review, for example the fruitful uses of the DRAWS sequence to study the distances in spin pairs, including spectrally coincident carbonyl pairs with substantial CSA values, and the complementary information on carbonyl tensor angles obtained through exchange spectroscopy (Weliky and Tycko, 1996).

For sidechain correlation experiments, it is desirable to achieve broadbanded efficient recoiling. Several of the sequences that are well suited for this task operate on double quantum recoupling conditions, i.e. matching conditions between the spin lock field strength and the rotor speed that lead to non-zero average of the double quantum dipolar Hamiltonian in the toggling frame. The conditions with the strongest recoupling achieve transfer efficiencies in excess of 50%, with average dipolar coupling typically much less than 50% of the static values. This remarkable efficiency is possible due to a matching condition and supercycling design principle for the RF in which the transfer pathway is filtered so that the efficiency depends only on the polar orientation of the bond vector in the rotor, and not on its azimuthal angle; furthermore the transfer efficiency is independent of chemical shift and shift anisotropy over a broad range. Strong recoupling conditions of this type can be obtained for various combinations of periods of RF subcycle elements over integer numbers of N rotor periods; many have a very broadbanded nature but with specific RF requirements that need to be arranged with the selection of spinning speed (e.g. C7 and Post C7, where $\omega_1 = 7 \, \omega_R$; SPC-5, where $\omega_1 = 5 \, \omega_R$; in all cases proton RF should be much higher than that of the recoupled nucleus). The high transfer efficiencies and broadbanded nature make these promising sequences for sidechain correlations. Very high transfer efficiencies have also been reported for DREAM, an adiabatic version of the HORROR condition, which involves rather mild RF field strengths conditions ($\omega_1^I + \omega_1^S = \omega_r$). DREAM achieves its high efficiency and bandwidth by virtue of the adiabatic nature of the transfer, and is also expected to be experimentally forgiving due to the use of a ramp. This sequence again appears to be highly suitable sequence for sidechain correlation experiments. For all of the double quantum sequences, filtering for coherence pathways is useful, since the spectral diagonal can be removed, and single step magnetization transfer processes can be readily distinguished from two-step processes by the sign of the cross peaks.

Others methods, such as rotational resonance and rotational resonance tickling, and RFDR operate on zero-quantum transfers. Although they lack some of the advantages mentioned above, they allow for bandwidth selectivity, which is important in the context of detecting tertiary contacts. They also can be optimised for backbone homonuclear transfers, where the spectral band can be well defined. These methods have the important advantage that many of them are highly robust with respect to pulse errors, spinning speed fluctuations, and poor decoupling, and therefore can be used in situations where sample heating must be minimized.

Heteronuclear ^{13}C-^{15}N correlation spectra play a crucial role for correlating backbone resonances and providing site-specific assignments. Hetero-nuclear methods include RFDRCP (Sun *et al.*, 1995a), TEDOR (Hing *et al.*, 1992), SPICP (Wu and Zilm, 1993), APHH-CP (Baldus *et al.*, 1996; Hediger *et al.*, 1995; Hediger *et al.*, 1994), and AMAP-CP (Hediger *et al.*, 1997). Transfer efficiencies about 25% have been reported on proteins and above 50% on model systems in some cases. Here again band selectivity vs. broadbanded transfers is an important distinguishing feature, to be discussed further below.

The physical mechanism supporting most of these sequences, the synchronization of spin and spatial degrees of freedom, can be understood using average Hamiltonian or Floquet theories; simulation programs based on numerical integration of finite time intervals have been available and are very useful for obtaining insight into experimental details (Bak *et al.*, 2000).

In the assignment efforts reported to date, adiabatic homonuclear and heteronuclear transfer sequences at moderate field strengths were featured (Pauli *et al.*, 2001), as well as spin diffusion and RFDR sequences (McDermott *et al.*, 2000). These selections partly reflect considerations of instrument stability and minimization of sample heating.

Heteronuclear and homonuclear shift correlation experiments have been combined to 3D ^{15}N-^{13}C-^{13}C correlation spectra, as has been reported previously (Sun *et al.*, 1997), and have been used to assign uniformly labeled small molecules in the solid state (Fujiwara *et al.*, 1995; Hong and Griffin, 1998; Michal and Jelinski, 1997, Rienstra, *et al.*, 1998) as well as uniformly labeled proteins (unpublished). In developing the three dimensional experiments for sequential backbone assignments via ^{15}N$_i$-^{13}Cα_i-^{13}Cβ_i , ^{15}N$_i$-^{13}Cα_i-^{13}CO$_i$, and ^{15}N$_i$-^{13}CO$_{i-1}$-^{13}Cα_{i-1} correlations, selective transfer methods (Baldus *et al.*, 1998b) have been used to optimize transfer and data collection efficiency. Directional transfers of the polarization (i.e. selecting ^{15}N to ^{13}CO vs. ^{15}N to ^{13}Cα transfer) allows for more efficient transfers and use of restricted spectral bandwidth for indirect dimensions. These demonstrations provide useful templates for the upcoming work on many solid state proteins. In summary, a protocol that has been implied and explicitly discussed in many publications would involve three experiments of roughly the following type: (1) ^{13}C homonuclear 2D experiments with broadband detection in both dimensions for sidechain assignments, (2) ^{15}N-^{13}Cα-^{13}CX 2D or 3D experiments, (3) ^{15}N-^{13}CO ^{13}Cα 2D or 3D experiments. These experiments have of course much in common thematically with analogous assignment protocols used popularly for solution NMR studies if isotopically enriched materials and those recently used for oriented solid state biopolymers.

2.1   With an Assigned Spectrum?

Partial assignments of the uniformly labeled material provide the opportunity for many kinds of biophysical studies, structural characterization being only one aspect. NMR shifts are sensitive probes of ligand binding and can be used to locate the important cavities in the protein: SAR by NMR might well have a solid state NMR analog wherein binding cavities and epitopes are characterized for membrane proteins. Characterization of sidechain ionization states and pH titration behaviour is another popular use of solution NMR and would in principle be possible with assigned NMR spectra in the solid state as well. It is of interest to compare the carbon shift values as measured in the solid state to those measured in solution state. Many aspects of the sample conditions necessarily differ in the two measurements (pH, ionic strength, presence of the precipitating agent, intermolecular contacts) but it is known for at least some systems that with the important exception of the ionizable residues, many of the carbon shifts are rather immune to these issues in solution state (Tuchsen and Hansen, 1988). Cole and Torchia (1991) observed that isotropic shifts tend to be similar in the solid and the solution state based on extensive comparative data on Staphylococcal Nuclease. More recently, detailed comparisons between solution and solid state chemical shifts have been made for α spectrin and BPTI, again illustrating that in the majority of cases the shifts measured in microcrystals mirror those measured in solution and reflect properties of a stable structural core. Thus these chemical shifts will be useful probes in the same way that they have been for selectively labelled solid state samples, and for solution NMR studies over the years.

3.   STRUCTURES IN THE NEAR FUTURE?

With the recent accomplishments of partial assignments of several proteins, the issues lying ahead concern the question of whether it is possible to solve the three dimensional structure of a protein. This objective can be divided into the problem of identifying secondary structure region, the question of identifying tertiary contacts, and the problem of computing structures.

Regions of secondary structure can be derived based on solid state NMR data; many measurements are sensitive to backbone torsional states. Isotropic shifts can be analyzed using the chemical shift index, a well calibrated indicator of secondary structure (Wishart and Nip, 1998). Chemical shift surfaces could also be used to utilize the torsional information in the shift and shift anisotropy more quantitatively (Heller et al., 1997). Heteronuclear local field spectroscopy, in which one probes the sum dipolar tensors for adjacent sites $D_{H-X} + D_{H-Y}$, can be used to measure X-Y bond torsions. The angular relationship between the two dipolar tensors has a strong effect on the sum dipolar tensor. This approach has been used already in many proteins, and is well suited for uniformly enriched proteins. Backbone torsion angles could also be measured with these methods; ϕ can be measured using the $^1H^{15}N^{13}C^1H$ spin system (Hong et al., 1997), and ψ can be measured using $^{15}N^{13}C^{13}C^{15}N$ systems (Costa et al., 1997a; Feng et al., 1997a), thus providing another reliable method for recognition of secondary structures. Sidechain torsion angles involving $^1H^{13}C^{13}C^1H$ spin systems can be similarly measured (Feng et al., 1996b). Other methods involve neighboring carbonyl lines that need not be resolved spectrally; relative CSA tensors and distances can be derived and offer additional backbone constraints for structure calculation (Burkoth et al., 2000; Weliky and Tycko, 1996). One of the most promising tools is the identification of non-local hydrogen bonding interactions such as main chain hydrogen bonding contacts (N-H ..CO).

What tertiary structural constraints can be measured in uniformly labeled systems? This important question is largely open at present. ^{13}C-^{13}C and ^{13}C-^{15}N long range contacts would be difficult to measure with broadbanded recoupling methods, because each ^{13}C is coupled to many directly neighboring ^{13}C nuclei. The couplings associated with direct neighbors are greater than 1kHz and those associated with useful tertiary contacts are less than 100 Hz; very little magnetization is expected to reach long-range tertiary contacts under a "total recoupling" scheme. On the other hand, band-selective homonuclear sequences (Bennett et al., 1998; Costa et al., 1997b) have a lot of promise for detecting relatively long range correlations. Analogously, REDOR experiments that selectively probe long range contacts have also been demonstrated in model systems such as tripeptides (Jaroniec et al., 2001). It is expected that couplings that exceed transverse relaxation rates could successfully be probed, so long as they are spectrally resolved well enough for the selective recoupling experiments. Thus we can expect to detect couplings as remote as 6 Å. How many contacts are needed to define the tertiary structure, and how many are expected in a protein structure? A cursory count of heteroatom pairs (i.e. pairs involving ^{13}C and ^{15}N rather than ^{1}H), at moderate distances (up to 5 Å), involving distant amino acids in the primary sequence (separated by more than four residues) indicates that several contacts per residue are expected. The current challenge is to develop protocols for detecting high percentage of these contacts. Protocols for analysis of tertiary constraints to define a three dimensional structure are expected to be highly analogous to those used for solution NMR-derived NOEs (Brunger et al., 1998).

4.   PROBES OF DYNAMICS FOR SOLID STATE NMR OF UNIFORMLY LABELED MATERIALS

NMR methods for probing protein motions in liquid state are extremely popular in the past decade; nearly every structure that is reported includes backbone order parameters. Newly developed methods for probing motion on several timescales have given important insights into protein function, binding, release, allostery, energetics and folding. Similarly, conformational dynamics of static protein solids have been studied with NMR. Although MAS spectra have an even greater sensitivity to conformational exchange, relatively less has been reported to date on experimental and computational methods or biological applications. Just as for solution or for static solids, motions on slow timescales (i.e., much slower than the relevant spectral splittings, and slower than the MAS rates), are expected to cause peak doubling. Fast motions (much faster than the relevant spectral splittings) are expected to cause a variety of relaxation phenomena. Both of these effects in MAS NMR have been utilized to characterize sidechain dynamics for a uniformly ^{13}C enriched cyclic decapeptide (Straus *et al.*, 1997). Functional groups in intermediate exchange typically show depressed NMR intensities and significantly broadened lines, for solution NMR or static solid NMR. In the case of MAS NMR the lines can be broad or weak over very broad temperature ranges (Heaton, 1997; Jørgen *et al.*, 1992; Long *et al.*, 1994) due to the many time-scales on which conformational dynamics might interfere with coherent evolution processes. BPTI has many amino acids that are known to be in or near intermediate exchange in solution near our experimental temperature (Wagner *et al.*, 1976) including ring flips (Wagner, 1983) and cysteine isomerization (Otting *et al.*, 1993). These residues were nonetheless detected using fast magic angle spinning and cross polarization (McDermott *et al.*, 2000), although some with depressed intensities or broadened lines. Thus characterization of dynamics of uniformly labeled proteins by magic angle spinning solid state NMR appears to be in principle possible, but additional methods development and data analysis tools are needed before these exciting studies will come to pass.

5.   THE FUTURE: STRUCTURAL GENOMICS, STRUCTURE–FUNCTION RELATIONS, MEMBRANE PROTEINS... AND SSNMR?

What are the likely application areas for this new method? Based on these preliminary results, it is likely that any small proteins that yields microcrystals (even low-quality microcrystals) are targets for full characterization by solid state NMR. For technical reasons some of these may not be amenable to solution NMR or X-ray crystallography. Membrane proteins remain the most open area in this sense; they are an important set of biological players that are not at all well characterized at the present. Several exciting studies of membrane systems by solid state NMR have already been reported. Partial structures at atomic detail are forthcoming for many fragments of membrane systems and fragments of insoluble proteins, including: gramicidin A (Cross and Quine, 2000; Fu *et al.*, 2000), a fragment M2 of the acetyl choline receptor (Kovacs *et al.*, 1998), phospholambin (Ying *et al.*, 2000), aspartate chemotaxis receptor (Murphy *et al.*, 2001), viral fusion peptides (Yang *et al.*, 2001), rhodopsin (Feng *et al.*, 1997b), and bacteriorhodopsin (Thompson *et al.*, 1992). Other medically interesting non-soluble systems have also been extensively studied by SSNMR because of their insolubility, including the β-amyloid peptides (Benzinger *et al.*, 1998; Lansbury *et al.*, 1995) and coat proteins of bacteriophage (Marassi *et al.*, 1997). These important accomplishments involved site-specific labels. Assignment of uniformly labeled materials might prove challenging, due to poor chemical shift dispersion in the all-helical membrane proteins. Nevertheless, initial efforts at 600–750 MHz are highly encouraging (Egorova-Zachernyuk *et al.*, 2001; McDermott, 2001). Thus far, domains involving 50-100 residues have been attempted, for both soluble and membrane systems, but there is as yet no clear indication of a molecular weight limit. Larger systems might be studied using labeled domains, possibly involving block labeling of portions of proteins (Muir *et al.*, 1998). Oligomeric systems are also probably well suited to solid state NMR, since the spectral complexity reflects only the symmetric unit. With the arrival of even higher field magnets, there is every reason for optimism in this field.

5.1   Acknowledgments

The author thanks her colleagues, particularly Chad Rienstra, Sharaon Rozovsky, Tatyana Polenova, and Tatyana Igumenova for many useful discussions. The previously unpublished one-dimensional spectra of ubiquitin were collected in collaboration with Kurt Zilm and Joshua Wand, and the BPTI spectra are based on the work previously reported (McDermott *et al.*, 2000), a collaboration with the groups of Kurt Zilm and Guy Montelione. This research was supported by NSF MCB 83581.

6.   REFERENCES

Bak, M., Rasmussen, J. T., and Nielsen, N. C., 2000, *J. Magn. Reson.* **147**: 296-330.
Baldus, M., Geurts, D. G., Hediger, S., and Meier, B. H., 1996, *J. Magn. Reson. (Ser. A)* **118**: 140-144.
Baldus, M., Geurts, D. G., and Meier, B. H., 1998a, *SSNMR* **11**: 157-168.
Baldus, M., Petkova, A. T., Herzfeld, J., and Griffin, R. G., 1998b, *Mol. Phys.* **95**: 1197 1207.
Baldus, M., Tomaselli, M., Meier, B. H., and Ernst, R. R., 1994, *Chem. Phys. Lett.* **230**: 329-336.
Bax, A., 1989, *Annu. Rev. Biochem.* **58**: 223-256.
Bennett, A. E., Ok, J. H., Griffin, R. G., and Vega, S., 1992, *J. Chem. Phys.* **96**: 8624-8627.
Bennett, A. E., Rienstra, C. M., Auger, M., Lakshmi, K. V., and Griffin, R. G., 1995, *J. Chem. Phys.* **103**: 6951-6958.
Bennett, A. E., Rienstra, C. M., Griffiths, J. M., Zhen, W., Lansbury, P. T., Jr., and Griffin, R. G., 1998, *J. Chem. Phys.* **108**: 9463-9479.
Benzinger, T. L. S., Gregory, D. M., Burkoth, T. S., Miller-Auer, H., Lynn, D. G., Botto, R. E., and Meredith, S. C., 1998, *Proc. Natl. Acad. Sci. USA* **95**: 13407-13412.

Bielecki, A., Kolbert, A. C., de Groot, H. J. M., Griffin, R. G., and Levitt, M. H., 1990, *Adv. Magn. Reson.* **14**: 111-124.

Bray, D., 1998, *Annu. Rev. Biophys. Biomol. Struct.* 59-75.

Brunger, A. T., Adams, P. D., Clore, G. M., DeLano, W. L., Gros, P., Grosse-Kunstleve, R. W., Jiang, J. S., Kuszewski, J., Nilges, M., Pannu, N. S., Read, R. J., Rice, L. M., Simonson, T., and Warren, G. L., 1998, *Acta Cryst. D* **54**: 905-921.

Burkoth, T. S., Benzinger, T., Urban, V., Morgan, D. M., Gregory, D. M., Thiyagarajan, P., Botto, R. E., Meredith, S. C., and Lynn, D. G., 2000, *J. Am. Chem. Soc.* **122**: 7883-7889.

Cole, H. B. R., and Torchia, D. A., 1991, *Chem. Phys.* **158**: 278-281.

Colombo, M. G., Meier, B. H., and Ernst, R. R., 1988, *Chem. Phys. Lett.* **146**: 189-96.

Costa, P. R., Gross, J. D., Hong, M., and Griffin, R. G., 1997a, *Chem. Phys. Lett.* **280**: 95.

Costa, P. R., Sun, B. Q., and Griffin, R. G., 1997b, *J. Am. Chem. Soc.* **119**: 10821-10830.

Cross, T. A., and Quine, J. R., 2000, *Concepts Mag. Res.* **12**: 55-70.

Dyson, H. J., and Wright, P. E., 1998, *Nat. Struct. Biol.*, **5 S**: 499-503.

Egorova-Zachernyuk, T. A., Hollander, J., Fraser, N., Gast, P., Hoff, A. J., Cogdell, R., de Groot, H. J. M., and Baldus, M., 2001, *J. Biomol. NMR* **19**: 243-253.

Ernst, R. R., Bodenhausen, G., and Wokaun, A. (1991). *International Series of Monographs on Chemistry* (Halpern, J. and Green, M. L. H., Eds.), **14**, Clarendon Press, Oxford.

Feng, X., Eden, M., Brinkmann, A., Luthman, H., Eriksson, L., Graslund, A., Antzutkin, O. N., and Levitt, M. H., 1997a, *J. Am. Chem. Soc.* **119**: 12006-12007.

Feng, X., Lee, Y., Sandström, D., Edén, M., Maisel, H., Sebald, A., and Levitt, M. H., 1996a, *Chem. Phys. Lett.* **257**: 314-320.

Feng, X., Lee, Y. K., Sandström, D., Edén, M., Maisel, H., Sebald, A., and Levitt, M. H., 1996b, *Chem. Phys. Lett.* **257**: 314-320.

Feng, X., Verdegem, P. J. E., Lee, Y. K., Sandström, D., Edén, M., Bovee-Geurts, P., de Grip, W. J., Lugtenburg, J., de Groot, H. J. M., and Levitt, M. H., 1997b, *J. Am. Chem. Soc.* **119**: 6853-6857.

Fernandez, C., Adeishvili, K., and Wüthrich, K., 2001, *Proc. Nat'l. Acad. Sci. USA* **98**: 2358-2363.

Fu, R. Q., Cotten, M., and Cross, T. A., 2000, *J. Biomol. NMR* **16**: 261-268.

Fujiwara, T., Sugase, K., Kainosho, M., Ono, A., Ono, A. M., and Akutsu, H., 1995, *J. Am. Chem. Soc.* **117**: 11351-11352.

Gardner, K. H., and Kay, L. E., 1998, *Annu. Rev. Biophys. Biomol. Struct.* 357-406.

Gregory, D. M., Mitchell, D. J., Stringer, J. A., Kiihne, S., Shiels, J. C., Callahan, J., Mehta, M. A., and Drobny, G. P., 1995, *Chem. Phys. Lett.* **246**: 654-663.

Griffin, R. G., 1998, *Nat. Struct. Biol. NMR Supp.* **5**: 508-512.

Griffiths, J. M., Bennett, A. E., Engelhard, M., Siebert, F., Raap, J., Lugtenburg, J., Herzfeld, J., and Griffin, R. G., 2000, *Biochemistry* **39**: 362-371.

Gu, Z., Drueckhammer, D. G., Kurz, L., Liu, K., Martin, D. P., and McDermott, A., 1999, *Biochemistry* **38**: 8022-8031.

Gullion, T., and Vega, S., 1992, *Chem. Phys. Lett.* **194**: 423-428.

Hajduk, P. J., Meadows, R. P., and Fesik, S. W., 1999, *Quart. Rev. Biophys.* **32**: 211-240.

Heaton, N. J., 1997, *Mol. Phys.* **92**: 251-263.

Hediger, S., Meier, B. H., and Ernst, R. R., 1995, *Chem. Phys. Lett.* **240**: 449-456.

Hediger, S., Meier, B. H., Kurur, N. D., Bodenhausen, G., and Ernst, R. R., 1994, *Chem. Phys. Lett.* **223**: 283-288.

Hediger, S., Signer, P., Tomaselli, M., Ernst, R. R., and Meier, B. H., 1997, *J. Magn. Reson.* **125**: 291-301.

Heller, J., Laws, D. D., Tomaselli, M., King, D. S., Wemmer, D. E., Pines, A., Havlin, R. H., and Oldfield, E., 1997, *J. Am. Chem. Soc.* **119**: 7827-7831.

Hill, R. B., Bracken, C., DeGrado, W. F., and Palmer, A. G. I., 2000, *J. Am. Chem. Soc.* **122**: 11610-11619.

Hing, A. W., Vega, S., and Schaefer, J., 1992, *J. Magn. Reson.* **96**: 205-209.

Hohwy, M., Jakobsen, H. J., Eden, M., Levitt, M. H., and Nielsen, N. C., 1998, *J. Chem. Phys.* **108**: 2686-2694.

Hohwy, M., Rienstra, C. M., Jaroniec, C. P., and Griffin, R. G., 1999, *J. Chem. Phys.* **110**: 7983-7992.

Hong, M., and Griffin, R. G., 1998, *J. Am. Chem. Soc.* **120**: 7113-7114.

Hong, M., Gross, J. D., Rienstra, C. M., Griffin, R. G., Kumashiro, K. K., and Schmidt-Rohr, K., 1997, *J. Magn. Reson.* **129**: 85-92.

Ishii, Y., Yesinowski, J. P., and Tycko, R., 2001, *J. Am. Chem. Soc.* **123**: 2921-2922.

Ishima, R., Freedberg, D. I., Wang, Y. X., Louis, J. M., and Torchia, D. A., 1999, *Struct. Fold. and Design* **7**: 1047-1055.

Jaroniec, C. P., Tounge, B. A., Herzfeld, J., and Griffin, R. G., 2001, *Biophys. J.* **80**: 1546.

Jørgen, H. K., Bildsøe, H., Jakobsen, H. J., and Nielsen, N. C., 1992, *J. Magn. Reson.* **100**: 437-443.

Kovacs, F. A., McNiel, S. M., Denny, J. K., Quine, J. R., and Cross, T. A., 1998, *Biophys. J.* **74**: A388-A388.

Lansbury, P. T., Jr., Costa, P. R., Griffiths, J. M., Simon, E. J., Auger, M., Halverson, K. J., Kocisko, D. A., Hendsch, Z. S., Ashburn, T. T., Spencer, R. G. S., Tidor, B., and Griffin, R. G., 1995, *Nat. Struct. Biol.* **2**: 990-998.

Lee, A. L., and Wand, A. J., 2001, *Nature* **411**: 501-504.

Lee, Y. K., Kurur, N. D., Helmle, M., Johannessen, O. G., Nielsen, N. C., and Levitt, M. H., 1995, *Chem. Phys. Lett.* **242**: 304-309.

LeMaster, D. M., 1994, *Prog. NMR Spec.* **26**: 371-419.

LeMaster, D. M., 1996, *J. Am. Chem. Soc.* **118**: 9255-9264.

Li, Y. K., Poliks, B., Cegelski, L., Poliks, M., Gryczynski, Z., Piszczek, G., Jagtap, P. G., Studelska, D. R., Kingston, D. G. I., Schaefer, J., and Bane, S., 2000, *Biochemistry* **39**: 281- 291.

Long, J. R., Sun, B. Q., Bowen, A., and Griffin, R. G., 1994, *J. Amer. Chem. Soc.* **116**: 11950-11956.

Marassi, F. M., Ramamoorthy, A., and Opella, S. J., 1997, *Proc. Natl. Acad. Sci. USA* **94**: 8551- 8556.

Matsuo, H., Kupce, E., Li, H., and Wagner, G., 1996, *J. Magn. Reson.* **B 113**: 91-96.

McDermott, A., 2001, unpublished.

McDermott, A., Polenova, T., Bockmann, A., Zilm, K. W., Paulsen, E. K., Martin, R. W., and Montelione, G. T., 2000, *J. Biomol. NMR* **16**: 209-219.

McDowell, L. M., Klug, C. A., Beusen, D. D., and Schaefer, J., 1996, *Biochem.* **35**: 5395-5403.

Michal, C. A., and Jelinski, L. W., 1997, *J. Am. Chem. Soc.* **119**: 9059-9060.

Minton, A. P., 2000, *Curr. Opin. Struct. Biol.* **10**: 34-39.

Muir, T. W., Sondhi, D. and Cole, P. A., 1998, *Proc. Natl. Acad. Sci. USA* **95**: 6705-6710.

Mulder, F. A. A., Hon, B., Muhandiram, D. R., Dahlquist, F. W., and Kay, L. E., 2000, *Biochemistry* **39**: 12614-12622.

Murphy, O. J., Kovacs, F. A., Sicard, E. L., and Thompson, L. K., 2001, *Biochemistry* **40**: 1358- 1366.

Otting, G., Liepinsh, E., and Wüthrich, K., 1993, *Biochemistry* **32**: 3571-82.

120   ANN MCDERMOTT

Pauli, J., Baldus, M., van Rossum, B., de Groot, H. J. M., and Oschkinat, H., 2001, *Chembiochem* **2**: 272-281.
Pauli, J., van Rossum, B., Forster, H., de Groot, H. J. M., and Oschkinat, H., 2000, *J. Magn. Reson.* **143**: 411-416.
Prestegard, J. H., Al-Hashimi, H. M., and Tolman, J. R., 2000, *Quart. Rev. Biophys.* 371-424.
Raleigh, D. P., Levitt, M. H., and Griffin, R. G., 1988, *Chem. Phys. Lett.* **146**: 71-76.
Ramakrishnan, V., and Moore, P. B., 2001, *Curr. Opin. Struct. Biol.* 144-154.
Rastogi, V. K., and Girvin, M. E., 1999, *Nature* **402**: 263-268.
Riek, R., Pervushin, K., and Wüthrich, K., 2000, *Trends Biochem. Sci.* **25**: 462-468.
Rienstra, C. M., Hatcher, M. E., Mueller, L. J., Sun, B.-Q., Fesik, S. W., Herzfeld, J., and Griffin, R. G., 1998, *J. Am. Chem. Soc.* **120**: 10602-10612.
Rosen, M. K., Gardner, K. H., Willis, R. C., Parris, W. E., Pawson, T., and Kay, L. E., 1996, *J. Mol. Biol.* **263**: 627-636.
Rozovsky, S., and McDermott, A. E., 2001, *J. Mol. Biol.* **310**: 259-270.
Sakai, H., and Tsukihara, T., 1998, *J. Biochem.* **124**: 1051-1059.
Schaefer, J., 1999, *J. Magn. Reson.* **137**: 272-275.
Serber, Z., Richter, C., and Dotsch, V., 2001, *Chembiochem* **2**: 247-251.
Straus, S. K., Bremi, T., and Ernst, R. R., 1996, *Chem. Phys. Lett.* **262**: 709-715.
Straus, S. K., Bremi, T., and Ernst, R. R., 1997, *J. Biomol. NMR* **10**: 119-128.
Straus, S. K., Bremi, T., and Ernst, R. R., 1998, *J. Biomol. NMR* **12**: 39-50.
Studelska, D. R., McDowell, L. M., Espe, M. P., Klug, C. A., and Schaefer, J., 1997, *Biochemistry* **36**: 15555-15560.
Sun, B. Q., Costa, P. R., and Griffin, R. G., 1995a, *J. Magn. Reson. (Ser. A)* **112**: 191-198.
Sun, B. Q., Costa, P. R., Kocisko, D., Lansbury, P. T., and Griffin, R. G., 1995b, *J. Chem. Phys.* **102**: 702-707.
Sun, B. Q., Rienstra, C. M., Costa, P. R., Williamson, J. R., and Griffin, R. G., 1997, *J. Am. Chem. Soc.* **119**: 8540-8546.
Thompson, L. K., McDermott, A. E., Raap, J., van der Wielen, C. M., Lugtenberg, J., Herzfeld, J. and Griffin, R. G., 1992, *Biochemistry* **31**: 7931-7938.
Tuchsen, E., and Hansen, P. E., 1988, *Biochemistry* **27**: 8568-8576.
Tycko, R., and Dabbagh, G., 1990, *Chem. Phys. Lett.* **173**: 461-465.
van Rossum, B.-J., Forster, H., and de Groot, H. J. M., 1997, *J. Magn. Reson.* **124**: 516-519.
Verel, R., Ernst, M., and Meier, B. H., 2001, *J. Magn. Reson.* **150**: 81-99.
von Heijne, G., 1999, *J. Mol. Biol.* **293**: 367-379.
Wagner, G., 1983, *Quart. Rev. Biophys.* **16**: 1-57.
Wagner, G., DeMarco, A., and Wüthrich, K., 1976, *Biophys Struct Mech* **2**: 139-58.
Wang, Y. X., Freedberg, D. I., Yamazaki, T., Wingfield, P. T., Stahl, S. J., Kaufman, J. D., Kiso, Y., and Torchia, D. A., 1996, *Biochemistry* **35**: 9945-9950.
Weliky, D. P., and Tycko, R., 1996, *J. Am. Chem. Soc.* **118**: 8487-8488.
Westler, W. M., Stockman, B. J., and Markley, J. L., 1988, *J. Am. Chem. Soc.* **110**: 6256-6258.
Wider, G., and Wüthrich, K., 1999, *Curr. Opin. Struct. Biol.* **9**:594-601.
Wishart, D. S., and Nip, A. M., 1998, *Biochem. Cell Biol.* **76**: 153-163.
Wishart, D. S., and Sykes, B. D., 1994, *J. Biomol. NMR* **4**: 171-180.
Wright, P. E., and Dyson, H. J., 1999, *J. Mol. Biol.* **293**: 321-331.
Wu, X., and Zilm, K. W., 1993, *J. Magn. Reson. (Ser. A)* **104**: 154-165.
Yang, J., Parkanzky, P. D., Khunte, B. A., Canlas, C. G., Yang, R., Gabrys, C. M., and Weliky, D. P., 2001, *J. Mol. Graph. Model.* **19**: 129-135.
Ying, W. W., Irvine, S. E., Beekman, R. A., Siminovitch, D. J., and Smith, S. O., 2000, *J. Am. Chem. Soc.* **122**: 11125-11128.

Chapter 6

NMR Spectroscopy of Encapsulated Proteins Dissolved in Low Viscosity Fluids

A. Joshua Wand, Charles R. Babu, Peter F. Flynn and Mark J. Milton
Johnson Research Foundation and Department of Biochemistry & Biophysics, University of Pennsylvania, Philadelphia, Pennsylvania USA 19104-6059

Abstract: The challenges to solution NMR spectroscopy originating from the slow tumbling of large proteins are reviewed and a novel solution based on a hydrodynamic approach is described. This solution is based on the effective decrease in tumbling time brought about by the encapsulation of the protein of interest in a protective reverse micelle assembly and its subsequent dissolution in a low viscosity fluid. The physical basis of the method is reviewed and a variety of technical issues are discussed. Recent advances that appear to validate the approach as a tool for structural biology are presented. Future applications and developments are also described.

1.   INTRODUCTION

Nuclear magnetic resonance (NMR) spectroscopy continues to play a central role in the characterization of the structure and dynamics of proteins, nucleic acids and their complexes. Over the past fifteen years there have been staggering developments in NMR techniques and supporting technologies such that the comprehensive structural characterization of 20 kDa proteins has become routine [for concise reviews, see Wagner, (1997); Clore and Gronenborn, (1997); Wider and Wüthrich, (1999)]. Second only to crystallography, NMR spectroscopy provides an unparalleled view of structure and it remains second to none in its ability to examine dynamic phenomena. NMR also provides a unique avenue to monitor the full structural and dynamic effects of changes in temperature, solution conditions and the binding of small and large ligands.

The size of protein structures that can be solved by modern NMR techniques has increased dramatically over the past decade. Coupled with the introduction of heteronuclear (Sørensen et al., 1987) and ultimately triple resonance (Kay et al., 1990) spectroscopy was the widespread use of recombinant technologies to introduce NMR-active isotopes into proteins and nucleic acids (LeMaster, 1994; McIntosh and Dahlquist, 1990). With the development of multinuclear and multidimensional capabilities, NMR is now able to confidently, efficiently and comprehensively deal with proteins of significant size and spectral complexity. However, increasing size brings with it several important limitations that unfortunately compound each other to significantly limit the size of a protein that can be efficiently and comprehensively approached by modern NMR techniques. Two obvious ones are:

- Increasing size leads to slower tumbling and correspondingly shorter spin-spin relaxation times. This causes the basic engine of NMR spectroscopy of proteins - the triple resonance technologies – to begin to fail. As lines broaden basic sensitivity also begins to become a limiting issue.
- Increasing size leads to increasingly complex spectra. Spectral degeneracy complicates the assignment process and renders assignment of NOEs and other structural restraints to parent nuclei a difficult and often impossible task

In our opinion, complexity is not currently the limiting factor and would in any event be dramatically reduced as more and more sophisticated spectroscopy can be applied by virtue of favourable relaxation behaviour. Thus if one is able to reduce the limitations presented by short spin-spin relaxation times the problems presented by complexity are also reduced.

Briefly stated, increasing size leads to slower tumbling of the macromolecule which in turn results in more efficient dipolar relaxation processes and shorter spin-spin relaxation times. The transfer of coherence underlying current triple resonance-based assignment strategies is time-dependent and begin to fail with proteins ~30 kDa and larger. Chemical approaches such as random partial or perdeuteration have been used successfully to reduce the dipolar field so that high-resolution ^{15}N-HSQC spectra can be obtained (LeMaster, 1994). Unfortunately, perdeuteration also drastically limits the structural information available from the NOE. Fractional deuteration has limited sensitivity and its applicability as a general solution to the dipolar broadening displayed by proteins above 35 kDa is uncertain. Spectroscopic solutions are also appearing. Some find their roots in the steady improvement in the use of the rotating frame to provide for more efficient isotropic mixing for coherence transfer (e.g. Brauschweiler and Ernst, 1983; Bax and Davis, 1985; Griesinger et al., 1988; Glaser and

Drobny, 1989; Mohebbi and Shaka, 1991). Two very recent advances are the selection of the narrow multiplet component arising due to the cross correlation of dipole-dipole coupling and chemical shift anisotropy in ^{15}N-^{1}H correlation experiments (Pervushin et al., 1997) and the use of slowly relaxing multiple quantum coherence (Larsson et al., 1999). These and other approaches are extremely helpful but they do not appear to resolve all of the issues facing the solution NMR spectroscopist. The difficulty of dealing comprehensively with large proteins in a general manner remains a significant limitation in the application of solution NMR methods to the rapidly growing list of proteins being discovered by the molecular biology community.

The need for additional techniques and approaches is quite clear - fully 25% of known open reading frame sequences appear to code for membrane proteins and over 50% code for proteins that are beyond the size accessible by current solution NMR methods. While solid state NMR methods continue to show great progress and recent triumphs like the determination of the gramicidin channel illustrate the potential of these approaches (Ketchem et al., 1996), solution NMR methods are clearly easier to employ (if they work) and a vast infrastructure of techniques is in place for their application.

This review describes our initial efforts to develop and prove the feasibility of a novel idea that seeks to address the long correlation time problem via a physical route: by the reduction of the effective tumbling time for a protein macromolecule by placing it, hydrated and without significant structural perturbation, within a reverse micelle rapidly tumbling in a suitable low viscosity fluid.

2. LIMITATIONS IMPOSED BY NMR RELAXATION

The spin-spin relaxation time (T_2) is often the dominant limitation with respect to the successful application of modern multinuclear and multidimensional NMR spectroscopy to a particular protein. For an isolated H-X spin pair, the transverse relaxation rate of the X-spin is given by:

$$T_2^{-1} = \frac{\hbar^2 \gamma_X^2 \gamma_H^2}{8 r_{HX}^6} \left[\begin{array}{l} 4J(0) + J(\omega_H - \omega_C) + 3J(\omega_C) \\ + 6J(\omega_H) + 6J(\omega_H + \omega_C) \end{array} \right] \tag{1}$$

For the sake of clarity, we will assume that the "model free" spectral density treatment of Lipari and Szabo (1982a,b) is sufficient to describe the motion of the internuclear H-X vector. In that case, the spectral density takes the form:

$$J(\omega) = \frac{2}{5}\left[\frac{S^2\tau_m}{1+\omega^2\tau_m^2} + \frac{(1-S^2)\tau}{1+\omega^2\tau^2}\right] \tag{2}$$

where $1/\tau = 1/\tau_m + 1/\tau_e$. Here S^2 is the square of the generalized order parameter and ranges between 0 and 1 (rigid), τ_e is the effective correlation time for internal motion, and, importantly, τ_m is the isotropic correlation time for global tumbling. Forms for anisotropic global tumbling are also available but are unnecessary to illustrate the point to be made. In the absence of unrestricted internal motions, as τ_m increases the transverse relaxation rate ($1/T_2$) will increase (Figure 1).

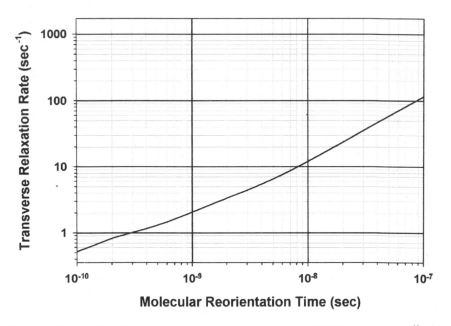

Figure 1. The relationship between the spin-spin relaxation rate ($1/T_2$) for an isolated ^{13}C-^1H spin pair and the correlation time for molecular reorientation (τ_m) is shown and illustrates the dominance of the J(0) term in Equation (1).

The effect of decreasing T_2 is at least two fold. On one hand, the signal-to-noise ratio (S/N) of a Lorenztian line rapidly degrades with decreasing T_2. This is illustrated in Figure 2.

On the other hand, the effectiveness of the current library of multidimensional and multinuclear NMR experiments begins to degrade as T_2 begins to approach the inverse of the coupling constant responsible for coherence transfer. To illustrate this consider the simplest case where coherence transfer is described by $sin(\pi J\tau)$ where J is the involved active coupling and τ is the time allowed for development of the coherence. Assuming a rigid molecule (i.e., $S^2 \sim 1$), we find the dependence of the transfer efficiency on the tumbling time shown in Figure 3.

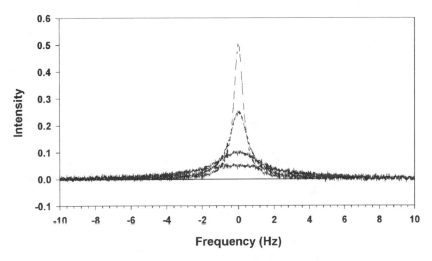

Figure 2. The influence of decreasing transverse relaxation time (T_2) on the S/N of a Lorentzian line of constant intensity. Shown are resonance lines corresponding to T_2 times of 0.5, 0.25, 0.10 and 0.05 seconds. Noise of 1% of the intensity has been applied to each curve.

Figures 1, 2 and 3 clearly show what is amply known in the protein NMR community: the standard triple resonance experiments begin to become unreliable at room temperature for proteins larger than ~30 kDa and fail significantly for proteins above 35 kDa in the absence of elevated temperature and/or extensive deuteration (Sattler *et al.*, 1999). We discuss several possible solutions to this apparent barrier to the application of modern triple resonance techniques to large proteins in the following section, with particular emphasis being placed upon the title approach: the protective encapsulation of proteins to allow their dissolution in low viscosity fluids.

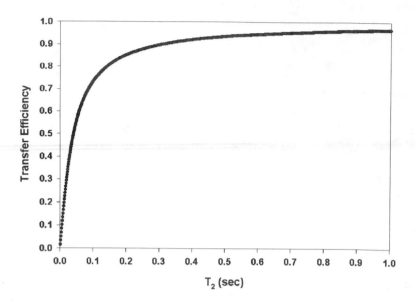

Figure 3. The effect of decreasing T_2 on the maximum transfer efficiency obtainable as a function of T_2. The calculation is based on a simple $\sin(\pi J\tau)\exp(-\tau/T_2)$ transfer function employing a 15 Hz coupling constant, typical of the interresidue N-Cα coupling constant.

3.   POTENTIAL SOLUTIONS TO THE RELAXATION BARRIER

The application of solution NMR techniques to systems of increasingly greater size has spawned several approaches for minimizing the negative influence of the rapid transverse (T_2) relaxation that is associated with slow molecular tumbling. These include extensive deuteration to reduce the "dipolar field" experienced by critical heteronuclei, use of the so-called "TROSY" effect and an active attempt to increase the tumbling time of large proteins by placing them in low viscosity fluids.

3.1   Deuteration

The gyromagnetic ratio of deuterium is almost 7 times smaller than that of hydrogen. As is apparent from Equation (1), substitution of deuterium for hydrogen at carbon sites reduces the dipolar contribution to transverse ^{13}C relaxation by a factor of about 45. The practical consequence of this substitution is that ^{13}C dipolar relaxation is virtually decoupled from molecular motion, leaving only contributions from CSA and R_{ex} relaxation mechanisms. This in turn leads to substantial improvement in many of the NMR triple-resonance assignment experiments, i.e., HNCA, HNCACB, CC(CO)NH, etc. Unfortunately, such perdeuterated samples are unusable in experiments that record nuclear Overhauser effects (i.e., ^{13}C-resolved and ^{15}N-resolved NOESY). Efforts have therefore been made to optimise relaxation while retaining varying levels of protonation. Random fractional deuteration at levels between 50 % and 75 % has been shown to provide a workable compromise in which the transverse relaxation is reduced while some level of side chain protonation is preserved. Efforts have also been made to selectively provide protonation in the isoleucine δ1, leucine δ and valine γ methyl groups in an otherwise perdeuterated molecule and to use the relatively sparse NOE information together with measurements of residual dipolar couplings to determine the global fold of a protein (Gardner and Kay, 1997; Goto et al., 1999; Mueller et al., 2000).

3.2   Transverse Optimised Spectroscopy

At high magnetic fields, chemical shift anisotropy (CSA) of ^{15}N and certain ^{1}H and ^{13}C nuclei can provide a significant contribution to transverse relaxation. A detailed analysis of transverse relaxation promoted by both CSA and dipolar relaxation mechanisms reveals that one component of the fine structure has unusual properties (Redfield, 1965, Farrar and Stringfellow, 1996; Pervushin et al., 1997). For an I-S spin pair, the equations of motion governing the components of the S-spin doublet are given as (Farrar and Stringfellow, 1996; Pervushin et al., 1997):

$$\frac{d}{dt}\begin{bmatrix} \left\langle S_{12}^{\pm}\right\rangle \\ \left\langle S_{34}^{\pm}\right\rangle \end{bmatrix} = \begin{bmatrix} c_{11} & c_{12} \\ c_{21} & c_{22} \end{bmatrix} \bullet \begin{bmatrix} \left\langle S_{12}^{\pm}\right\rangle \\ \left\langle S_{34}^{\pm}\right\rangle \end{bmatrix}$$

$$c_{11} = \pm i\omega_s^{12} + R_{1212} + \frac{1}{T_{2S}} + \frac{1}{2T_{1I}}$$

$$c_{12} = 3\left(p^2 - \delta_I^2\right)J\left(\omega_I\right) - \frac{1}{2T_{1I}} \qquad (3)$$

$$c_{21} = 3\left(p^2 - \delta_I^2\right)J\left(\omega_I\right) - \frac{1}{2T_{1I}}$$

$$c_{22} = \pm i\omega_s^{34} + R_{3434} + \frac{1}{T_{2S}} + \frac{1}{2T_{1I}}$$

where the subscripted $\pm$ denotes the orientation of the transition and the resonance frequencies of the individual components are defined as $\omega_S^{12} = \omega_S + \pi J_{IS}$ and $\omega_S^{34} = \omega_S - \pi J_{IS}$ and J_{IS} is the direct scalar coupling constant between the I and S spins.

In the slow tumbling limit, the $3(p^2 - \delta_I^2)J(\omega_I)$ term goes to zero and the individual transverse relaxation rates can be approximated as (Pervushin et al., 1997):

$$R_{3434} = (p + \delta_S)^2 \, 4J(0)$$
$$R_{1212} = (p - \delta_S)^2 \, 4J(0)$$

$$(4)$$

The key is to take advantage of the fact that the dipole-dipole relaxation term (p) is field independent while that due to CSA (δ_S) is field dependent:

$$p = \frac{1}{2\sqrt{2}} \gamma_I \gamma_S \left/ r_{IS}^3 \right. \qquad \delta_S = \frac{1}{3\sqrt{2}} \gamma_S B_0 \Delta \sigma_S \tag{5}$$

This allows one tune in favourable relaxation by nulling the (p-δ_S) term of Equations 4. It is predicted then the R_{1212} relaxation rate will reach a minimum at around ~ 1 GHz. To obtain optimal performance, however, dipolar coupling with remote hydrogen spins should be eliminated by deuteration. This, combined with the requirement for a high chemical shift anisotropy of the S spin appears to restrict this approach largely to spectroscopy of the backbone.

3.3   Reduced Macromolecular Tumbling

A potential physical approach to overcoming the barrier arising from the slow tumbling of a large macromolecule is to find some way to actively decrease its effective tumbling time. Considerable insight can be obtained from a simple hydrodynamic analysis. Approximating a protein as an isotropically reorienting molecule (i.e., as a sphere), one can relate its correlation time for global tumbling to the solvent viscosity via the Stokes-Einstein relation:

$$\tau_m = \frac{1}{6 D_{rot}} = \frac{\eta V_h}{kT} \tag{6}$$

where D_{rot} is the rotational diffusion constant for a sphere, η is the viscosity, V_h is the volume of the sphere, k is the Boltzmann constant and T is the absolute temperature. More complicated forms are available but are unnecessary here.

One sees, then, that the correlation time for tumbling is linearly related to the viscosity of the solvent. This is the key to the approach.

The radius (r_h) of a spherical protein of molecular weight M_r having a specific volume (v) and a hydration shell corresponding to a given frictional ratio of (f/f_o) will be:

$$r_h = \sqrt[3]{\frac{3V_h}{4\pi}} = \sqrt[3]{\frac{3\left(\frac{v \bullet M_r}{N_o}\right) \bullet \left(\frac{f}{f_o}\right)}{4\pi}} \tag{7}$$

The (f/f_o) term is used here to scale the volume of the protein in order to reflect a hydration shell. Values of ~1.2 are typically used. Consider, for example, that a spherical protein with molecular weight of 50 kDa and an average specific volume (v, ~0.75 mL/g) and a typical (f/f_o) ratio of 1.2 will therefore have a radius of ~26 Å and will tumble in water at 300K ($\eta \sim 850$ μPa•s) with a correlation time of ~15 ns. To give this protein the spectroscopic performance of a small protein like ubiquitin (8.5 kDa), the protein would have to tumble with a correlation time of ~ 3 ns. From the point of view of viscosity, this would require tumbling in a solvent with a viscosity of ~170 μPa•s or less. This is clearly not obtainable in water over a reasonable range of temperature and pressure. Clearly, another solvent is required.

However, short chain alkanes whose intermolecular interactions are restricted to London forces show viscosities of the range necessary. Ethane, propane and butane are gases at room temperature and pressure but are liquefied by modest pressures (Table 1).

Table 1. Viscosity of Short Chain Alkane Fluids at 300 K

	Ethane	Propane	Butane	Pentane	Water
Viscosity (μPa•s)	35	97	158	~220	850
Pressure (MPa)	4.7	1.05	0.40	0.10	0.10

Note: 1 MPa = 10 bar = 145 psi ~ 10 atm; 1 μPa•s = 10^{-3} centipoise. The figure for pentane results from an interpolation.

The challenge then is to figure out a way to place a protein in a hydrophobic solvent such as liquid propane. Here we have turned to a technology that was extensively investigated in the 1980s - encapsulation of molecules such as proteins within the water cavity formed by so-called reverse micelles.

4.   REVERSE MICELLE TECHNOLOGY

Reverse micelles will spontaneously form as transparent solutions in a low polarity liquid and are thermodynamically stable assemblies of surfactant molecules organized around a water core. Reverse micelles were the subject of extensive attention in the 1980s as potential devices for a range of applications including separations, chromatography and reaction processes (Goklen and Hatton, 1985). They have also been used in the context of hosting chemical reactions in solvents with low environmental impact such as supercritical carbon dioxide (Johnston *et al.*, 1996).

The size and stability of reverse micelles depends on many factors, most critically on the shape and charge of the surfactant and the amount of water loading. Sodium bis(2-ethylhexyl)sulfosuccinate (AOT) has been extensively studied in a variety of organic solvents. The chemical structure of AOT is shown in Figure 4 and illustrates the common structural theme for reverse micelle surfactants: small polar or charged headgroups with (ideally) short branched aliphatic tails. Such a chemical structure favors close packing in the inner shell of the reverse micelle while simultaneously providing a large volume tail to occupy the outer shell of the reverse micelle.

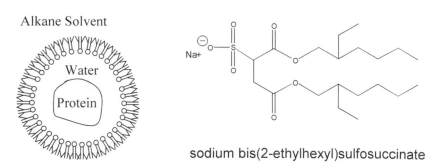

Figure 4.   The chemical structure of AOT, the surfactant that is commonly used to construct reverse micelle systems having a significant internal aqueous core. On the left is a schematic illustration of an AOT reverse micelle containing a water-soluble protein.

Water loading (w_o) is commonly defined as the molar ratio of water to surfactant. Water loadings have been described that yield stable reverse micelles of AOT in a variety of long and short chain alkanes large enough to accommodate proteins (e.g., Frank and Zografi, 1969; Gale et al., 1987; Fulton and Smith, 1988; Fulton et al., 1989). Indeed, there have been reports that all of the three short chain alkane solvents will support formation of a single phase containing reverse micelles. Ethane requires relatively elevated pressures, 50 to perhaps up to 200 bar, to obtain optimal behavior while butane requires only a few bar (Fulton et al., 1989; Smith et al., 1990).

4.1   Potential NMR Performance

In principle, spherical reverse micelles will have a radius determined by the width of the surfactant shell (15 Å for AOT) and the volume of the water pool enclosed within. The radii of "empty" (i.e. containing only water) reverse micelles formed by AOT in long chain alkane solvents such as isooctane have been shown to increase in a roughly linear fashion with the molar ratio of water to surfactant (w_0) (reviewed by Luisi, 1985). This is important as the tumbling time of the spherical reverse micelle particle depends on the cube of its radius (see Equations 6 and 7).

$$r_{mic} \approx 1.5w_0 + 15 \text{ Å} \tag{8}$$

Loading with small proteins such as ribonuclease (M_r 13.5 kDa) or larger proteins such as liver alcohol dehydrogenase (M_r 82 kDa) results in a limiting micelle radius at low w_0 corresponding to the simple sum of the hydrated protein's effective radius and the length of the surfactant (Luisi, 1985).

Given this general behavior one can calculate the anticipated hydrodynamic performance of reverse micelles containing a single protein molecule and a minimum hydration shell (here assumed to correspond to 20% of the protein volume). The results of such calculations for the short chain alkane solvents are presented in Figure 5 and illustrate several important points. As illustrated graphically in Figure 4, the addition of a layer of surfactant around the surface of a hydrated protein introduces a significant volume penalty and accordingly will increase the effective tumbling time of the protein. This volume penalty is very significant and is most dramatic for smaller proteins and is responsible for the non-linear dependence of the effective tumbling time on the molecular weight of the encapsulated protein (Figure 5). Nevertheless all five alkane solvents are predicted to provide low enough viscosity to overcome this volume penalty and to provide faster effective tumbling of a reverse micelle containing a single protein molecule of M_r greater than 50 kDa than the protein would have in free aqueous solution. Though these simple calculations give absolute values for the tumbling times that are somewhat shorter than actually observed (even for the free solution situation), one anticipates that they provide an accurate means of comparison. Indeed, as will be shown below, this is the case. Importantly, these simulations predict that the molecular tumbling time of a ~30 kDa protein in free aqueous solution can be obtained by a 100 kDa protein encapsulated in AOT and dissolved in propane. As noted above, 30 kDa is the approximate limit of uniformly reliable application of triple resonance and total correlation spectroscopy to proteins without benefit of deuteration. In summary, these analyses indicate that the approach may indeed provide the necessary decrease in effective molecular reorientation time.

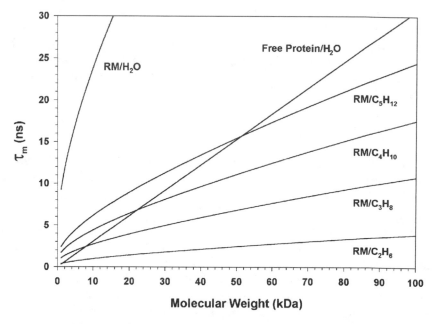

Figure 5. Anticipated hydrodynamic performance for AOT reverse micelles containing a single spherical hydrated protein molecule. Shown are the predicted dependencies of effective molecular tumbling time of a single protein molecule of a given molecular weight in free aqueous solution (free protein/H_2O), encapsulated in a reverse micelle dissolved in H_2O (RM/H_2O), pentane (RM/C_5H_{12}), butane (RM/C_4H_{10}), propane (RM/C_3H_8) and ethane (RM/C_2H_6). A twenty percent hydration volume ($f/f_o = 1.2$), a specific volume of 0.75 mL/g, a surfactant length of 15 Å, a temperature of 300 K, and the viscosities listed in Table 1 were employed in the calculation.

4.2   Preparation of Encapsulated Proteins

A number of approaches have been developed to prepare solutions of encapsulated proteins dissolved in a variety of organic solvents. Unfortunately, the restraint on viscosity imposed by the requirement for rapid molecular tumbling severely restricts the number of fluids with sufficiently low viscosity. Suitable fluids include the short chain alkanes, supercritical carbon dioxide, and various halocarbon refrigerants. All of these require the application of significant pressure to be kept liquefied at room temperature. The preparation of stable well-behaved solutions of reverse micelles in low viscosity fluids is anticipated to generally require the maintenance of significant sample pressure. This will be particularly true for the short chain alkanes such as propane and ethane (Smith *et al.*, 1990) and supercritical carbon dioxide (Johnston *et al.*, 1996).

4.2.1   Pressurized NMR Samples

In developing this method, we have designed an NMR tube suitable for the preparation of solutions of proteins encapsulated in reverse micelles dissolved in fluids that require pressures approaching 200 bar (~3,000 psi) to achieve optimal solution behaviour (Ehrhardt *et al.*, 1999). To safely reach potentially necessary pressures approaching 200 bar, the NMR tube is made from "hot isostatically pressed" zirconium oxide which has a relatively high tensile strength. In this context, zirconium oxide has several advantages over sapphire including lower cost. Collars to hold the tube were constructed from BeCu and machined to accept two standard stainless steel HPLC fittings (Ehrhardt *et al.*, 1999). The tube is sealed to the collar without benefit of glue by the action of the embedded o-ring at low pressures and the boron nitride washer at high pressures. It should be emphasized that glass NMR tubes that can safely operate up to 200 p.s.i. are commercially available. Many applications using samples of encapsulated proteins dissolved in propane or butane are readily prepared and used in such tubes.

4.2.2   Sample Preparation

The transfer of protein from a free aqueous solution environment to the restricted aqueous core of a reverse micelle dissolved in an organic solvent is a poorly understood process. Two variations of the same basic method of preparing samples are routinely used in our laboratory to successfully encapsulate proteins in their native state. In both cases the solvent is first liquefied within a pressure cell in the presence of surfactant. Concurrent stirring using a small conventional magnetic stir-bar is sufficient to ensure preparation of a homogeneous solution of reverse micelles in the liquefied solvent. In the first scheme the solution of reverse micelles is transferred into a second mixing cell that has been preloaded with the hydrated protein (Ehrhardt *et al.*, 1999). The solution of reverse micelles and the hydrated protein are then mixed until the desired loading of protein is achieved. In our hands, preparation of protein-containing reverse micelles is best done in a two step passive phase transfer process whereby reverse micelles are prepared and then subsequently loaded with protein. This is illustrated schematically in Figure 6.

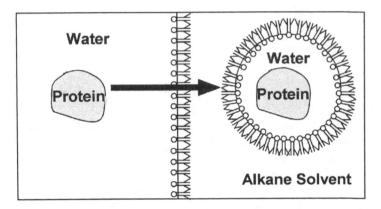

Figure 6. Illustration of the phase transfer method for solublization of water-soluble proteins in the water core of reverse micelles formed in organic solvent. The "infinite" water method is illustrated. This approach allows the creation of two bulk phases with protein in the aqueous phase and surfactant in organic phase. The partitioning of the protein between the two phases is strongly influenced by the pH and ionic strength of the aqueous phase. For the preparation of encapsulated proteins with defined and low amounts of water, lyophilized protein is hydrated and the resulting "solution" shaken in a solution of surfactant.

This procedure requires mixing under defined conditions of temperature and pressure. In our experience, solutions as high 0.3 mM in protein-loaded reverse micelles can be prepared. Protein loading can be followed directly by absorbance or fluorescence since the surfactant solution is often transparent in near UV and visible region of the optical spectrum. The protein-solvent-surfactant system is then transferred into the NMR cell. In a second scheme, the solvent-AOT phase is delivered directly into the NMR cell, which has been preloaded with hydrated protein. In this variation, the hydrated protein is combined with the solvent-surfactant phase using gentle agitation that is aided by a glass capillary (which also carries D_2O used to lock the spectrometer). In principle the former scheme provides additional control over the process. However, experience suggests that the results are generally identical and the second method avoids a transfer step, which is often the source of failure when preparing these samples.

Liquefaction of the solvent at a given temperature is achieved by simply raising the pressure of the system above that specified at the gas-liquid phase boundary of the pressure-temperature phase diagram for the solvent of interest. Elevated pressure may be generated using a liquid pressure generator or through the use of pressurized inert gas. Transfer of the solution of reverse micelles to a second mixing cell or to the NMR cell is accomplished by creating a small pressure difference (~10 p.s.i) between the vessels. In all transfer steps the solvent must be kept from boiling which leads to sample loss. Much of the difficulty associated with making and manipulating solutions of encapsulated proteins dissolved in the low viscosity solvents required for NMR spectroscopy arises from the need to have precise control over the pressure and temperature of the solutions at all times. Current technology for dealing with this is crude and this Achilles' heel of the method is ripe for further development.

4.2.3   Sample Composition

There are a number of competing considerations when defining the precise composition of a preparation of encapsulated protein. The fundamental driving force for the transfer of protein from an aqueous phase to the water core of a reverse micelle appears to be electrostatic in nature and arises from the interaction between the charged head groups of the surfactant and the protein. Accordingly, the degree of transfer can be modulated by variation of the charge of the protein (via the pH of the aqueous phase), by the amount of water available to the reverse micelles and by electrostatic screening of the interaction (via the ionic strength of the aqueous phase). A quantitative understanding of the role of each of these environmental variables is currently lacking and optimal conditions must be defined empirically for each protein. Fortunately, the integrity of the protein can be easily assessed using ^{15}N-HSQC spectra.

A critical issue is the amount of water that is to be made available since excess water will unnecessarily increase the radius of the reverse micelle particle with disastrous consequences on the obtained rotational correlation time. Ideally, one would like to restrict the amount of water to the minimum required to maintain its native aqueous phase structure. Hopefully, this minimum water corresponds to only one to three solvation layers.

Water enters into consideration not only from the point of view of the final size of the reverse micelle particle but also with respect to the type of reverse micelle assembly that is formed. The AOT-water-propane phase diagram provides an illustrative example of the fine line that must be delineated (Figure 7). It also points to a potential general solution. Low water loading and low surfactant concentration favours the formation of a monodisperse solution of well-formed reverse micelles having near ideal hydrodynamic behaviour. Increasing water content and/or increasing surfactant concentration favours the formation of non-spherical ribbon-like surfactant assemblies or even phase separation. Neither condition is appropriate for rapid molecular reorientation. Fortunately, the phase diagram can be manipulated with hydrostatic pressure. Application of significant but not unreasonable pressure, up to 200 bar (3000 psi), broadens the region of the ternary phase diagram that is capable of supporting homogenous solutions of spherical reverse micelle assemblies (Figure 7).

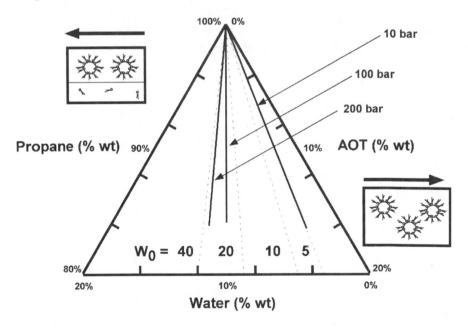

Figure 7. The ternary phase diagram for liquid propane, water and AOT. The phase diagram partitions into homogeneous solutions of well defined reverse micelles (to the right of the indicated pressure lines) and inhomogeneous mixtures of reverse micelles and both smaller and larger assemblies of surfactant (to the left of the indicated pressure lines). Redrawn with permission from Smith *et al.* (1990) *J. Phys. Chem.* **94**: 781-787. Copyright (1990) the American Chemical Society.

It is important to note that not much is known about how the ternary phase diagrams of the short chain alkanes respond to the presence of protein molecules. However, initial success described below seems to indicate that the same trends are followed at least semi-quantitatively.

4.2.4   Reverse Micelle Surfactants

The dominant force driving the encapsulation of proteins by reverse micelle assemblies is the electrostatic interaction between the headgroup of the surfactant and the surface charges of the protein. Therefore to broaden the range of proteins that can be encapsulated, a range of cationic and anionic surfactants may be required. A number of surfactants that form reverse micelles have been reported and include CTAB (Mantalvo *et al.*, 1995; Zhu *et al.*, 1999), DOLPA (Goto *et al.*, 1990; Ono *et al.*, 1996; Goto *et al.*, 1997), TOMAC (Drekker *et al.*, 1986), DODMAC (Wang *et al.*, 1994) and DEPTA, DEHPA, and PC-88A (Rong *et al*, 1998, 1999). Some natural lipids will also form reverse micelles (Walde *et al.*, 1990). The chemical structures of many of these surfactants are shown in Figure 8.

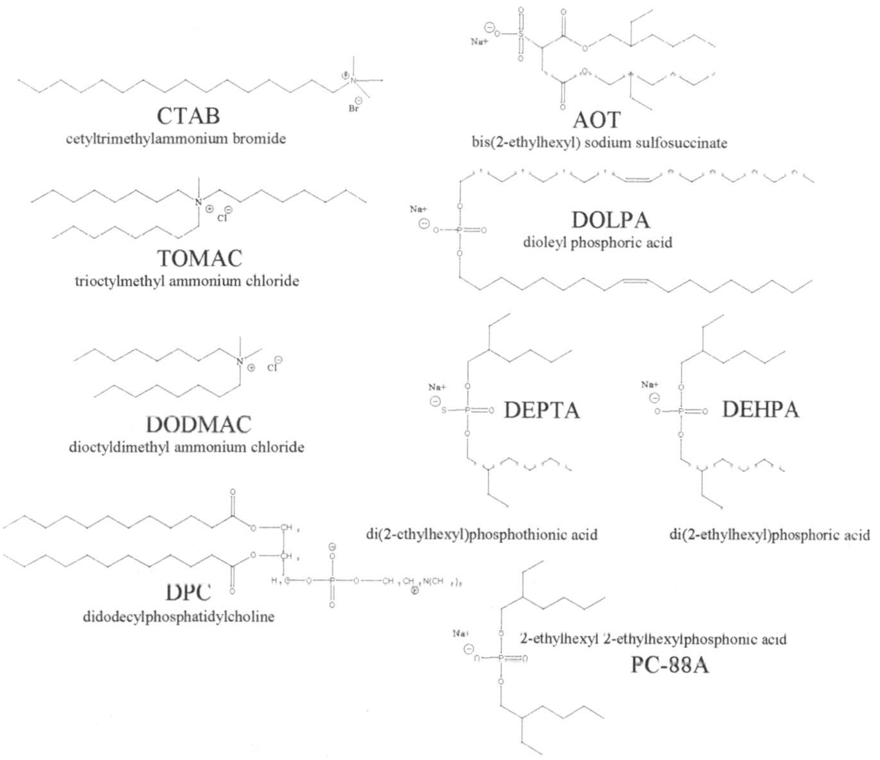

Figure 8. Chemical structures and common nomenclatures of various reverse micelle forming surfactants.

5.   NMR SPECTROSCOPY

A key goal in the development of the reverse micelle strategy is to preserve the capabilities of modern triple resonance NMR spectroscopy. In other words, one would like to avoid the need for extensive modification or further development of the current state-of-the-art. There are only two serious issues that distinguish preparations of encapsulated proteins dissolved in low viscosity fluids from simple aqueous solutions of proteins. These are the relatively low concentrations, typically 0.1 to 0.2 mM, that can be prepared and the presence of numerous coupled solvent and surfactant lines that dwarf the resonance lines of interest (Figure 9).

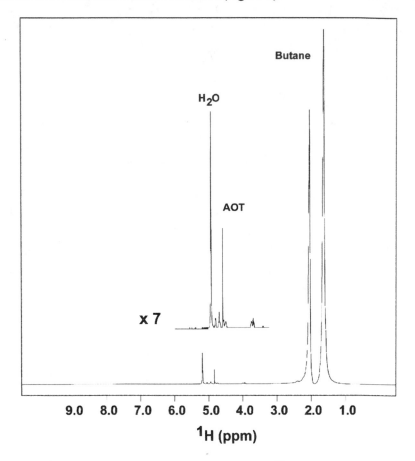

Figure 9. ^{1}H NMR spectrum of a preparation of 0.2 mM ^{15}N-labeled recombinant human ubiquitin in 75 mM AOT surfactant, 750 mM H_2O in liquefied butane (~10 M). This spectrum illustrated the unusual dynamic range problem presented by these systems. Deuteration of the alkane solvent is a viable and reasonably cost effective option while deuteration of the surfactant usually is not.

5.1   Solvent and Solute Suppression

A typical preparation of an encapsulated protein might be composed of 0.2 mM protein, ~10 M butane, ~1 M H_2O and 100 mM AOT. This creates a huge dynamic range problem for ^1H-detected NMR spectroscopy (Figure 9). Indeed, direct one-and two-dimensional ^1H NMR spectroscopy is largely unworkable. Fortunately, the general H_2O solvent suppression strategies that are inherent to modern NMR experiments are largely capable of suppressing unwanted ^1H resonances, mostly by virtue of heteronuclear filtration. This is particularly true for experiments that resolve information on the amide ^1H resonance via correlation with the amide ^{15}N. Especially good examples are the HN-resolved carbon TOCSY (Montelione *et al.*, 1992), ^{15}N-resolved NOESY and the HNHA (Vuister *et al.*, 1993) (see Figures 10, 11 and 12). Nevertheless, one is able to obtain high quality ^{13}C-resolved NOESY (Figure 13) and HCCH-TOCSY experiments in deuterated solvents (Figure 14).

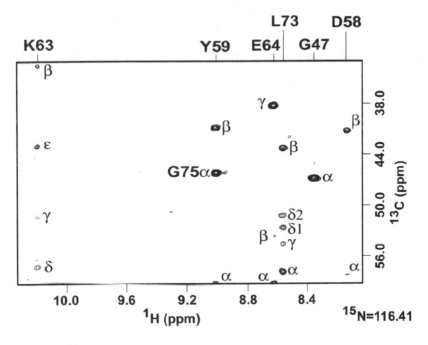

Figure 10. ^{15}N-slice through a three dimensional CC(CO)NH-TOCSY spectrum of recombinant human ubiquitin (0.25 mM) in liquid butane at 20 C. The spectrum was collected at 750 MHz (^1H). Total recording time was 103 hrs with the field strength of 8 kHz and the DIPSI mixing time of 20.4 ms.

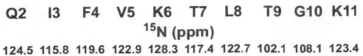

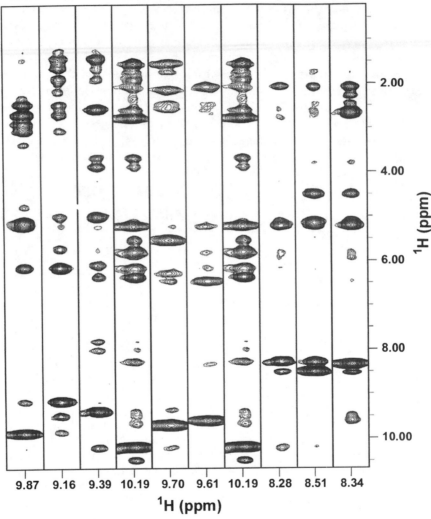

Figure 11.   Strip plot of ^{15}N-slices from a three-dimensional ^{15}N-resolved NOESY of recombinant human ubiquitin encapsulated in AOT reverse micelle dissolved in liquid d_{12}-pentane. A 90 ms NOE mixing period was used and the total recording time was 67 hrs.

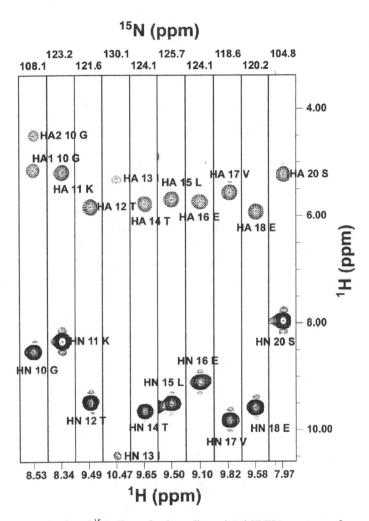

Figure 12. Strip plot of ^{15}N-slices of a three dimensional HNHA spectrum of recombinant human ubiquitin (0.25 mM) in liquid d_{12}-pentane at 20 C. The spectrum was collected at 750 MHz (^{1}H). The dotted peaks are HN/HA cross peaks with negative intensities. The total recording time was 118 hrs.

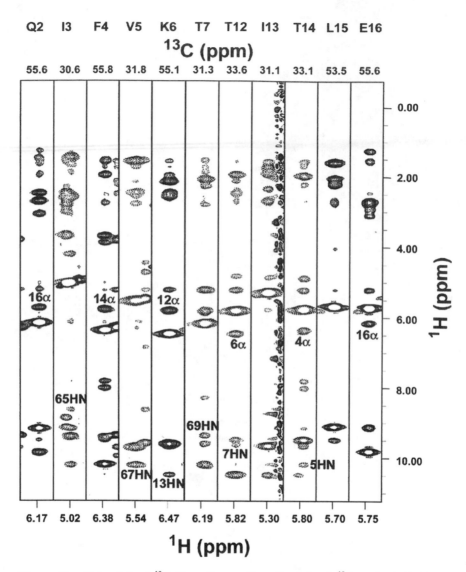

Figure 13. Strip plot of ^{13}C-slices from a three-dimensional ^{13}C-resolved NOESY of recombinant human ubiquitin encapsulated in AOT reverse micelles dissolved in liquid d$_{12}$-pentane at 20 C. A 90 ms NOE mixing period was used. The total recording time was 93 hrs. The indicated NOEs correspond to inter-strand H$_\alpha$/H$_\alpha$ and H$_\alpha$/H$_N$ NOEs arising in parallel and anti-parallel beta sheets.

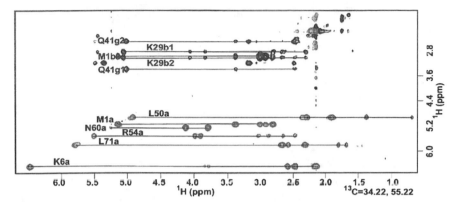

Figure 14. ^{13}C-slice through a three dimensional HCCH-TOCSY spectrum of recombinant human ubiquitin (0.25 mM) in liquid d_{12}-pentane at 20 C. The spectrum was collected at 750 MHz (^{1}H). Total recording time was 91 hrs with a B_1 field strength of 8.9 kHz and the DIPSI mixing time of 18.3 ms.

5.2   Sensitivity

Unfortunately NMR spectroscopy is an inherently insensitive technique and this insensitivity is often a fundamental limitation to its application in a variety of contexts in physics, chemistry and biology. This is particularly true in the context of NMR spectroscopy of encapsulated proteins dissolved in low viscosity fluids. This is because the region of phase space available to well-behaved reverse micelle solutions is quite limited with respect to their concentration. The effective concentrations of encapsulated proteins that can be prepared is accordingly limited to ~0.3 mM and often significantly less. Thus, a further extension of the sensitivity of the NMR method to allow for routine acquisition of triple resonance data in the sub-mM sample concentration regime is highly desirable. Such an extension is now on the near horizon and is particularly well suited for application to NMR studies of encapsulated proteins dissolved in low viscosity fluids like the short chain alkanes.

It has long been realized that cooling the receiver coil to cryogenic temperatures would significantly reduce the noise voltage associated with signal detection. To efficiently explain this let us represent the sensitivity or signal-to-noise (S/N) of an NMR probe as a function of its filling factor (η), the effective quality factor of the probe (Q_{eff}), the effective noise-temperature of a room-temperature sample in a coil (T_{eff}), and the noise-temperature of the receiver (T_A). Then we have (Hoult and Richards, 1976; Hoult and Lauterbur, 1979; Gadian and Robinson, 1979):

$$T_{eff} = \frac{T_c Q_s + T_s Q_c}{Q_c + Q_s} \; ; \; \frac{S}{N} \propto \left[\frac{\eta Q_{eff}}{T_{eff} + T_s} \right]^{1/2} \; ; \; Q_{eff} = \frac{Q_c Q_s}{Q_c + Q_s} = \frac{\omega_0 L}{R_c + R_s} \qquad (9)$$

Q_c is the directly measured quality factor of the empty coil and Q_s is the contribution of the sample to the effective quality factor, ω_0 is the resonance frequency, L is the inductance of the circuit, R_c and R_s are the equivalent resistances respectively of the coil circuit and sample, T_s is the sample temperature and T_c is the coil temperature. Equations 9 indicate that cooling of the receiver coil to cryogenic temperatures should, in principle, dramatically increase the signal-to-noise of the an NMR probe.

Styles *et al.* (1984) made the first practical demonstration of a cryogenic probe showing substantial gain in sensitivity. Unfortunately, the performance of the cryogenic probe is exquisitely sensitive to conductivity of the sample rendering their utility in the context of samples containing high (~100 mM) concentrations of salts problematic. This is particularly pertinent in the context of multinuclear multidimensional NMR studies of proteins and other biopolymers in aqueous solution. Inductive losses result from the dissipation of power due to the induction of current in the conducting sample (Hoult and Richards, 1976; Hoult and Lauterbur, 1979; Gadian and Robinson, 1979). Dielectric losses result from the passage through the conducting sample of the electrical lines of force arising from the distributed capacitance of the **rf** coil. This dependence of Q_s on the conductivity of a cylindrical sample of radius r_s and conductivity σ is conveniently expressed by Equation 10.

$$Q_s \propto \left[\eta \omega_0 r_s^2 \sigma \right]^{-1} \qquad (10)$$

The high Q of a cryogenic probe is rapidly degraded by the effective resistance caused by inductive and dielectric losses being introduced into a circuit of otherwise very low resistance. Accordingly, the conventional probe is relatively insensitive to conductive samples as compared to a cryogenic probe. This is illustrated in Figure 15 for a 400 MHz cryogenic probe and its room temperature counterpart. A simple solution to this unsatisfactory loss of performance comes with the realization that the losses arise from the bulk properties of the solvent and have little to do with the protein solutes themselves. Ideally, then, one would simply like to replace the ~99% of the volume of the sample occupied by bulk solvent and dissolved ions with a solvent of low conductivity. This is precisely the situation that one has with the solutions of encapsulated proteins where the majority of the bulk aqueous phase has been discarded in favour of a low dielectric non-conducting organic solvent.

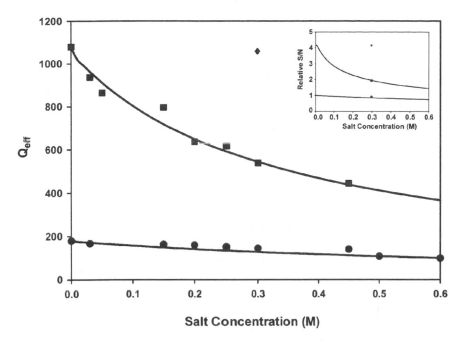

Figure 15. Effective probe quality factor (Q_{eff}) as a function of salt concentration. Measured Q_{eff} values for a conventional 400 MHz probe (solid circles) and a cryogenic 400 MHz probe (solid squares) are shown. Solid lines indicate the fitted theoretical curves (see Equations 9 and 10). The effective probe quality factor obtained for ubiquitin dissolved in 50 mM acetate buffer and 250 mM NaCl encapsulated in AOT reverse micelles dissolved in liquid pentane is shown with a solid diamond. The inset shows the fitted theoretical curves expressed as signal-to-noise. These data indicate that the full performance of the cryogenic probe is recovered for lossy protein samples when encapsulation in low dielectric solvent is employed. Adapted with permission from Flynn *et al.* (2000) *J. Am. Chem. Soc.* **122**: 4823-4824. Copyright (2000) the American Chemical Society.

Thus the combination of cryogenic probe technology with encapsulation of proteins within reverse micelles dissolved in organic solvents is perhaps a perfect marriage of methods. On the one hand, the apparent limitation on protein loading to ~0.25-0.5 mM solutions of reverse micelles is overcome by the use of the high sensitivity cryogenic probe. On the other hand, the enormous sensitivity of the cryogenic probe is made accessible to lossy samples such as protein solutions of high ionic strength by use of the reverse micelle approach. Indeed, essentially the full performance of the probe is recovered (Figure 15). This allows very high quality two-dimensional [15]N-

HSQC spectra to be obtained in as little as 7 minutes at a field strength of 7 T (400 MHz ^{1}H) (Figure 16).

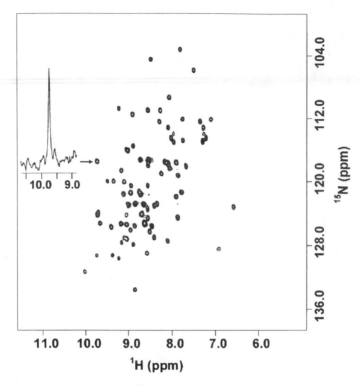

Figure 16. Contour plot of the ^{15}N-HSQC spectrum of recombinant human ubiquitin encapsulated in AOT reverse micelles dissolved in n-pentane. The protein was encapsulated at a concentration of ~ 0.25mM in sodium acetate buffer (50 mM) containing 0.25M NaCl. The spectrum was recorded at 400 MHz (^{1}H) using a cryogenic probe with the sample chamber temperature equilibrated at 20°C. The spectral width in the direct ^{1}H-dimension was 8000 Hz (20 ppm) digitized into 512 complex points (64 ms acquisition time); the spectral width in the indirect ^{15}N-dimension was 2200 Hz (54.3 ppm) using 76 complex points (34.5 ms acquisition time). Total acquisition time for the experiment was approximately 7 minutes. The inset shows a ^{1}H cross section through the cross peak indicated by the arrow and illustrates the typical S/N of the spectrum. Reprinted with permission from Flynn *et al.* (2000) *J. Am. Chem. Soc.* **122**: 4823-4824. Copyright (2000) the American Chemical Society.

6.   VALIDATION

A great deal of effort has been put forth over the past several decades in the development of reverse-micelle-forming solvents and surfactants and in the characterization of the reverse micelles and the molecules encapsulated within them (Luisi, 1985; De Gomez-Puyou and Gomez-Puyou, 1998). Remarkably however, it is only recently that comprehensive structural information has been obtained for an encapsulated protein. The absence of such information raised the issue of whether reverse micelle encapsulation is a viable tool for the determination of biologically relevant protein structures. Furthermore the basic proposition of the general approach – that the relaxation properties of protein encapsulated within a reverse micelle can be modulated by simply changing the bulk solvent viscosity – remains to be demonstrated. Finally, one would hope that the full battery of modern triple resonance NMR spectroscopy techniques can be applied to encapsulated protein systems. Here we briefly summarize recent results that demonstrate that all these criteria can be met.

6.1   Uniqueness of Encapsulated Protein Structure

As a first step towards validation of using reverse micelle encapsulation to facilitate characterization of proteins by solution NMR methods, we have examined the behaviour of recombinant human ubiquitin in AOT reverse micelles prepared in a variety of short chain alkane solvents (Wand et al., 1998; Babu et al., 2001). Conditions were easily found which resulted in an ^{15}N-HSQC spectrum typical of a unique folded structure and, importantly, closely similar to that obtained for ubiquitin in aqueous buffer. Equally important, the ^{15}N-HSQC spectrum of encapsulated ubiquitin is, apart from resonance line widths, independent of the alkane solvent (Figure 17). Obviously, the ^{15}N-HSQC spectrum is by far the easiest, most direct and most sensitive way to assess the NMR performance and integrity of an encapsulated protein.

These results indicate that the protein is folded into a unique structure. In a sense this is remarkable, since the volume of a reverse micelle is so small it precludes the concept of a bulk pH. This is to say that the existence of a single unique structure (spectrum) implies that a homogenous population of protein molecules is present. This in turn requires that either every protein molecule exists in the same ionisation state or that the population of protein molecules is highly averaged. Because of the small volume of the reverse micelle, the latter possibility is most likely and there is indeed extensive documentation of the dynamic exchange of contents between reverse micelles that is fast on the NMR chemical shift time scale (Carlstrom and Halle, 1988). This then leads to the averaged spectra observed.

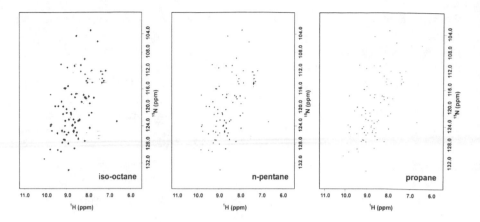

Figure 17. ^{15}N-HSQC spectra obtained for encapsulated human ubiquitin in AOT reverse micelles dissolved in propane, pentane and octane. The protein was prepared in 50 mM acetate buffer (pH 5) and 0.25 KCl to a w_o of 10. Spectra were obtained at 25 C at 750 MHz (^{1}H).

6.2   Assessment of Hydrodynamic Performance

A key component of the approach is the ability to obtain homogenous preparations of proteins encapsulated in reverse micelles containing only a small amount of water. The intent is to keep the volume of the reverse micelle assembly as small as possible. This serves to maximize the rate of molecular reorientation by through both the volume and viscosity terms of Equation 6. The hydrodynamic performance of a reverse micelle preparation is most easily assessed by measurement of translational diffusion via by use of pulsed field gradient diffusion measurements (Altieri *et al.*, 1995).

$$D_{trans} = kT/6\pi\eta r_h \tag{11}$$

For example, one can compare the apparent diffusion rates of ubiquitin in water (M_r 8.5 kDa) to the diffusion rates of "empty" AOT micelles in pentane and cytochrome c loaded AOT micelles in pentane (Figure 18). At 20 C, ubiquitin in water and the "empty" AOT reverse micelles in pentane have diffusion rates differing by a factor of about 4. The "empty" AOT reverse micelles and cytochrome c loaded AOT micelles in pentane differ in diffusion rates about a factor of 2.

At this temperature, water and pentane differ in viscosity by ~4 fold. This suggests that the effective radius of the "empty" AOT micelle in pentane is about that of ubiquitin in water. A more interesting result is the observation that the "empty" AOT reverse micelle has a diffusion constant approximately twice that of an AOT reverse micelle loaded with cytochrome c which is completely consistent with one cytochrome c molecule per micelle. These kinds of experiments are very sensitive and are therefore very quick to carry out. In a very real sense, they provide a "proof of the pudding" test of the appropriateness of a given preparation. Both the diffusion of the surfactant (by simple ^{1}H NMR) or the protein (by ^{15}N-filtered NMR) can be followed.

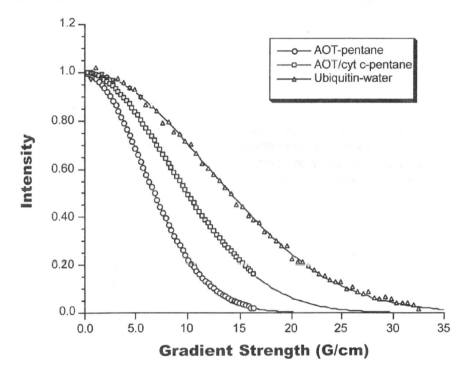

Figure 18. sLED pulsed field gradient diffusion profiles of ubiquitin in water (upper curve), cytochrome c in AOT reverse micelles in pentane (middle curve) and empty AOT reverse micelles in pentane. The fitted diffusion constants are 1.9×10^{-6} cm^2/sec, 4.0×10^{-6}, 7.3×10^{-6} respectively. The fact that ubiquitin in water diffuses more slowly than cytochrome c in AOT reverse micelles in pentane is an indirect proof of the principle of the reverse micelle approach.

6.3   Towards Validation as a Tool for Structural Biology

It is only recently that comprehensive structural information has been previously obtained for an encapsulated protein. Such information is vital to the acceptance of reverse micelle encapsulation is a viable tool for the determination of biologically relevant protein structures. As a first step to evaluate whether an encapsulated protein adopts its native (i.e. free solution) structure we have assigned the spectrum of recombinant human ubiquitin encapsulated in AOT reverse micelles and compared the obtained chemical shifts to those of the protein in free solution (Wand *et al.*, 1998).

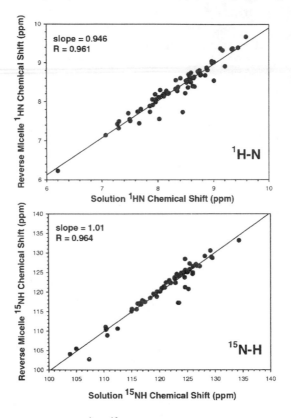

Figure 19. Correlation of amide ^{1}H - ^{15}N chemical shifts of recombinant human ubiquitin in aqueous buffer with those of the protein encapsulated in a reverse micelle dissolved in pentane. In both cases the protein was solvated by aqueous buffer containing 50 mM sodium acetate, pH 5, and 250 mM NaCl. ^{15}N-HSQC spectra were obtained at 25 C at 750 MHz. The upper panel shows the comparison of amide ^{1}H chemical shifts and the lower panel shows the comparison of ^{15}N chemical shifts. Reprinted with permission from Wand *et al.*, (1998) *Proc. Natl. Acad. Sci. USA* **95**: 15299–15302. Copyright (1998) National Academy of Sciences USA.

Comparison of the chemical shifts of amide ^{1}H and ^{15}N in the AOT reverse micelle and those in water reveals minimal differences (Figure 19). The standard deviations between amide ^{1}H and ^{15}N chemical shifts of ubiquitin in the two states are 0.17 and 1.6 ppm, respectively. Two localized regions display chemical shift perturbations. These include the C-terminal residues R72 and R74 and residues 45 to 48 that form a tight turn.

As a second step towards validation that an encapsulated protein adopts its native (i.e. free solution) structure we have determined the structure of human ubiquitin encapsulated in AOT reverse micelles (Babu *et al.*, 2001). Distance restraints were derived from NOESY-^{15}N-HSQC and NOESY-^{13}C-HSQC spectra acquired with mixing times of 90 ms as illustrated above. Dihedral angle (ϕ) restraints were calculated from the HNHA quantitative J correlation experiment. A comprehensive array of structural restraints was obtained: 1805 unique NOEs (873 intraresidue, 336 short-range, 239 medium-range, and 357 long-range) and 63 phi torsion angle restraints derived from J-coupling.

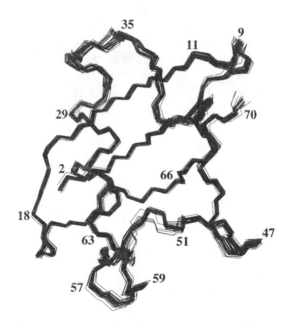

Figure 20. Structure of encapsulated human ubiquitin. The family of 32 structures were superimposed on backbone atoms for residues 2–70. Residues 71 through 76 are disordered and are not shown. The structures have been deposited in the PDB under code 1G6J. Reprinted with permission from Babu *et al.* (2001) *J. Am. Chem. Soc.* **123**: 2691-2692. Copyright (2001) the American Chemical Society.

Structures were calculated in the usual way using the torsion angle dynamics structure refinement program DYANA (Guntert *et al.*, 1997). The final family of structures has an average rmsd of 0.26 ± 0.05 Å for backbone atoms and 0.77 ± 0.04 Å for heavy atoms to the mean structure (residues 2–

70) (Figure 20). The target function values ranged from 0.05 to 0.23 Å² with an average of 0.16 ± 0.05 Å². These structural statistics indicate that the structure is determined to high precision. Residues 9, 10, 35, 36, 52, and 53 have slightly higher local rmsds in comparison to the average. These regions are suggested to be flexible according to generalized order parameters obtained for N–H vectors for ubiquitin in free solution (Schneider *et al.*, 1992). Ramachandran plot analysis using the Procheck criteria (Laskowski *et al.*, 1993) indicates that 87% of residues fall in the "most favoured" and 13% in "additional allowed" regions when glycines and prolines are excluded.

To identify potential differences, the family of structures for encapsulated ubiquitin was compared to the crystal structure (Vijay-Kumar *et al.*, 1987) and to the structure determined in free solution (Cornilescu *et al.*, 1998). These quantitative comparisons are summarized in Table 2. Superpositions were generated using residues 2 through 70, i.e., omitting the C-terminus that is known to be unstructured in solution (Di Stefano and Wand, 1987) and to have large thermal motion in the crystal (Vijay-Kumar *et al.*, 1987). Earlier work (Clore and Gronenborn, 1998) and the precision of the model determined here (0.26Å) suggests that the structure obtained for encapsulated ubiquitin and those obtained for the protein in free solution (Cornilescu *et al.*, 1998) and in the crystal (Vijay-Kumar *et al.*, 1987) are in quantitative agreement.

Table 2. Comparison of crystalline, free solution and reverse micelle encapsulated human ubiquitin structures[a]

Comparison	Backbone rmsd (Å)	Heavy atom rmsd (Å)
Crystal *vs* rev. micelle	0.77 ± 0.03	1.63 ± 0.05
Free solution *vs* rev. micelle	0.79 ± 0.04	1.57 ± 0.07
Crystal *vs* free solution	0.35 ± 0.01	1.15 ± 0.05

[a]rmsds were calculated for residues 2–70.

Minor variations in the structure of encapsulated ubiquitin can be identified in the first reverse turn between residues 8 and 10, in the region between residues 32 and 34 and near residue 62. Most of these differences probably arise as a result of the relative scarcity of NOEs derived for these regions. One apparent exception, however, involves residues 62–64 where short distance interactions (NOEs) are seen in the structure of the encapsulated protein but are not observed in the free solution structure. These distance restraints cause a minor localized variance (< 1.3 Å) for residue 62. The origin of these minor variations is unclear. Nevertheless, these structural comparisons demonstrate that the structure of ubiquitin remains largely undisturbed upon encapsulation. The results ubiquitin therefore represent a significant first step in the validation of the reverse micelle approach as a tool for determining structures of proteins using standard triple-resonance based solution NMR methods.

6.4 Modulation of Relaxation Properties

The central goal of the reverse micelle encapsulation strategy is to reduce the rate of spin-spin relaxation in a protein, relative to its rate for the protein in free solution. In accordance with the Stokes-Einstein relation (Equation 6), one anticipates then that effective tumbling time of the encapsulated protein should be directly related to the <u>bulk</u> solvent viscosity. One could imagine that this need not be the case and so it is important to demonstrate that this is indeed true. In an initial effort to study line narrowing we encapsulated ^{15}N-ubiquitin in AOT reverse micelles prepared in iso-octane, n-heptane, n-hexane, n-pentane, n-butane and propane.

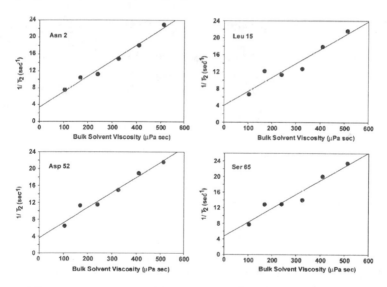

Figure 21. Spin-spin relaxation rates ($1/T_2$) for the amide nitrogens of example residues are plotted against the bulk solvent viscosity. Data points, in order of decreasing viscosity are for iso-octane, n-heptane, n-hexane, n-pentane, n-butane, and propane. All data was collected a 25 C. The butane and propane samples were prepared at pressures of 7 and 12 bar, respectively. All other samples were prepared at ambient pressure. Reprinted with permission from Wand *et al.* (1998) *Proc. Natl. Acad. Sci. USA* **95**: 15299–15302. Copyright (1998) National Academy of Sciences USA.

Since the main chain dynamics of ubiquitin in water have been characterized previously by NMR relaxation methods (Schneider *et al.*, 1992), regions of the protein where T_2 is dominated by the influence of global tumbling are known and can be used as monitors of molecular tumbling. The expected roughly linear dependence of $1/T_2$ on solvent viscosity is clearly revealed (Figure 21). As hoped, the spin-spin nitrogen relaxation times of ubiquitin in water are largely recovered when ubiquitin encapsulated in AOT reverse micelles is dissolved in liquid propane. In this range of viscosity, the reverse micelles remain in the slow tumbling limit ($\omega\tau_m > 1$). Extrapolation from this roughly linear region to zero viscosity should ideally yield a limiting ($1/T_2$) value of ~2 sec^{-1}, about one half of that obtained. This deviation presumably reflects differences between the effective microviscosity experienced by the reverse micelle and the viscosity of the bulk solvent as well as changes in the volume of the reverse micelle with solvent.

7.   FUTURE PROSPECTS

7.1   Application to Large Soluble Proteins

The original motivation behind the development of the reverse micelle strategy was to overcome the slow tumbling problem presented by large soluble proteins. There have been extensive literature reports of the encapsulation of large soluble enzymes with maintenance of significant activity, which implies successful conservation of native structure. Many large enzymes are oligomeric proteins and an important issue is whether the appropriate oligomer can be encapsulated. Initial studies with triose phosphate isomerase (TIM) from T. brucei are most encouraging in this regard. TIM is a 56 kDa homodimer, has a basic pI and is encapsulated by AOT reverse micelles (Figure 22).

7.2   Extension to Membrane Proteins

The reverse micelle carries with it other important potential applications. In contrast to surfactants in water, the reverse micelle in an organic solvent can potentially offer two solvents to the solublized protein. In principle, this provides a route to the solubilization of integral membrane proteins where the aqueous phase would solubilize the hydrophilic component while the alkane phase to be used here would solublize the integral membrane component. The surfactant would act as a bridge between the two phases. This is schematically illustrated below (Figure 23). There is precedence in the literature for this solubilization scheme for membrane proteins (Darszon et al., 1979; Schonfeld et al., 1980; Warncke and Dutton, 1993).

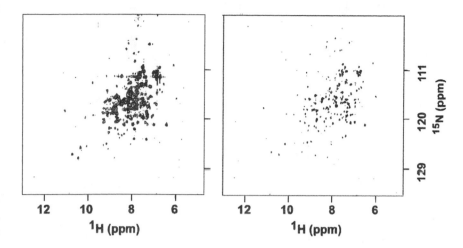

Figure 22. ^{15}N-HSQC spectra of triose phosphate isomerase from T. Brucei in water (left panel) and encapsulated in AOT reverse micelles dissolved in liquid propane. The protein concentration in both cases was ~0.1 mM. Spectra were obtained at 20 C at 750 MHz (^{1}H).

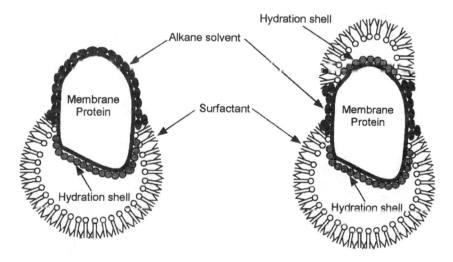

Figure 23. Schematic illustration of the solubilization scheme envisaged for membrane proteins. The reverse micelle surfactant would form an inverted "shower cap" over the aqueous face(s) of the integral membrane protein while the membrane spanning portion would be bathed in the alkane solvent.

7.3 Future Technical Advances

Solutions to several technical issues loom on the near horizon. These
include exploring the viability of the ethane/surfactant/water ternary phase
diagram to support encapsulated proteins, examination of other low viscosity
solvents such as fluoroform (CHF_3), and development of tools for obtaining
partial alignment of encapsulated proteins and thereby recover the full range
of experimental solution NMR techniques in this context.

7.3.1 Ethane and Fluoroform – Ultimate Solvents?

It has already been established that supercritical ethane will support well-
behaved AOT/water reverse micelles (Smith *et al.*, 1990). Unfortunately,
ethane will require substantially higher pressures than that needed by
preparations using propane. The phase diagram for ethane/AOT/water is
shown in Figure 24 and indicates that pressures on the order of 100-200 bar
may be required to maintain stable solutions of reverse micelles.

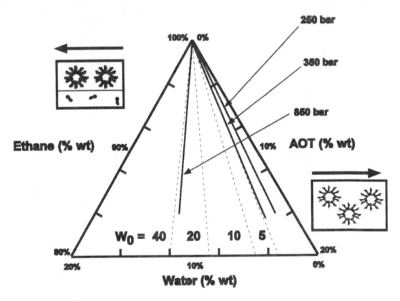

Figure 24. The ternary phase diagram for liquid ethane, water and AOT. The phase diagram
partitions into homogeneous solutions of well defined reverse micelles (to the right of the
indicated pressure lines) and inhomogeneous mixtures of reverse micelles and both smaller and
larger assemblies of surfactant (to the left of the indicated pressure lines). Increasing pressure
stabilizes the desired homogeneous solutions. Pressures as high as 200 bar may be required to
obtain well behaved solutions of AOT reverse micelles containing sufficient levels of water.
Redrawn with permission from Smith *et al.* (1990) *J. Phys. Chem.* **94**: 781-787. Copyright
(1990) the American Chemical Society.

It must be emphasized that *the pressures required are far below those that perturb the <u>structure</u> of a native-state protein* (Robinson and Sligar, 1995). At pressures somewhat above those anticipated to be required for ethane (200 bar), the chemistry of enzyme activity is affected, due to effects in the transition state not to structural effects. Similarly, CF_3H has a very attractive viscosity ($\sim$ 25 μPa s at 25 C). Though both ethane and CF_3H are very attractive solvents, the requirement for sample preparation and maintenance under significant pressure presents a considerable challenge. In closing, we should mention that supercritical carbon dioxide also has an attractive viscosity ($\sim$ 95 μPa s) but it has proven quite difficult to find suitable surfactants (see Johnson *et al.* 1996). Bax has recently encapsulated BPTI using a fluorocarbon surfactant (Gaemers *et al.*, 1999). However, in the presence of water (as will be the case here), the resulting aqueous solution becomes quite acidic which significantly limits the lifetime of proteins encapsulated in this solvent.

7.3.2   Partial Alignment of Encapsulated Proteins

It is probably accurate to say that the recently introduced ability to obtain residual dipolar couplings arising from the partial alignment of macromolecules in solution has provided access to an array of structural restraints of unprecedented breadth and accuracy. The original partial alignment strategy of Tjandra and Bax using bicelles as steric alignment tools (Tjandra and Bax, 1997) has more recently been supplemented by use of phage (Hansen *et al.*, 1998; Clore *et al.*, 1998) and purple membrane (Koenig *et al.*, 1999). Obviously, one would very much like to see this capability in the context of the reverse micelle approach. Unfortunately, none of the current approaches will work in the context of the organic solvent/surfactant systems. A "back of the envelope" calculation seems to suggest that electric fields can be used to align the reverse micelle/protein assemblies. For many of the same reasons that the cyrogenic probe performs so much better with a reverse micelle preparation, one anticipates that the very large dipole moments of proteins, generally on the order of hundreds of Debye, can be used for alignment. The cavity field in the region occupied by the reverse micelle but outside the protein is estimated by:

$$E_c = \frac{3\varepsilon_0}{2\varepsilon_0 + \varepsilon_i} \bullet E_{ext} \tag{13}$$

We wish to obtain $\sim$ 0.1 % alignment. Treating the protein as a Langevin dipole and using the dipole moment of ubiquitin, for example, we find that E_c

should reach ~100 V/cm.   In the absence of salt in the aqueous layer surrounding the protein, we therefore anticipate that the applied field should be ~ 1500 V/cm. Considering the presence of 0.5 M salt, the applied field must reach ~9,000 V/cm.   Fortunately, the protein itself provides an additional downward correction such that a 0.1 % alignment would be achieved at ~6,000 V/cm of applied electric field. This avenue is being actively pursued in our laboratory.

8.   SUMMARY

We have described the basic ideas underlying the use of encapsulated proteins dissolved in low viscosity fluids to decrease the effective molecular tumbling of large soluble proteins. It has been shown that the technical difficulties presented by the use of pressurized fluids are easily overcome and that the desired results are obtained. It has also been shown that for the proteins thus far examined conditions can be found where encapsulation results in the maintenance of the native structure of the protein. Thus the general approach appears to be valid technique to be used in the context of structural biology.

8.1.   Acknowledgments

This work is supported by operating grants from the National Institutes of Health, a grant from the Culpeper Biomedical Pilot Program and by equipment grants from the NIH and the Army Research Office.

9.   REFERENCES

Altieri, A. S., Hinton, D. P., and Byrd, R. A., 1995, *J. Am. Chem. Soc.* **117**: 7566-7567.
Babu, C. R., Flynn, P. F., and Wand, A J., 2001, *J. Am. Chem. Soc.* **123**: 2691-2692.
Bax, A. and Davis, D. G., 1985, *J. Magn. Reson.* **65**: 355-360.
Brauschweiler, L., and Ernst, R. R., 1983, J. Magn. Reson. **53**: 521-528.
Carlström, G. and Halle, B., 1988, *Mol. Phys.* **64**: 659-678.
Clore, G. M., and Gronenborn, A. M., 1997, *Nat. Struct. Biol.* **4**: 845-848.
Clore, G. M., and Gronenborn, A. M., 1998, *Proc Natl Acad Sci U S A* **95**: 5891–5898.
Clore, G. M., Starich, M. R., and Gronenborn, A. M., 1998, *J. Amer. Chem. Soc.* **120**: 10571-10572.
Cornilescu, G., Marquardt, J. L., Ottiger, M., and Bax, A., 1998, *J. Am. Chem. Soc.* **120**: 6836–6837.
Drekker, M., Van 'T Riet, K., and Weijers, S. R., 1986, *Chem. Eng. J.* **33**: B27-B33.
Darszon, A., Strasser, R. J., and Montal, M., 1979, *Biochemistry* **18**:5205-13.

De Gomez-Puyou, M. T., and Gomez-Puyou, A., 1998, *Crit. Rev. Biochem. Mol. Biol.* **33**: 53–89.

Di Stefano, D. L., and Wand, A. J., 1987, *Biochemistry* **26**: 7272–7281.

Ehrhardt, M. R., Flynn, P. F., and Wand, A. J., 1999, *J. Biomol. NMR* **14**: 75-78.

Farrar, T.C., and Stringfellow, T.C., 1996, In *Encyclopedia of Nuclear Magnetic Resonance Spectroscopy*, (Grant, D.M. and Harris, R.K., eds.), Wiley and Sons, New York, Vol. 6, 4101-4107.

Flynn, P. F., Mattiello, D. L., Hill H. D. W., and Wand, A. J., 2000, *J. Am. Chem. Soc.* **122**: 4823-4824.

Frank, S. G., and Zografi, G., 1969, *J. Colloid Interface Sci.* **29**: 27-35.

Fulton, J. L., and Smith, R. D., 1988, *J. Phys. Chem.* **92**: 2903-2907.

Fulton, J. L., Blitz, J. P., Tingey, J. M., and Smith, R. D., 1989, *J. Phys. Chem.* **93**: 4128-4204.

Gadian, D. G., and Robinson, F. N. A., 1979, *J. Magn. Reson.* **34**: 449-455.

Gaemers, S., Elsevier, C. J., and Bax, A., 1999, *Chem. Phys. Lett.* **301**: 138-144

Gale, R. W., Fulton, J. L., and Smith, R. D., 1987, *J. Am. Chem. Soc.* **109**: 920-921.

Gardner, K.H., and Kay, L.E., 1997, *J. Am. Chem. Soc.* **119**: 7599-7600.

Glaser, S. J., and Drobny, G. P., 1989, *Chem. Phys. Lett.* **164**: 456-462.

Goklen, K. E., and Hatton, T. A., 1985, *Biotechnology Progess* **1**: 69-74.

Goto, M., Kondo, K., and Nakashio, F., 1990, *J. Chem. Eng. Japan* **23**: 513-515.

Goto, M., Ono, T., Nakashio, F., Hatton, T. A., 1997, *Biotech. Bioeng.* **54**: 26-32.

Goto, N., Gardner, K.H., Mueller, G.A., Willis, R.C., and Kay, L.E., 1999, *J. Biomol. NMR* **13**: 369-374.

Griesinger, C., Otting, G., Wüthrich, K., and Ernst, R. R., 1988, *J. Am. Chem. Soc.* **110**: 7870-7872.

Guntert, P., Mumenthaler, C., and Wüthrich, K., 1997, *J. Mol. Biol.* **273**: 283–298.

Hansen, M. R., Mueller, L., and Pardi, A., 1998, *Nature Struct. Biol.* **5**: 1065-1074.

Hoult, D. I., and Richards, D. E., 1976, *J. Magn. Reson* **24**: 71-85.

Hoult, D. I., and Lauterbur, P. C., 1979, *J. Magn. Reson.* **34**: 425-433.

Johnston, K. P., Harrison, K. L., Clarke, M. J., Howdle, S. M., Heitz, M. P., Bright, F. V., Carlier, C., and Randolph, T. W., 1996, *Science* **271**: 624-626.

Kay, L. E., Ikura, M., Tschudin, R., and Bax, A., 1990, *J. Magn. Reson.* **89**: 495-514.

Ketchem R. R., Lee K. C., Huo S., and Cross T. A ., 1996, *J. of Biomolecular NMR* **8**: 1-14.

Koenig, B. W., Hu, J. S., Ottiger, M., Bose, S., Hendler, R. W., and Bax, A., 1999, *J. Amer. Chem. Soc.* **121**: 1385-1386.

Larsson, G., Wijmenga, S., and Schleucher, J., 1999, *J. Biomolecular NMR* **14**: 169-174.

Laskowski, R. A., MacArthur, M. W., Moss, D. S., and Thornton, J. M., 1993, *J. Appl. Cryst.* **26**: 283–291.

Vijay-Kumar, S., Bugg, C. E., and Cook, W. J., 1987, *J Mol Biol.* **194**: 531–544.

LeMaster, D. M., 1994, *Progr. NMR Spectroscopy* **26**: 371-419.

Lipari, G., and Szabo, A., 1982a, *J. Amer. Chem. Soc.* **104**: 4546-4559.

Lipari, G., and Szabo, A., 1982b, *J. Amer. Chem. Soc.* **104**: 4559-4570.

Luisi, P. L., 1985, *Angew. Chem. Int. Ed.* **24**: 439-528.

McIntosh, L. P., and Dahlquist, F. W., 1990, *Q. Rev. Biophys.* **23**: 1-38.

Mantalvo, G., Valiente, M., and Rodenas, E., 1995, *J. Colloid Interface Sci.* **172**: 494-501.

Mohebbi, A., and Shaka, A. J., 1991, *Chem. Phys. Lett.* **178**, 374-378.

Montelione, G. T., Lyons, B. A., Emerson, S. D., and Tashiro, M., 1992, *J. Am. Chem. Soc.* **114**: 10974–10975.

Mueller, G.A., Choy, W.Y., Forman-Kay, J.D., Venters, R.A., and Kay, L.E., 2000, *J. Mol. Biol.* **300**: 192-212.

Ono, T., Goto, M., Nakashio, F., and Hatton, T. A., 1996, *Biotech. Prog.* **12**: 793-800.
Pervushin, K., Riek, R., Wider, G., and Wüthrich, K., 1997, *Proc. Nat. Acad. Sci. USA* **94**: 12366-12371.
Redfield, A.G., 1965, *In* Advances in Magnetic Resonance, (Waugh, J.S., ed.), Academic Press, New York, Vol. 1, 1-32.
Robinson, C. R., and Sligar, S. G., 1995, *Methods Enzym.* **259**: 395-427.
Rong, L., Yamane, T., and Takeuchi, H., 1999, *J. Chem. Eng. Japan* **32**: 530-534.
Rong, L., Yamane, T., and Takeuchi, H., 1998, *J. Chem. Eng. Japan* **31**: 434-439.
Sattler, M., Schleucher, J., and Griesinger, C., 1999, *Prog. NMR Spectr.* **34**: 93-158.
Schneider, D. M., Dellwo, M. J., and Wand, A. J., 1992, *Biochemistry* **31**: 3645-3652.
Schonfeld, M., Montal, M., and Feher G., 1980, *Biochemistry* **19**:1535-1542.
Smith, R. D., Fulton, J. L., Blitz, J. P., and Tingey, J. M., 1990, *J. Phys. Chem.* **94**: 781-787.
Sørensen, O. W., Eich, G. W., Levitt, M. H., Bodenhausen, G., and Ernst, R. R., 1987, *Progr. NMR Spectroscopy* **16**: 163-192.
Styles, P., Soffe, N. F. J., Scott, C. A., Cragg, D. A., White, D. J., and White, P. C. J., 1984, *J. Magn. Reson.* **60**: 397-404.
Tjandra, N., and Bax, A., 1997, *Science* **278**: 1111-1114.
Vuister, G. W., and Bax, A., 1993, *J. Am. Chem. Soc.* **115**: 7772-7777.
Wagner, G., 1997, *Nat. Struct. Biol.* **4**: 841-844.
Walde, P., Giuliani, A. M., Boicelli, C. A., and Luisi, P. L., 1990, *Chem. Phy. Lipids* **53**: 265-288.
Wand, A. J., Ehrhardt, M. R., and Flynn, P. F., 1998, *Proc. Natl. Acad. Sci. U. S. A.* **95**: 15299-15302.
Wang, W., Weber, M. E., and Vera, J. H., 1994, *J. Colloid Inter. Sci.* **168**: 422-427.
Warncke, K., and Dutton, P. L., 1993, *Proc. Nat. Acad. Sci. USA* **90**: 2920-2924.
Wider, G., and Wüthrich, K., 1999, *Curr. Opin. Struct. Biol.* **9**: 594-601.
Zhu, H., Fan, Y-X., Shi, N., and Zhou, J-M., 1999, *Arch. Biochem. Biophys.* **368**: 61-66.

II

Structure Refinement

Chapter 7

Angular Restraints from Residual Dipolar Couplings for Structure Refinement

Christian Griesinger, Jens Meiler[1], Wolfgang Peti[2]
Max Planck Institute of Biophysical Chemistry, Group 030,
Am Fassberg 11
37077 Göttingen, Germany

Abstract: Dipolar couplings have proven to provide long range restraints before not available in high resolution NMR. This review presents an overview of recent techniques for the measurement of dipolar couplings. A special focus on side chain and proton-proton dipolar couplings can be found. The second part centers on the use of dipolar couplings in structure calculations of proteins and carbohydrates. Their usage in a 3D homology search and structural genomic efforts using NMR are also discussed.

1. INTRODUCTION AND THEORY

Nuclear magnetic resonance (NMR) spectroscopy is a standard technique used for the characterization of structure and dynamics of a wide range of compounds in all branches of chemistry and biology. NMR spectroscopy applied in solution has established its place as a major technique to determine structures of proteins, oligonucleotides, and other biologically relevant macromolecules and their complexes (Wüthrich, 1986; Croasmun and Carlson, 1994; Cavangh *et al.*, 1996; Clore and Gronenborn, 1998; Clore and Gronenborn, 1998). In addition, dynamics on time scales ranging from picoseconds to seconds and covering molecular events from

---

[1]*Current address: University of Washington, Department of Biochemistry, Box 357350, Seattle, Wa, 98195, USA*
[2]*Current address: The Scripps Research Institute, Department of Molecular Biology, MB-44 10550 North Torrey Pines Road, La Jolla, Ca, 92037, USA*

vibrations to chemical reactions can be monitored by NMR (Palmer *et al.*, 1996; Kay, 1998; Meiler *et al.*, 2001; Peti *et al.*, 2002). A suite of multidimensional NMR experiments have become routine experiments for the spectroscopic characterization of molecules (Sattler *et al.*, 1999). Experiments have been developed for the determination of distance restraints derived from quantification of NOESY and ROESY spectra, as well as for the determination of torsion angle restraints from measurement of scalar ^{3}J coupling constants (Griesinger *et al.*, 1998; Vuister *et al.*, 1998). These traditional NMR parameters provide short- to medium-range structural restraints.

The traditional NMR parameters have been augmented recently by the measurement and interpretation of anisotropic parameters (Prestegard, 1998). In this review we will concentrate on the usage of dipolar couplings measured in weakly aligned systems (Tolman *et al.*, 1995; Tjandra and Bax, 1997; Prestegard, 1998; Tjandra, 1999). We will briefly introduce the theory of dipolar couplings, discuss means of alignment as well as methods to determine dipolar couplings, and discuss the usage of dipolar couplings for structure refinement and their impact on 3D homology searching.

The dipolar coupling, as well as the CSA or a quadrupolar coupling, are given by the following Hamiltonian:

$$H_{kl} = b_{kl} \sum_{q=-2}^{2} F_{kl}^q(\theta_{kl}, \phi_{kl}) A_{kl}^q(I_k, I_l) \tag{1}$$

with the terms given in Table 1.

In the high field approximation, these equations can be simplified for the dipolar coupling and the chemical shift anisotropy to yield:

$$D_{kl} = b_{kl} F_{kl}^0(\theta_{kl}) \left(I_{kz} I_{lz} - \frac{1}{2}(I_{kx} I_{lx} + I_{ky} I_{ly}) \right) \tag{2}$$

and

$$\sigma_k = \frac{\gamma B_0}{3} \left[\left(\sigma_{zz} - \frac{1}{2}(\sigma_{xx} + \sigma_{yy}) \right) F_{kl}^0(\theta_{kl}) + \sqrt{\frac{3}{8}} (\sigma_{xx} - \sigma_{yy})(F_{kl}^2(\theta_{kl}) + F_{kl}^{-2}(\theta_{kl})) \right] I_{kz}$$

$$\tag{3}$$

Table 1. Tensor operators in the rotating frame and modified spherical harmonics for the dipolar and CSA Interaction. The calibration has been chosen such that $\int F^{(q)}(\theta,\phi)F^{(-q)}(\theta,\phi)d(\cos\theta d\phi)$ is independent of θ.

	Tensor operators for the dipolar interaction $b_{kl} = -\mu_0 \dfrac{\gamma_k\gamma_l\hbar}{4\pi r_{kl}^3}$	Tensor operators for the CSA Interaction $b_k = \frac{1}{3}(\sigma_\parallel - \sigma_\perp)\gamma_k B_0$	Modified spherical harmonics	Frequency
	$\hat{A}_{kl}^{(q)}(\hat{I}_k,\hat{I}_l)$	$\hat{A}_k^{(q)}(\hat{I}_k)$	$F_k^{(q)}(\theta,\phi), F_{kl}^{(q)}(\theta,\phi)$	ω_q
2	$\sqrt{\frac{3}{8}}\,\hat{I}_k^-\hat{I}_l^-$		$\sqrt{\frac{3}{2}}\sin^2\theta\exp(+2i\phi)$	$\omega(\hat{I}_k)+\omega(\hat{I}_l)$
1	$\sqrt{\frac{3}{8}}\,\hat{I}_{k,z}\hat{I}_l^-$		$\sqrt{6}\sin\theta\cos\theta\exp(+i\phi)$	$\omega(\hat{I}_l)$
1	$\sqrt{\frac{3}{8}}\,\hat{I}_k^-\hat{I}_{l,z}$	$\sqrt{\frac{3}{8}}\,\hat{I}_k^-$	$\sqrt{6}\sin\theta\cos\theta\exp(+i\phi)$	$\omega(\hat{I}_k)$
0	$\hat{I}_{k,z}\hat{I}_{l,z}$	$\hat{I}_{k,z}$	$3\cos^2\theta-1$	0
0	$\frac{1}{4}\left(\hat{I}_k^+\hat{I}_l^- + \hat{I}_k^-\hat{I}_l^+\right)$		$3\cos^2\theta-1$	$\omega(\hat{I}_k)-\omega(\hat{I}_l)$
1	$\sqrt{\frac{3}{8}}\,\hat{I}_{l,z}\hat{I}_k^+$	$\sqrt{\frac{3}{8}}\,\hat{I}_k^+$	$\sqrt{6}\sin\theta\cos\theta\exp(-i\phi)$	$\omega(\hat{I}_k)$
1	$\sqrt{\frac{3}{8}}\,\hat{I}_{k,z}\hat{I}_l^+$		$\sqrt{6}\sin\theta\cos\theta\exp(-i\phi)$	$\omega(\hat{I}_l)$
2	$\sqrt{\frac{3}{8}}\,\hat{I}_k^+\hat{I}_l^+$		$\sqrt{\frac{3}{2}}\sin^2\theta\exp(-2i\phi)$	$\omega(\hat{I}_k)+\omega(\hat{I}_l)$

An anisotropic distribution of orientations of molecules can be described by a probability function in the molecular frame of the molecule as:

$$P(\beta,\gamma) = A_p F_{kl}^0(\theta_{kl}) + A_r \sqrt{\frac{3}{8}}\left(F_{kl}^2(\theta_{kl}) + F_{kl}^{-2}(\theta_{kl})\right) \qquad [4]$$

With the Euler rotation connecting the molecular frame and the alignment tensor $R(\phi,\theta,\psi)$ and integrating over all orientations, we arrive at the following results for the dipolar coupling and the chemical shift:

$$\langle D_{kl}\rangle = \frac{1}{4\pi}\int_0^{2\pi} d\gamma \int_0^{\pi} d\cos\beta\, P(\beta,\gamma) b_{kl} F_{kl}^0(\theta_{kl})$$

$$= \frac{1}{4\pi}\int_0^{2\pi} d\gamma \int_0^{\pi} d\cos\beta\, P(\beta,\gamma) b_{kl} \sum D_{m0}(\phi_{kl},\theta_{kl},\psi_{kl}) F_{kl}^m(\beta,\gamma)$$

$$= \frac{b_{kl}}{10}\left(A_p D_{00}(\phi_{kl},\theta_{kl},\psi_{kl}) + \sqrt{\frac{3}{8}}A_r\left(D_{20}(\phi_{kl},\theta_{kl},\psi_{kl}) + D_{-20}(\phi_{kl},\theta_{kl},\psi_{kl})\right)\right)$$

$$= \frac{b_{kl}}{10}\left(A_p\frac{3\cos^2\theta_{kl}-1}{2} + \frac{3}{4}A_r\sin^2\theta_{kl}\cos 2\phi_{kl}\right)$$

$$[5]$$

From similar formulae and with $\sigma_{zz} - \frac{1}{2}(\sigma_{xx}+\sigma_{yy}) = \sigma_p$ and $\sigma_{xx} - \sigma_{yy} = \sigma_r$ we get the following expression for the chemical shift:

$$\langle \sigma_k\rangle = \frac{1}{4\pi}\int_0^{2\pi} d\gamma\int_0^{\pi} d\cos\beta P(\beta,\gamma)b_k\left(\sigma_p F_k^0(\theta_{kl}) + \sqrt{\frac{3}{8}}\sigma_r\left(F_k^2(\theta_{kl}) + F_k^{-2}(\theta_{kl})\right)\right)$$

$$= \frac{1}{4\pi}\int_0^{2\pi} d\gamma\int_0^{\pi} d\cos\beta P(\beta,\gamma)b_k\left(\begin{array}{l}\sigma_p\sum_{m=-2}^{2}D_{m0}(\phi_k,\theta_k,\psi_k)F_k^m(\beta,\gamma)\\[2mm] + \sqrt{\frac{3}{8}}\sigma_r\left(\sum_{m=-2}^{2}D_{m2}(\phi_k,\theta_k,\psi_k)F_k^m(\beta,\gamma) + \sum_{m=-2}^{2}D_{m-2}(\phi_k,\theta_k,\psi_k)F_k^m(\beta,\gamma)\right)\end{array}\right)$$

$$= \frac{b_k}{10}\sigma_p\left[A_p\frac{3\cos^2\theta_k-1}{2} + \frac{3}{4}A_r\sin^2\theta_k\cos 2\phi_k\right]$$

$$+ \frac{3b_k}{40}\sigma_r\left[A_p\sin^2\theta_k\cos 2\psi_k + A_r\left(\sin^4\frac{\theta}{2}\cos 2(\phi+\psi) + \cos^4\frac{\theta}{2}\cos 2(\phi+\psi)\right)\right]$$

$$[6]$$

Following is a summary of several methods used to obtain weak alignment, as well as methods used to measure dipolar couplings constants in the most optimal way. Examples for structure refinement, as well as the analysis of dynamical information from dipolar couplings, are also provided.

2.   MEASUREMENT OF DIPOLAR COUPLINGS

Due to the fast reorientation and isotropic orientation distribution in an isotropic solution, dipolar couplings are usually not visible in liquid state NMR spectroscopy. The small amount of order can be introduced either by the molecule itself (paramagnetic proteins, DNA, RNA due to their high magnetic susceptibility) at high magnetic fields, or anisotropic solutions consisting of, for example, dilute liquid crystal media reintroduce dipolar couplings. While the still fast reorientation of the molecules prevents the occurrence of a powder pattern for each resonance due to the dipolar coupling, a weak alignment makes the molecule behave as if all molecules were in a single crystal, with a dramatic scaling of all anisotropic interactions by a factor of approximately 1000. Therefore, spectra of molecules measured in anisotropic media have properties similar to high resolution NMR-spectra, but include the dipolar coupling information. The different methods to achieve weak alignment are summarized in Table 2.

Alignment due to magnetic field was the first method for measuring dipolar couplings applied to biomacromolecules (Kung et al., 1995; Tolman, et al., 1995; Beger et al., 1998). This method was shown to work with paramagnetic proteins or larger oligonucleotides because of their considerable anisotropic magnetic susceptibility. Experiments on ubiquitin showed that the alignment induced due to the aromatic side chains is very small and the dipolar couplings do not exceed ± 0.3Hz (Tjandra et al., 1996).

As previously mentioned, more general methods for the measurement of dipolar couplings have been introduced which allow the alignment of a large variety of molecules. The alignment relies on the steric interaction between the highly aligned bicelles or other alignment media and the biomacromolecules. This induces a weak alignment of the biomacromolecule. In general, the bonding interactions which can be either hydrophobic or electrostatic in nature should be weak in order to prevent shortening of the transverse relaxation time of the protein.

Table 2. Different methods of alignment. The source of alignment and the properties of the biomolecules to be oriented are shown with the phases.

Aligning medium	Source of alignment	Recommended Biomolecules	Published
Bicelles	steric interactions	Non hydrophobic molecules	(Sanders and Schwonek, 1992; Ottiger and Bax, 1998)
Doped Bicelles	steric + electrostatic interactions	Non hydrophobic molecules	(Losonczi and Prestegard, 1998)
Phages	steric + electrostatic interactions	pI ≤ 5, proteins, RNA, DNA	(Clore, et al., 1998; Hansen, et al., 1998a; Hansen, et al., 1998b; Barrientos et al., 2001)
Purple membrane fragments	steric + electrostatic interactions	proteins	(Sass, et al., 1999; Koenig, et al., 1999)
Crystallites of Cellulose	steric + electrostatic interactions	proteins	(Flemming, et al., 2000)
Helfrich phases	steric + electrostatic interactions	proteins	(Prosser, et al., 1998; Barrientos, et al., 2000)
n-alkyl-polyethylenglycol /n-Alkylalcohol/H_2O	steric interactions	proteins	(Ruckert and Otting, 2000)
polyacrylamide gels	steric + electrostatic interactions	proteins	(Tycko, et al., 2000; Sass, et al., 2000)
Electric Field	electric dipole moment	proteins	(Hilbers and MacLean, 1968; Hilbers and MacLean, 1970; Hilbers and MacLean, 1971; van Zijl et al., 1984; Ruessink and MacLean, 1987; Peshkovsky and McDermott, 1999; Riley and Augustine, 2000)
Magnetic Field	Magnetic susceptibility	DNA,RNA, paramagnetic proteins	(Tolman, et al., 1995; Kung, et al., 1995)

This concept has been implemented by bicelles, which consist of "long arm" glycolipids (DMPC dimyristoyl phosphatidylcholine, DLPC dilauroyl phosphatidylcholine) and "short arm" glycolipids (DHPC dihexanoyl phosphatidylcholine) (Sanders and Schwonek, 1992; Sanders II et al., 1994; Ottiger and Bax, 1998) or CHAPSO (3-[(3-cholamidopropyl) dimethyl ammonio]- 2-hydroxy-1-propanesulfonate) (Sanders and Schwonek, 1992; Tjandra and Bax, 1997; Wang et al., 1998a). Many different bicelle mixtures for different sample conditions have been proposed (e.g. different temperatures (Cavagnero et al., 1999) or different pH values (Ottiger and

Bax, 1999)). Additives for stabilization and different charging (CTAB, SDS) of bicelles have also been introduced (Losonczi and Prestegard, 1998; Zweckstetter and Bax, 2000). The use of bicelles in high pressure systems has been shown by Brunner *et al.*, (2001). Ruckert and Otting (2000) recently presented new uncharged bicelles made of n-Alkyl-poly(ethylene glycol)/n-alkyl alcohol and glucopone/n-hexanol mixtures. The uncharged poly(ethylene glycol)-based systems are insensitive with respect to pH. Most of the interactions are purely from spatial interactions between the bicelles and the biomacromolecule. This phase is also quickly prepared and stable for months.

Other methods use phages (Clore *et al.*, 1998; Hansen *et al*, 1998a; Hansen *et al.*, 1998b; Ojennus *et al.*, 1999) or purple membranes (Koenig *et al.*, 1999; Sass *et al.*, 1999) for the alignment. These methods are useful for oligonucleotides or acidic proteins. In these media, the alignment is not only caused by steric interactions, but also by either weak transient binding or large electrostatic interactions of the protein with the aligning medium. This leads to a decrease of the effective transverse relaxation times that can be severely reduced compared to the isotropic solution.

Recently proposed alignment media are cellulose crystallites (Flemming *et al.*, 2000), surfactant liquid crystals (Helfrich phases) (Prosser *et al.*, 1998; Barrientos *et al.*, 2000) and anisotropic aqueous polymeric gels (Sass *et al.*, 2000; Tycko *et al.*, 2000; Ishii *et al.*, 2001) that can be used in high resolution NMR spectroscopy. Inorganic materials (Desvaux *et al.*, 2001) have also been presented for the alignment of molecules.

The application of different alignment media for a single molecule has proven to be beneficial for structure determination since ambiguities of the translation of dipolar couplings into orientations are reduced (Ramirez and Bax, 1998). Another application relies on the measurement of dipolar couplings in at least five different alignment media in order to determine the dynamics from dipolar couplings (Meiler, *et al.*, 2001). Applications of the use of dipolar couplings for structure elucidation is discussed in Section 3.

2.1.   Measured Parameter

N-H^N dipolar couplings were the first measured dipolar couplings in proteins. Many different dipolar couplings in the protein backbone have been measured thereafter. Heteronuclear dipolar couplings in RNA/DNA can also be easily accessed. Size and sign of heteronuclear dipolar couplings can be measured by the difference of the splitting observed in partially aligned molecules, in comparison with the scalar couplings measured in isotropic solutions. The dipolar coupling can be directly extracted, provided the amount of alignment is weak enough that the absolute value of the dipolar coupling does not exceed the absolute value of the scalar coupling. This requirement is fulfilled for directly bound H-C and H-N pair's (Tjandra, et al., 1996; Tjandra and Bax, 1997; Yang et al., 1998).

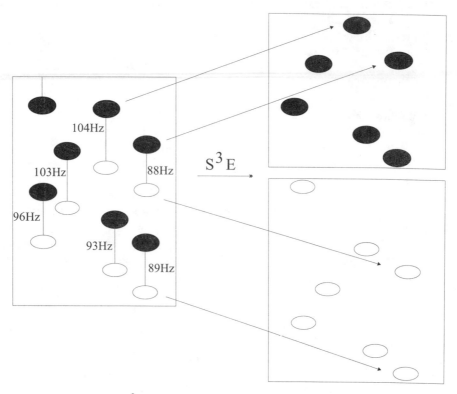

Figure 1. Schematic S³E spectrum. The resolution is improved by separating the upfield from the downfield doublets.

Dipolar couplings are measured from the splitting in a non-decoupled experiment, leading to more peak overlap. Thus it is desirable to record each line of the doublet in separate spectra (Figure 1). In particular, the S³E (Meissner *et al.*, 1997) (spin state selective excitation) and the related IPAP (in-phase, anti-phase) methodology (Ottiger *et al.*, 1998a) remove overlap and allow recording of each doublet line individually (Meissner, *et al.*, 1997; Ottiger, *et al.*, 1998a; Yang, *et al.*, 1998; Yang *et al.*, 1999; Chou *et al.*, 2000a; Kontaxis *et al.*, 2000; Permi and Annila, 2000).

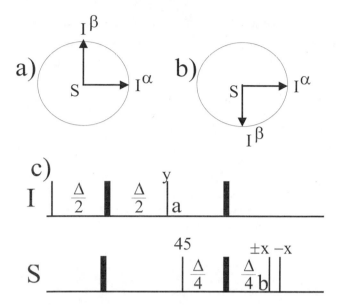

Figure 2. Effect of the S^3E element (c) with phase x (a) and –x (b) in a IS spin system. The number 45 refers to the phase, not the flip angle of the pulse. $\Delta = (2J)^{-1}$.

In Figure 2 the most important part of the S^3E pulse sequence is displayed. After the first INEPT transfer of the proton magnetization to the attached heteronucleus the $2I_zS_z$ is generated at **a**. The next 90°(S) pulse has the phase 45° and creates the following coherence:

$$2I_zS_z \xrightarrow{\left(\frac{\pi}{2}\right)^{45}} 2I_z\left(S_x \cos45 - S_y \cos45\right) \rightarrow 2I_z\frac{1}{\sqrt{2}}\left(S_x - S_y\right)$$

After the following delay of $\dfrac{\Delta}{2} = \dfrac{1}{4^1J_{NH}}$ the coherence at point b is:

$$\left(S_x \sin\pi J\frac{\Delta}{2} - 2S_yI_z \cos\pi J\frac{\Delta}{2} + \left(S_y \sin\pi J\frac{\Delta}{2} + 2S_xI_z \cos\pi J\frac{\Delta}{2}\right)\right)\frac{1}{\sqrt{2}} =$$

$$= \left(S_x + 2S_xI_z + S_y - 2S_yI_z\right)\frac{1}{2} =$$

$$- S_xI^\alpha + S_yI^\beta$$

Application of a 0 or $180^\circ_x(S)$ pulse leads to the following situations:

$$S_xI^\alpha + S_yI^\beta \xrightarrow{0^\circ} S_xI^\alpha + S_yI^\beta$$

$$S_xI^\alpha + S_yI^\beta \xrightarrow{180^\circ} S_xI^\alpha - S_yI^\beta$$

Addition of the two spectra leads to $2S_x I^\alpha$, while subtraction leads to $2S_y I^\beta$. This allows separation of the two peaks into two different spectra. The S^3E element can be combined with sensitivity-enhanced HSQC spectra in combination with gradient echo selection. The sensitivity of the experiment is ½ of a sensitivity-enhanced and fully decoupled HSQC (Schleucher et al., 1993). Nevertheless there are drawbacks in using the S^3E techniques. The amount of the evolution of the couplings is no longer uniform because the J+D dipolar couplings are not uniform. The dipolar couplings vary because of the different orientations of the I-S vectors in the protein. Additionally, water suppression is critical in diluted samples.

A second option for the separation of the two doublet components in two spectra is based on the TROSY technique (Pervushin et al., 1997) which causes the following simultaneous transfers:

$$S_- I^\alpha \longrightarrow I_- S^\alpha \text{ antiecho}$$
$$S_+ I^\beta \longrightarrow I_- S^\beta \text{ echo}$$

Table 3. Phase cycle for the pulse sequence in Figure 3. The first four phases and the second four phases need to be stored at different locations. After summation the two FIDS for the α-line contain the echo and the antiecho part of the spectrum and are processed accordingly (AE flag on Bruker spectrometers).

ϕ	ψ	α-line	β-line
x	x	x	x
y	x	y	-y
-x	x	-x	-x
-y	x	-y	y
x	-x	x	x
y	-x	-y	y
-x	-x	-x	-x
-y	-x	y	-y

It is obvious that one of the two doublet components is in the echo, whereas the other component is in the anti-echo spectrum. The two components can be separated using well established phase cycling techniques as summarized in Table 3 and shown in Figure 3. The drawback of this method is that gradient echo selection can only be applied at the expense of a loss in signal to noise since one of the components is lost.

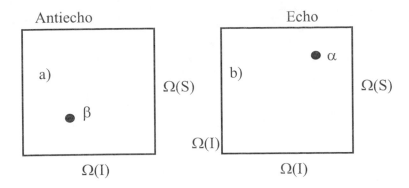

Figure 3. The TROSY pulse sequence (c) to separate the two lines indicated in a) and b) is an alternative way of recording simultaneously and separately two submultiplets of the IS-HSQC multiplet.

Other techniques are based on the HNCO spectrum and allow the measurement of dipolar coupling in the protein backbone (Yang, *et al.*, 1998; Yang, *et al.*, 1999; Permi and Annila, 2000).

In addition to the backbone, dipolar couplings derived for the side chains are useful. Therefore, methods for the measurement of CH dipolar couplings in CH_2 and CH_3 groups have been developed. Sums of these couplings can be determined from the modulation pattern of carbon coherence by the heteronuclear couplings (Kontaxis and Bax, 2001) that is similar to the HSQC-J approach already used for the measurement of backbone dipolar couplings (Kung *et al.*, 1995). Dipolar couplings for individual proton carbon resonances can be measured using the method of Carlomagno *et al.*, (2000). The **SPITZE-HSQC** (**SP**In state selec**T**ive **Z**Ero overlap) approach allows the simultaneous measurement of all three dipolar couplings in a CH_2 group ($^1D_{CH1}$, $^1D_{CH2}$ and $^1D_{H1H2}$). The **SPITZE-HSQC** in combination with S^3E (Meissner, *et al.*, 1997) achieves selective transfer between the following operators (shown in Figure 4 in a CH_2 group):

$$C^+ H_1^\alpha H_2^\alpha \rightarrow H_1^- H_2^\beta C^\alpha \qquad \text{peak 1} \qquad [7]$$

$$C^+ H_1^\beta H_2^\beta \rightarrow H_1^- H_2^\alpha C^\beta \qquad \text{peak 8}$$
or:
$$C^+ H_1^\alpha H_2^\alpha \rightarrow H_1^- H_2^\alpha C^\alpha \qquad \text{peak 2} \qquad [8]$$

$$C^+ H_1^\beta H_2^\beta \rightarrow H_1^- H_2^\beta C^\beta \qquad \text{peak 7}$$

This reduces spectral overlap and allows the measurement of all three involved couplings. Four spectra are obtained, each containing only one peak per CH_2 group. In large molecules where peaks 1 and 2 are not resolved in general, this approach allows the measurement of $J_{H_1 H_2} + D_{H_1 H_2}$ coupling constants, since peaks 1 and 2 appear in two different spectra. Due to the selectivity of the transfer described in Eq. 7 and 8, the magnetization corresponding to the $C^+ H_1^\alpha H_2^\alpha$ and the $C^+ H_1^\beta H_2^\beta$ terms is also not distributed on four lines, as in a coupled HSQC, but concentrated on one single line. This leads to a factor 4 gain in signal-to-noise with respect to a coupled HSQC in molecules for which the $J_{H_1 H_2} + D_{H_1 H_2}$ coupling is resolved, and a factor 2 gain in those molecules for which the $J_{H_1 H_2} + D_{H_1 H_2}$ coupling is not resolved.

Using a different approach (DQ/ZQ HMQC spectra) the measurement of the same three couplings dipolar couplings in NH_2 side chain groups in paramagnetic protein was presented (Bertini et al., 2000). The extraction of the couplings is performed in the ω_1 dimension where the resolution is typically worse.

Recently two new methods for the extraction of dipolar couplings in CH_3 groups were published. These dipolar couplings are important for the structure elucidation of large proteins, particularly if deuteration schemes are applied that replace protons by deuterons except for the methyl groups (Gardner and Kay, 1998). CH_3 group signals are well resolved, intense lines and are therefore accessible in large proteins. The method of Kontaxis and Bax (2001) uses the IPAP approach (Ottiger, et al., 1998a) to separate subspectra for the methyl-$^{13}C^1$H-multiplet components in a ^1H,^{13}C-HSQC spectrum. This method allows the quick measurement of ^1H-^{13}C scalar and dipolar couplings.

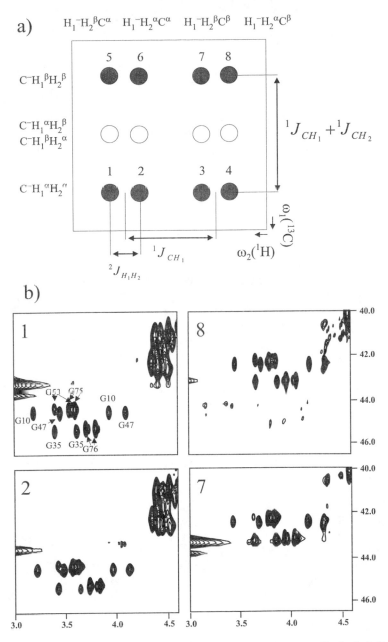

Figure 4. Scheme of a fully coupled HSQC peak of a CH_2 group a). Peaks 1, 2, 7 and 8 contain the desired information about the CH and HH dipolar couplings The SPITZE-HSQC allows to measure them. The four separated spectra corresponding to peaks 1, 2, 7 and 8 are given in b).

A second method was presented by Kaikkonen and Otting (2001). This approach enables the extraction of the magnitude and the sign of ^{1}H-^{13}C dipolar couplings in methyl groups. Their DiM-sequence (dipolar couplings in methyl groups) was designed to measure the separation of the two outmost multiplet components of the methyl proton triplet and to suppress the central line simultaneously. This yields the $^2D_{HH}$ dipolar couplings. The sign is measured relative to a large heteronuclear one-bond scalar coupling. The measurement of methyl dipolar couplings might be particularly useful in very large, fully deuterated proteins, where only the methyl groups are protonated.

2.2.   Proton-Proton Dipolar Couplings

Since the density of hydrogen atoms is notably high in proteins, proton-proton dipolar couplings provide additional structural information. First reports concentrated on the detection of dipolar couplings between protons that had not given rise to cross peaks via scalar couplings (Bolon and Prestegard, 1998; Hansen, et al., 1998). However, these methods did not quantify the dipolar couplings. COSY and TOCSY techniques (DCOSY) were used for measurement of the dipolar couplings where only the COSY experiments allow the quantification of dipolar couplings using a new tool for spectral analysis (Delaglio et al., 2001). New DCOSY based on MOCCA (modified phase-cycled Carr-Purcell-type) sequences (Kramer et al., 2001) improve the efficiency of the dipolar transfer dramatically (Kramer, et al., 2001) compared to DIPSI, MLEV or WALTZ sequences that are generally used for coherence transfer via J couplings. An example spectrum derived from a MOCCA sequence is shown in Figure 5 for ubiquitin. All H^N,H^N cross peaks that are only visible in anisotropic media originate from dipolar couplings.

Quantitative determination of the size of the dipolar homonuclear couplings constants is more difficult than for heteronuclear couplings. While the sign of heteronuclear dipolar couplings between directly bound nuclei can be easily determined due to the known sign of the heteronuclear ^{1}J coupling, and the fact that this ^{1}J coupling is normally larger than the dipolar coupling, this is not the case for proton-proton dipolar couplings. Depending on the amount of alignment, proton-proton dipolar couplings can be larger than the respective scalar coupling. This is always the case for dipolar couplings between protons that are more than three bonds apart in the bonding network. Therefore, experiments that allow measurement of size and sign of proton-proton dipolar couplings require special design. H,H dipolar couplings also depend on the interproton distance according to $\frac{1}{r^3}$.

ZMOCCA-XY-16

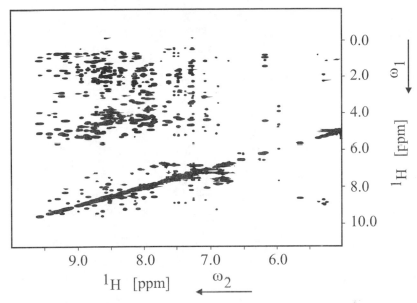

Figure 5. MOCCA correlation spectrum of ubiquitin. The cross peaks are generated from dipolar couplings between the amide protons and other protons. This is especially obvious for the H^N,H^N cross peaks.

The first measurement of H,H dipolar couplings was performed using the quantitative-J approach (Tjandra and Bax, 1997) using a 3D HNHA pulse sequence (Vuister and Bax, 1993). This method delivers the absolute value of J+D allowing the sign determination for only sufficiently large scalar H^N,H_α coupling constants.

The idea of using COSY (Aue *et al.*, 1976; Brunner *et al.*, 2000) measurement for the determination of dipolar couplings was also exploited in the James H. Prestegards lab. They showed that one can measure qualitatively H,H dipolar couplings in carbohydrate using COSY (Bolon and Prestegard, 1998). CT-COSY (constant time COSY) (Girvin, 1994) provides a quantitative measure of the absolute sign of the dipolar coupling, provided that there is knowledge about the size ratio of the scalar and the dipolar coupling (Tian *et al.*, 1999). The intensities of the cross- and the auto-peaks are proportional to the measured coupling: $\left(\dfrac{I_{cross}}{I_{auto}}\right) = k \tan\left(\pi\left(J + D\right)\Delta\right);$

Δ is the constant time delay used in the measurement. The main drawback is that the cross peaks in the CT-COSY have an anti-phase or phase-shifted character and thus are difficult to integrate. This can be partly overcome by using many different constant time delays Δ. The observed ratios $\left(\dfrac{I_{cross}}{I_{auto}} \right)$

can be plotted against Δ; this graph can be fit to extract the active coupling for the cross peak. Tian et al., (1999) first used this methodology for dipolar coupling measurements in carbohydrates and also recently showed the applicability of CT-COSY to the structure determination of proteins (Tian et al., 2000). Particularly in cases with defined secondary structure $^1H^N$-$^1H^N$ (α-helix) and $^1H^\alpha$-$^1H^N$ (β-strand) these constraints were shown to be very useful. They also introduced an ^{15}N-edited version of the CT-COSY that resolves overlap problems observed for larger proteins. Using CT-COSY methodology, the size of the dipolar coupling is accessible but the sign is not. This is not as important when dipolar couplings are only measured in regular secondary structure elements.

Measurements of sign and size of proton-proton dipolar couplings is possible with E.COSY type (Griesinger et al., 1985; Griesinger et al., 1986; Griesinger et al., 1987) techniques irrespective of the existence of a sizeable J coupling between the two protons. As introduced for scalar coupling between $^1H^N$, $^1H^\alpha$, the HNCA-E.COSY (Weisemann et al., 1994) experiment generates an E.COSY pattern with the $^1J(C^\alpha,H^\alpha)$ coupling associated with the desired H^N,H^α coupling (Cai et al., 1999). This methodology allows the measurement of the sign and size of $^1H^N(i)$,$^1H^\alpha(i)$ and $^1H^N(i)$,$^1H^\alpha(i$-$1)$ dipolar couplings in ^{13}C, ^{15}N labeled proteins. Similarily, the measurement of dipolar couplings in proteins that are only ^{15}N labeled is possible (Pellecchia et al., 2000; Peti and Griesinger, 2000) for ^{15}N-H_1,H_2 moieties (Montelione and Wagner, 1989; Gemmecker and Fesik, 1991; Schulte-Herbrüggen et al., 1991; Willker and Leibfritz, 1992). In a J_{HH} NOESY experiment that is built from a ^{15}N edited NOESY spectrum with planar mixing before and after the NOESY step, the desired H,H coupling is associated with the ^{15}N,H splitting (J+D) whose sign is known. Closely related to the J_{HH} NOESY sequence is the "signed COSY" (Otting et al., 2000). It is based on the E.COSY principle and associates the desired H,H dipolar coupling with the $^1J_{CH}$ couplings constant.

Although not published for dipolar couplings, J_{HH} type experiments can use MOCCA (Kramer, et al., 2001) sequences rather than NOESY for the H,H transfer. This has the potential advantage of a shorter mixing time than in NOESY and adverse effects from short ^{15}N T_1 times are avoided. Using the MOCCA rather than NOESY could be beneficial for medium sized proteins where the ^{15}N T_1 are rather short and the proton line widths are narrow.

The different techniques for measurement of dipolar couplings summarized in this section are listed in Table 4.

Table 4. Methods and underlying principles for the measurement of dipolar couplings together with original literature.

Dipolar coupling	Method	Principle	Published
H^N, $H_\alpha C_\alpha$	CT-J-HSQC	J-modulation	(Tjandra, *et al.*, 1996)
	S3E-HSQC	Line position from two spectra	(Meissner, *et al.*, 1997)
$H^N C'$	S3E/IPAP-HNCO	E.COSY type splitting	(Wang *et al.*, 1998b)
$H_2 C$	CT-J-HSQC	J-modulation	(Ottiger, *et al.*, 1998a)
	SPITZE-HSQC	Line positions from four spectra	(Carlomagno, *et al.*, 2000)
$H_3 C$	CT-J-HSQC	J-modulation	(Ottiger, *et al.*, 1998a)
	IPAP-HSQC	Line position from two spectra	(Kontaxis and Bax, 2001)
H,H	DiM	H,H coupling as splitting	(Kaikkonen and Otting, 2001)
	COSY	ACME amplitude-constrained multiplet evaluation	(Delaglio, *et al.*, 2001)
	CT-COSY	Intensity modulation	(Tian, *et al.*, 1999; Tian, *et al.*, 2000; Wu and Bax, 2001)
	E.COSY	E.COSY extraction	(Griesinger, *et al.*, 1985; Griesinger, *et al.*, 1986; Griesinger, *et al.*, 1987; Neubauer, *et al.*, 2001)
	Signed COSY		(Otting, *et al.*, 2000)
$^{15}N^H$-H	J_{HH}-NOESY	E.COSY extraction	(Peti and Griesinger, 2000)
H^H-H_α	HNHA	J modulation	(Tjandra and Bax, 1997)
	HNCA-E.COSY	E.COSY extraction	(Cai *et al.*, 1999)

3.   DIPOLAR COUPLINGS AS A STRUCTURAL RESTRAINT FOR PROTEINS

The character of dipolar couplings as a structural parameter is distinct from the character of the NOE which is the classical NMR parameter for determining three dimensional structures. While the NOE only reflects pure distance information and orientation information for severely anisotropically tumbling molecules, dipolar couplings reflect distance and orientation information. For a pair of nuclei with fixed distances the dipolar coupling value only depends on the orientation of the internuclear vector. While the NOE only reflects short range distances, the dipolar coupling introduces information about the orientation relative to the molecular frame for all nuclei making this a local, as well as long range, structural parameter.

To use dipolar couplings for structure elucidation, incorporation into the current protocols for the determination of structures is required (also see Chapter 8 in this volume). The most common tool to generate structures that fulfil NMR restraints is the application of a restrained molecular dynamics that employs a simulated annealing protocol (Brünger, 1992). The extended protein structure is heated to approximately 1000 K. The NMR parameters are translated into additional force field parameters whose weights are steadily increased in this high temperature phase. During this period the global fold is established, mainly using these experimental observables. After the high temperature phase the structure is slowly cooled down to a low temperature and all additional force field parameters (molecular geometry, VDW interactions, electrostatics) are introduced or increased in weight, respectively. A three dimensional structure is established that fulfils all force field parameters together with the information obtained from NMR spectroscopy such as NOE's, hydrogen bonds, or scalar coupling constants.

Residual dipolar couplings depend on the orientation of the intermolecular dipole in the alignment frame according to Eq. 5. While it is easily possible to translate an orientation into a dipolar coupling once the size and the orientation of the alignment tensor in the molecular frame are known, the dipolar couplings measured for an unknown structure are more difficult to translate back since there are two continua of orientations compatible with one dipolar coupling. Even if several alignment media are used, the inversion of Eq. 5 has four discrete solutions because it is not sensitive to the sign of either θ or ϕ. These multiple solutions lead to a complex energy surface upon which optimization algorithms often fail.

Several approaches were introduced to use residual dipolar couplings in the structure elucidation process. Direct implementation of Eq. 5 requires the introduction of a tensor coordinate system in which the orientation of a vector can be easily expressed (Tjandra *et al.*, 1997). The tensor is represented by four atoms located far away from the molecule of interest to avoid interactions. The relative position of these four atoms is fixed by distance restraints with strong force constants. Expression of the relative orientation of the dipolar vectors in this artificial coordinate system allows the calculation of dipolar coupling value if the size and the rhombicity of the tensor is known. While the orientation is not known a priori, the size of the tensor and its rhombicity can be extracted by analyzing the distribution of a set of dipolar couplings obtained for one biomolecule (Clore *et al.*, 1998). These calculated dipolar coupling values can be compared with the experimentally determined values, and the deviation is used to calculate forces that influence the orientation of the dipolar vector in order to decrease this deviation. This approach has been successfully applied to a number of structure calculations (Clore and Gronenborn, 1998; Ottiger *et al.*, 1998b; Bayer *et al.*, 1999; Fischer *et al.*, 1999; Olejniczak *et al.*, 1999).

However, this approach does not overcome the problem resulting from the ambiguous character of the dipolar coupling information in combination with the introduced tensor coordinate system. The complexity of the energy hypersurface remains, which may cause convergence problems. These convergence problems can be overcome by using very small force constants that are slowly increased over a long timeframe, or by only using dipolar couplings in subsequent refinement protocols. In both cases, the duration of the calculation is increased and, more importantly, the dipolar coupling information is not used to its full potential during the global folding process in the case of structure elucidation of proteins.

A solution to this problem is the translation of residual dipolar couplings into intramolecular structure restraints (Meiler *et al.*, 2000a). The possible range for the projection angle between two dipolar vectors can be decreased from the range [0°,180°] to a smaller range using the dipolar coupling obtained for the two dipolar vectors. The advantages of this approach are: (1) The introduction of the reference coordinate system for the tensor is no longer necessary; and (2) the complex energy surface can be simplified (to be shown later). Therefore the dipolar couplings can be used from the start of the structure elucidation process to fold the protein structure.

The methods described thus far do not allow the building of a structure without a considerable number of NOE's. It has been shown that by measuring several dipolar couplings for a rigid molecular fragment (e. g. N–H^N / N–C' / C'–H^N / C'–C^α for a peptide plane) in at least two sufficiently different alignments, the backbone fold can be established without using NOE's. This method constitutes an important step towards development of a fast method for the determination of a low resolution structure in a short time. Other approaches to this problem are discussed in detail in section 5.

In the following an approach is reviewed for the introduction of dipolar couplings as structural restraints that minimizes the convergence problems mentioned above, and allow the dipolar couplings to assist the folding of the protein from the start of the structure calculation. This is achieved by transforming dipolar couplings restraints into a purely intramolecular projection restraint (Meiler, *et al.*, 2000a), that does not require the knowledge of the orientation of the molecule with respect to the alignment tensor. The following two examples demonstrate that with this implementation the convergence of the simulated annealing protocol with and without dipolar couplings is almost identical. On a model system, as well as using experimental data, the accuracy and the precision of the structures is enhanced by the use of dipolar couplings.

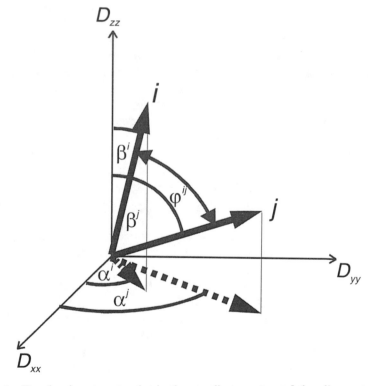

Figure 6. Two bond vectors *i* and *j* in the coordinate system of the alignment tensor (D_{xx}, D_{yy}, D_{zz}). The angles α^{ij} and β^{ij} determine the projection of the vector onto the z-axis and of the x,y-component onto the x-axis of the alignment tensor, respectively. φ^{ij} determines the angle between the two vectors *i* and *j*. Reproduced from *J. Biomol. NMR.*, **16**:245 (2000) with kind permission of Kluwer Academic Publishers.

To derive equations for the projection angles of interatomic vectors as a substitute for the orientation restraints, a coordinate system is introduced (Figure 6), in which the two vectors i and j are defined in the coordinate system of the alignment tensor with the values D_{xx}, D_{yy} and D_{zz}.

Defining an axially symmetric $D_{\parallel}$ and a rhombic part $D_{\perp}$ of the tensor according to:

$$D_{\parallel} = \frac{1}{3}\left(D_{zz} - \frac{D_{xx} + D_{yy}}{2}\right)$$

$$D_{\perp} = \frac{2}{3}\left(\frac{D_{xx} - D_{yy}}{2}\right)$$

[9]

one obtains:

$$D^i\left(\beta^i \alpha^i\right) = D_{\parallel}\left(3\cos^2 \beta^i - 1\right) + \frac{3}{2}D_{\perp}\left(\cos 2\alpha^i \sin^2 \beta^i\right)$$

[10]

With an experimental set of dipolar couplings the eigenvalues of the tensor D_{xx}, D_{yy}, D_{zz} can be determined from a histogram of the experimental couplings.

The following equations allow the use of the dipolar couplings as restraints without the need for defining the orientation of the alignment tensor will be derived. This is accomplished by calculating the projection angle φ^{ij} between all pairs of internuclear vectors i and j for which dipolar couplings have been measured.

$$\cos \varphi^{ij} = \begin{pmatrix} \cos \alpha^i \sin \beta^i \\ \sin \alpha^i \sin \beta^i \\ \cos \beta^i \end{pmatrix}^T \begin{pmatrix} \cos \alpha^j \sin \beta^j \\ \sin \alpha^j \sin \beta^j \\ \cos \beta^j \end{pmatrix}$$

[11]

Applying Eq. 10 to Eq. 11 one can eliminate two angles β^i, β^j and arrive at:

$$\cos \varphi^{ij} = \sqrt{\frac{2(3D^i - 2D_{zz} + D_{xx} + D_{yy})}{3(\frac{3}{2}(D_{xx} - D_{yy})\cos \alpha^i - 2D_{zz} + D_{xx} + D_{yy})}}$$

$$\bullet \sqrt{\frac{2(3D^j - 2D_{zz} + D_{xx} + D_{yy})}{3(\frac{3}{2}(D_{xx} - D_{yy})\cos \alpha^j - 2D_{zz} + D_{xx} + D_{yy})}} \cos(\alpha_2 \pm \alpha_1)$$

$$\pm \sqrt{1 - \frac{2(3D^i - 2D_{zz} + D_{xx} + D_{yy})}{3(\frac{3}{2}(D_{xx} - D_{yy})\cos \alpha^i - 2D_{zz} + D_{xx} + D_{yy})}}$$

$$\bullet \sqrt{1 - \frac{2(3D^j - 2D_{zz} + D_{xx} + D_{yy})}{3(\frac{3}{2}(D_{xx} - D_{yy})\cos \alpha^j - 2D_{zz} + D_{xx} + D_{yy})}}$$

[12]

In addition, the possible range for angle α^i is sometimes reduced by the measured coupling values:

$$if \left| \frac{6D^i + 2D_{zz} - D_{xx} - D_{yy}}{3(D_{xx} - D_{yy})} \right| \leq 1$$

$$\begin{cases} \alpha^i_{\min} = \frac{1}{2}\arccos\left(\frac{6D^i + 2D_{zz} - D_{xx} - D_{yy}}{3(D_{xx} - D_{yy})}\right); \\ \\ \alpha^i_{\max} = \pi - \frac{1}{2}\arccos\left(\frac{6D^i + 2D_{zz} - D_{xx} - D_{yy}}{3(D_{xx} - D_{yy})}\right) \end{cases}$$

[13]

$$else \begin{cases} \alpha^i_{\min} = 0; \\ \alpha^i_{\max} = \pi \end{cases}$$

Depending on the size of the measured coupling values, the angle φ^{ij} is no longer allowed to vary in the whole interval from 0 to π. Eq. 13 states that the extreme values of φ^{ij} will always be found at the extreme values for α^i and α^j. Additionally the allowed range for φ^{ij} is always symmetric on $\varphi^{ij} = \pi/2$ (Figure 7).

Two general possibilities are found:

one allowed range: $\varphi^{ij} \in [\varepsilon_1, \pi - \varepsilon_1]$
two allowed ranges: $\varphi^{ij} \in [\varepsilon_1, \pi/2 - \varepsilon_2]$ or $\varphi^{ij} \in [\pi/2 + \varepsilon_2, \pi - \varepsilon_1]$ with $\varepsilon_{1,2} \in [0, \pi/2]$.

The symmetry of the allowed φ^{ij} ranges is directly related to the geometric symmetries of the dipolar couplings (see Eq. 10). Figure 8 shows a range for one angle φ^{ij} that has been calculated in this way.

Using this approach from n dipolar couplings $n(n-1)/2$, ranges for angles φ^{ij} can be calculated. These ranges do not include any information about tensor orientation. The translation of the dipolar couplings into intervector projection angles, i.e., the scalar product is the simplest pair wise relation between two vectors and any more complex vector relation yielding a scalar value can be derived from it. At the same time, the scalar product conserves all information and more complex intervector relations are not required.

To use these angle ranges φ^{ij} for structure determination, a new restraint was introduced in the X-PLOR program (Brünger, 1992). The energy function used is displayed in Figure 7 and given in Eq. 14:

$$E^{ij}_{0 \to \varphi_{ext1}} = k_1 \left(\varphi^{ij} - \varphi^{ij}_{ext1} \right)^2$$

$$E^{ij}_{\varphi_{ext1} \to \varphi_{ext2}} = 0$$

$$E^{ij}_{\varphi_{ext2} \to \varphi_{ext3}} = k_2 \cos^2 \left(\pi \left(\frac{\varphi^{ij} - \varphi^{ij}_{ext2}}{\varphi^{ij}_{ext3} - \varphi^{ij}_{ext2}} - \frac{1}{2} \right) \right) \qquad [14]$$

$$E^{ij}_{\varphi_{ext3} \to \varphi_{ext4}} = 0$$

$$E^{ij}_{\varphi_{ext4} \to 180°} = k_1 \left(\varphi^{ij} - \varphi^{ij}_{ext4} \right)^2$$

k_1 and k_2 are the energy constants used in this implementation. Note that k_2 directly gives the height of the barriers between the two allowed φ^{ij} ranges, while k_1 scales the square of the deviation from the extreme values. The two energy constants must be scaled separately during the simulated annealing protocol. Although the energy surface is simplified by excluding the tensor, the barrier between the two allowed ranges again introduces multiple minima into the energy surface. However, the translation into relative projections allows the temporary introduction of a smooth energy function by setting k_2 to 0. This does not exclude the equator part of the projection angle, but does exclude the polar ranges if this is derived from the dipolar couplings. This potential is very simple and smooth, and uses the "unambiguous" part of the dipolar couplings from the start of the structure calculation.

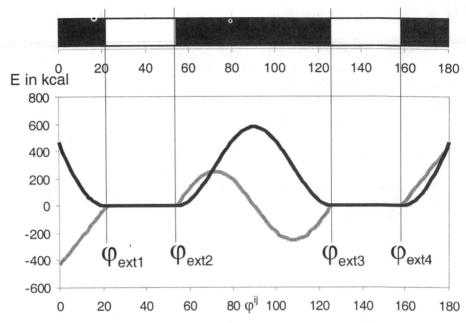

Figure 7. Potential employed to confine φ^{ij} within the allowed range (white) and exclude it from the forbidden range (black). A flat bottom potential is used for the allowed region, a parabolic potential for the margins close to 0 and π, and a $cos^2\Delta\varphi^{ij}$ function for the inner forbidden part. The energy term used is given by the black line and its derivative (negative force) by the gray line. Reproduced from *J. Biomol. NMR.*, **16**:245 (2000) with kind permission of Kluwer Academic Publishers.

The protocol has been applied to the protein Rhodniin, which has 103 amino acid and contains two similar folded domains of 46 amino acids and a flexible linker of 11 amino acids. A set of dipolar couplings between amide nitrogen and amide hydrogen was calculated from the known NMR structure (Maurer and Griesinger, 1998) of the protein, assuming a specific orientation of the alignment tensor. The eigenvalues were set to be $D_{zz} = 20.0$ Hz, $D_{yy} = 17.5$ Hz and $D_{xx} = 2.5$ Hz amounting to a rhombicity of 0.5. These dipolar couplings are used as an "experimental" test set of data. The Rhodniin structure we used for their calculation is called "target structure". All NOE's used in the following calculations were within three groups: strong NOE's < 2.5 Å, medium NOE's < 3.5 Å and weak NOE's < 5.0 Å. A flat bottom potential of zero was used for all values smaller than the mentioned distances and quadratically increased for larger distances.

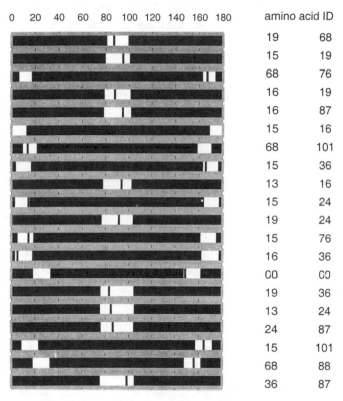

0 20 40 60 80 100 120 140 160 180	amino acid ID	
	19	68
	15	19
	68	76
	16	19
	16	87
	15	16
	68	101
	15	36
	13	16
	15	24
	19	24
	15	76
	16	36
	00	00
	19	36
	13	24
	24	87
	15	101
	68	88
	36	87

Figure 8. Allowed and forbidden ranges for the angles φ^{ij} between those N-H^N vectors that are most restricted by the dipolar couplings derived from one alignment tensor. The black bars are the φ^{ij} values of the "target" structure of Rhodniin. Reproduced from *J. Biomol. NMR.*, **16**:245 (2000) with kind permission of Kluwer Academic Publishers.

The eigenvalues of the alignment tensor were back calculated from the histogram of the dipolar couplings yielding $D_{zz} = 19.6\ Hz$, $D_{yy} = 17.4\ Hz$ and $D_{xx} = 2.2\ Hz$. From the dipolar couplings more than 4000 possible φ^{ij} – angle ranges were calculated using these eigenvalues. Figure 8 shows the 20 most restricted ranges. Table 5 reports the number of φ^{ij} – angle restraints that exclude a given percentage of the possible range from $[0,\pi]$. Since only a small part of the restraints contains most of the structural information, only 30% to 50% of the most restricted ranges (depending on the data) need to be applied in the calculations.

Table 5. Number of restraints that exclude a defined percentage of the possible φ^{ij} ranges.

percentage of φ^{ij} range excluded	number of restraints	percentage of restraints
0% – 10%	802	26.18%
10% – 20%	781	22.75%
20% – 30%	736	18.92%
30% – 40%	624	14.25%
40% – 50%	494	8.51%
50% – 60%	304	5.41%
60% – 70%	213	2.86%
70% – 80%	163	0.91%
80% – 90%	67	0.38%
90% – 100%	2	0.00%
sum	4186	100.00%

k_1 and k_2 varied between 0 $kcal/Mol$ and 200 $kcal/Mol$ and gave the best results with $k_1 = 40$ $kcal/Mol$ rad^{-2} and $k_2 = 10$ $kcal/Mol$. Figure 9 shows the scaling of the force constants used during the simulated annealing X-PLOR protocol. To keep the energy surface simple, only k_1 is ramped up together with the force constant for the unambiguous NOE's in the high temperature phase. k_2 is ramped up in the first cooling phase together with force constant of the ambiguous NOE's. k_2 defines the height of the barrier between the two possible ranges φ^{ij}. It is essential to choose the right period in the simulated annealing protocol when k_2 is ramped up. The convergence deteriorates when k_2 is switched on too early, namely, in the high temperature phase. In contrast, the number of violated angle restraints increases when k_2 is ramped up too late, namely, during the second cooling phase.

With the scaling of force constants as depicted in Figure 9, two ensembles of 100 structures were calculated with and without all angle restraints, all experimentally determined NOE's and J-coupling restraints. The N-terminal domain of Rhodniin was focused upon for the convergence. Figure 10 shows the rmsd of the heavy atoms in the N–terminal domain of the best 10 of 100 calculated structures to the target structure during the simulated annealing protocol, with and without the angle restraints, respectively. The rmsd decreases from *1.2 Å* to *0.5 Å* by introducing the information derived from dipolar couplings, which is in-line with previous results obtained for ubiquitin (Bax and Tjandra, 1997).

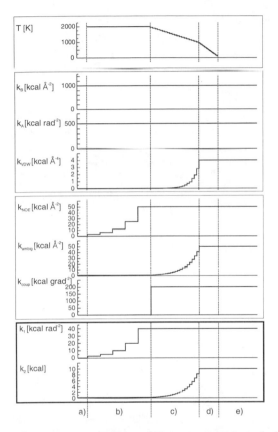

Figure 9. Time scale, temperature and size of the energy constants during the simulated annealing as used for the example Rhodniin in X-PLOR (Brünger, 1992). Section a) Initial energy minimization of 50 steps is performed. Section b) High temperature phase lasting for 32.5 ps the energy constants used for dipolar couplings k_1 is switched together on with the unambiguous NOE's. Section c) First cooling phase lasting for 25.0 ps: k_2 is switched on together with the rest of the NOE. Temperature is decreased from 2000 K to 1000 K. Section d) Second cooling phase lasting 10.0 ps. Temperature is decreased from 1000 K to 100 K. Section e) Final energy minimization of 200 steps are performed using all restraints and their final force constants.   Reproduced from *J. Biomol. NMR..*, **16**:245 (2000) with kind permission of Kluwer Academic Publishers.

In the best 10 of these structures, no NOE's and between zero to three of
the angle ranges are violated. Table 6 reports the energy distribution of the
100 structures. The introduction of the φ^{ij} restraints increases the mean
energy of the ensemble. However, only 5 of 100 structures do not fold to a
meaningful NMR structure and have a very high energy (beyond *5000
kcal/mol*); however 83 have an energy lower then *650 kcal/mol*. Without the
use of the couplings, 96 structures have an energy below *650 kcal/mol*. The
agreement between the dipolar couplings and the "experimental" restraints is
very good (see Figure 11) for 5 of 100 structures with the lowest energy.

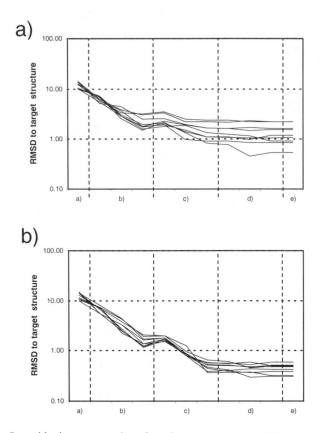

Figure 10. Logarithmic representation of rmsd to target structure during simulated annealing
without (a) and with (b) the use of dipolar couplings for N-domain of Rhodniin are performed
to the section of the SA explained in Figure 4. Best 10 out of an ensemble of 100 calculated
structures. The precision and accuracy of the structure with dipolar couplings (0.5 Å) is
higher than without (1.2 Å). Reproduced from *J. Biomol. NMR.*, **16**: 245 (2000) with kind
permission of Kluwer Academic Publishers.

Table 6. Energy distribution of an ensemble of 100 structures after the simulated annealing protocol without angle restraints, with restraints, and with restraints for two different tensors. With the use of the dipolar restraints five structures do not converge at all. One converges, however, at 990 kcal/mol and 1100 kcal/mol, respectively .

energy range	number of structures		
	without dipolar restraints	with one set of dipolar restraints	with two sets of dipolar restraints
350kcal – 450kcal	47	31	12
450kcal – 550kcal	43	38	30
550kcal – 650kcal	6	14	28
650kcal – 750kcal	3	6	13
750kcal – 850kcal	0	2	9
850kcal – 950kcal	0	3	2
> 950kcal	1	5 + 1	5+1

After performing the simulated annealing protocol, the orientation of the tensor was optimized (Losonczi *et al.*, 1999) using the program *DipoCoup* (Meiler *et al.*, 2000b). All deviations are given as dipolar Q-factor (Cornilescu *et al.*, 1999). This factor is larger than the R-factor (Clore and Garrett, 1999) by $\sqrt{2}$. The dipolar Q-factor is 0.19 with the use of the restraints, and 0.66 without angle restraints. Small deviations can be explained with the slightly incorrect alignment tensor eigenvalues used for calculating the restraints, as well as with the violation of angle ranges in special cases. The orientation of the tensor is reproduced within $\pm 5°$.

In another experiment only part of the NOE's was used to test the possibility of replacing NOE information by dipolar couplings. All NOE's containing at least one amide hydrogen were selected first. A varying percentage of 20% to 100% of this NOE subgroup was randomly selected for use in the calculation. Amide hydrogens were chosen since their NOE's are the first and easiest to obtain during evaluation of the spectra and are the only NOE's available in deuterated proteins. Table 7 shows the rmsd values to the target structure with respect to the amount of NOE's used for N-domain of the protein. The structures can be determined more accurately by using the angle restraints. The largest improvement is observed when 60% to 80% of the amide-to-all-proton NOE's are used. Differences are smaller if more or less than 60% to 80% of the NOE's are used. Thus, based on a minimal amount of NOE's that define the fold, dipolar couplings improve the structure. If the number of NOE's exceeds a certain number, they define the structure so well that dipolar couplings no longer considerably improve

the accuracy of the structure. If the number of NOE's is too low the fold is
no longer determined and the dipolar couplings cannot remedy this situation.

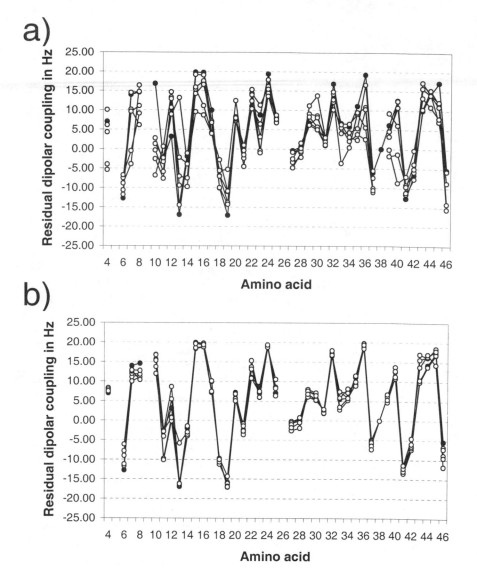

Figure 11.   Couplings recalculated for the resulting structures without (a) and with (b) the
restriction of the angle ranges. The normalized standard deviation (Q) decreases from 0.66
without (a) the use of the angle restraints to 0.19 with (b) the use of the angle restraints.
Filled circles represent the dipolar couplings of the target structure, open circles of the five
best calculated structures.   Reproduced from *J. Biomol. NMR.,* **16**:245 (2000) with kind
permission of Kluwer Academic Publishers.

Table 7. RMSD to target structure with and without the use of angle restraints. 20%, 40%, 60%, 80% and of the NOE's including at least one H^N-atom are used. All other NOE's including ambiguous NOE's are excluded.

percentage of NOE's used	RMSD to target structure in (Å)	
	with dipolar restraints	without dipolar restraints
20%	5.39	5.75
40%	4.20	4.40
60%	2.90	4.13
80%	1.78	3.50
100%	1.73	2.26

A third calculation was performed to test this implementation's effectiveness in determining the global structure of Rhodniin. Fifty structures were generated and the orientation of the two Rhodniin domains with respect to each other was investigated. For this calculation the 500 most restricted interdomain φ^{ij} ranges, as well as the 500 most restricted intradomain φ^{ij} ranges for the N-domain and the C-domain are used. Figure 12 shows the five resulting structures with lowest energy. Three of these five structures show the same relative orientation of the domains identical to the target structure. In the other two cases, one domain is turned around the x-axis of the alignment tensor by 180°. Due to the symmetry of the tensor, this is a valid solution. In all cases, not more than five φ^{ij} ranges are slightly outside the allowed range. Recalculating dipolar couplings from all five structures with an optimized tensor orientation shows that all five structures fulfill the dipolar coupling values.

Figure 12 (overleaf). The relative orientation of the N- and C-domains of Rhodniin with respect to each other. The tensor a) and the target structure b) in the coordinate system of the tensor are shown. c) and d) display the best five out of 50 structures from the simulated annealing protocol of Figure 4. c) The three structures with the same orientation of the domains as in the target structure are shown. d) Two structures are obtained in which one domain is rotated by *180°* about the D_{xx} axis with respect to the target structure. Reproduced from *J. Biomol. NMR.,* **16**: 245 (2000) with kind permission of Kluwer Academic Publishers.

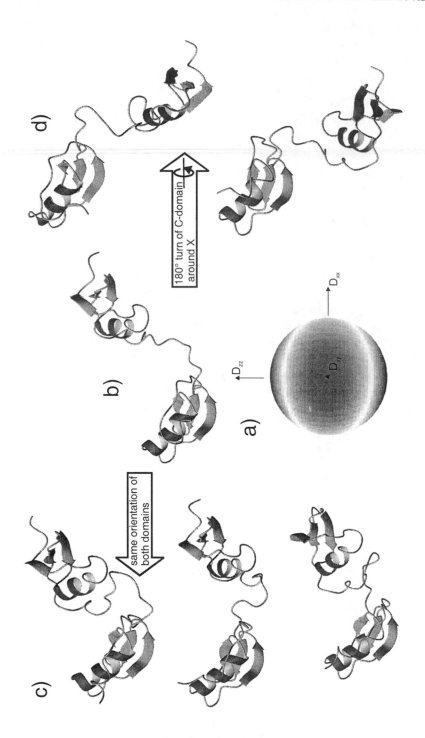

To check the ability of the program to use two sets of dipolar couplings obtained with two different alignment tensors (Ramirez and Bax, 1998) a second tensor was defined $D_{zz} = 10.0\,Hz$, $D_{yy} = 6.5\,Hz$ and $D_{xx} = 3.5\,Hz$ with a rhombicity of 0.2. The orientation of the tensor with respect to the original PDB file of the target structure was changed from $(\alpha,\beta,\gamma) = (60°,135°,0°)$ for the first alignment tensor to $(\alpha,\beta,\gamma) = (120°,90°,133°)$ for the second alignment tensor. A second set of dipolar couplings was calculated and from these values derived φ^{ij} ranges were used, together with the ranges depicted from the first set of dipolar couplings, to recalculate the N-domain of Rhodniin.

With the use of both sets of angle restraints and the previously used protocol 70 of 100 calculated structures have an energy lower than 650 $kcal/Mol$. For the 10 structures with the lowest energy no NOE is violated, and only 0 to 10 of the angle restraints are slightly violated. The dipolar Q-factor is 0.16 for the first set and 0.14 for the second set of dipolar couplings. The rmsd of the structures to the target structure also decrease to $0.3\,\mathring{A}$.

The use of dipolar couplings as φ^{ij} restraints includes these experimental data from the start of the simulated annealing protocol. Moreover, with this implementation the dipolar couplings can be directly translated into intramolecular restraints for the first time without orienting the alignment tensor during the simulated annealing protocol. The convergence of the simulated annealing protocol is almost unchanged as compared to calculations without dipolar couplings. The calculation time increases only slightly (below 5% for these examples) when dipolar couplings are taken into account.

To show the "experimental verification" the protocol has been successfully applied to the structure calculation of trigger factor from M. genitalium using a standard molecular dynamics protocol (Parac et al., 2001). Using 959 interresidual NOEs, 35 hydrogen bonds, 72 $^3J_{HN-H\alpha}$ coupling constants, 68 dihedral angle restraints and 2628 dipole-dipole projection restraints derived from 68 amide dipolar couplings, an ensemble of 200 structures was calculated. Thus, there were on average 11 inter-residual NOEs per residue. Using the dipolar coupling restraints a tight ensemble of structures was obtained. The 12 of the 200 calculated structures with the lowest energy display a backbone rmsd of 0.23 Å and an rmsd of 0.72 Å for all heavy atoms. No NOE violations over 0.5 Å and maximal one NOE violation between 0.2 and 0.5 Å were found. Figure 13 shows an overlay of the 12 best structures fit for minimal backbone rmsd of the (29-113) regions.

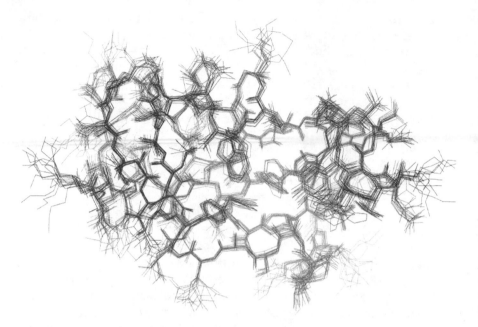

Figure 13. Superposition of the eight lowest energy TF$_{PPIase}$ structures: The high precision of the structure is clearly visible. Only the folded parts of the structure (residues 29-113) are shown.

Since the TF$_{PPIase}$ structure was restrained using residual dipolar couplings, evaluation of the impact of those additional restraints on accuracy and precision of the resulting model was possible. The same data set without any dipolar couplings leads to a distinctly higher backbone rmsd of 0.3 Å. The Q value without dipolar couplings is 0.5, which indicates several large deviations between back-calculated RDCs using our structural model and measured dipolar coupling data (Figure 14). After applying dipolar coupling restraints the Q value drops to 0.28, hence, the agreement of the couplings with the structure is greatly improved. Concomitantly, the dipolar coupling restraints result in a significant improvement of the rmsd value (0.23 Å versus 0.3 Å). The 'flap' region (amino acids between Pro90 and Pro102), in particular, became better defined with the use of dipolar couplings.

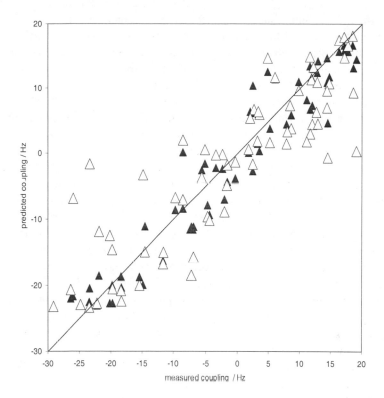

Figure 14. Correlation between measured and theoretically predicted and measured residual dipolar couplings. Open symbols indicate the result of a conventional restrained MD run without dipolar couplings, resulting in a backbone rmsd of 0.3 Å. For this case, the measured dipolar couplings are not very well fulfilled (Q=0.5). Filled symbols indicate the result of a restrained MD run with identical parameters but including residual dipolar coupling restraints. Both the precision (rmsd=0.23 Å) and the quality (Q=0.29) are significantly improved.

Recent examples of dipolar couplings used to enhance the structure of biomolecules include NMR on fully deuterated proteins where the density of protons is very low. In this case, dipolar couplings were very important to describe the fold with higher precision and accuracy (Huang *et al.*, 2000; Mueller *et al.*, 2000a; Mueller *et al.*, 2000b; Choy *et al.*, 2001). Without these new parameters, structure determination of large deuterated proteins was critical due to lack of structural constraints. Clore and co-workers described the impact of dipolar couplings in structure calculation (Clore *et al.*, 1999). They showed the effect on three proteins: B1 domain of streptococcal protein G (GB1) (Clore, *et al.*, 1998); barrier-to-autointegration factor BAF (Cai *et al.*, 1998) and cyanovirin-N (Bewley *et al.*, 1998). Detailed calculations were performed on GB; two different sets of experimental dipolar couplings were used (bicelles and phages). In GB1 approximately 90% of the residues are in secondary structure elements. This is particularly important because dynamics are much smaller in secondary structure elements as shown in the literature (Meiler, *et al.*, 2001; Peti, *et al.*, 2002). Inclusion of the experimental dipolar couplings in the calculations improved the structure dramatically when compared with a structure calculated based on only H-Bond restraints. The rmsd value dropped from 4.3Å to 1.2Å. Using the second alignment the precision was again increased by approximately 25% to a rmsd value of 0.9Å. GB1 is a unique protein due to its extremely low mobility. Therefore, the overall effect of the dipolar couplings should be more pronounced than in less rigid proteins.

Homonuclear dipolar couplings between protons constitute a valuable source of structural information (see Section 2). To check whether a structure is compatible with the H,H dipolar couplings, back calculations were performed on the H,H dipolar couplings for ubiquitin, based on the alignment tensor back-calculated from the known X-ray and NMR structure (Vijay-Kumar *et al.*, 1987; Cornilescu *et al.*, 1999a) and the $^1D_{NH}$ dipolar coupling constants (Clore, *et al.*, 1998). The results showed that the scatter of the H,H dipolar couplings was much larger than observed for the NH dipolar couplings and, on average, the back-calculated couplings were smaller than the experimental couplings (Peti and Griesinger, 2000). Quantitatively, the Q-value of H,H dipolar couplings was generally much larger than of H-N^H and other backbone dipolar couplings. This was not just recognized for only H,H dipolar couplings. C,H dipolar couplings measured in the side chains of proteins, mentioned above (Ottiger, *et al.*, 1998b, Carlomagno, *et al.*, 2000), also showed higher Q-values. However, it is observed that the dipolar H,H and the side chain dipolar C,H couplings fit the calculated NMR structure better than the X-ray structure (backbone dipolar couplings were used for refinement of the NMR structure). Similar results on carbonyl chemical shifts were previously observed (Cornilescu, *et al.*, 1999b). The percentage of decrease in the Q-factors for the H,H couplings from the X-ray structure to the NMR structure is even higher when compared with the N-H^N couplings. The larger discrepancies (Q-factors) of experimental and structure derived dipolar couplings for the side chain can be attributed to larger dynamics of the side chains when compared to the protein backbone.

As mentioned above, all calculated ^1H-^1H dipolar couplings are smaller, on average, than the experimental couplings due to a dynamically scaled alignment tensor derived from the N-H^N couplings used for the calculations. A slope of 0.75 was found for ^1H-^1H dipolar couplings in ubiquitin of the curve that correlated experimental and back-calculated couplings (Peti and Griesinger, 2000). As recently published, a large amount of overall motion is absorbed in the alignment tensor derived from the NH dipolar couplings (Meiler, *et al.*, 2001; Tolman *et al.*, 2001; Peti, *et al.*, 2002). The decrease of the Q-factor and thus the agreement between experimental and calculated dipolar couplings has also been realized by other groups (Brunner and Kalpitzer, 2000).

Only one work (Tjandra *et al.*, 2000) describes the usage of these dipolar couplings for direct refinement of macromolecules. Because the distance of H,H is not fixed as in backbone N,H^N vectors, difficulties arise when using H,H dipolar couplings. This means that one reveals not only orientational but also distance restraints in one measured parameter. This non-fixed geometry makes the usage of these dipolar couplings difficult. Due to the above-mentioned facts the alignment tensor cannot be directly calculated as it is routine for backbone dipolar couplings. Tjandra *et al.*, (2000) used ubiquitin, where they measured H^N,H^α dipolar couplings using the HNHA experiment (Tjandra and Bax, 1997) as a test case. Quantitative J approaches deliver no sign. To increase the number of proton proton dipolar couplings for structure calculations Tjandra also used so-called "null" H,H dipolar couplings which couldn't be measured in the spectra. After refinement the agreement between experimental and theoretical dipolar couplings increased dramatically. This again shows that the precision and the accuracy of structures can be increased using H,H dipolar couplings.

4.   OLIGOSACCHARIDE AS A MULTIDOMAIN MOLECULE

The following example shows how dipolar couplings are used for orienting structurally well defined domains with respect to each other.

An oligosaccharide was chosen since carbohydrates orient well in alignment media that were already established for proteins (Kiddle and Homans, 1998; Rundlöf *et al.*, 1998; Landersjo *et al.*, 2000; Martin-Pastor and Busch, 2000a; Martin-Pastor and Busch, 2000b; Tian *et al.*, 2001). Carbohydrates can either be in the free form or bound to proteins or other biomacromolecules (Bolon *et al.*, 1999; Olejniczak, *et al.*, 1999; Al-Hashimi *et al.*, 2000; Medek *et al.*, 2000; Thompson *et al.*, 2000) in solution. They often play important roles in many biological processes. Most deposited structures of carbohydrates were determined using X-ray crystallography or neutron diffraction spectroscopy. Because of packing interactions differences between solution (NMR) and X-ray structures are expected. Additional parameters should be very useful to get much more accurate solution structures of oligosaccharides. As for protein domains that show few mutual NOE's, structure determination of oligosaccharides has always been difficult due to the lack of NMR parameters across the glycosidic bond. Usually only few distances can be measured between sugar rings. The coupling constants are mainly heteronuclear and experiments for the measurement are not very sensitive (Halbeek, 1994). Therefore, solution structures of carbohydrates have been less accurate due to the lack of NOE restraints.

However, carbohydrates pose some specific problems when compared to proteins. While the measurement of one set of dipolar couplings e.g. NH is possible and sufficient in protein domains is sufficient to orient the protein domains with respect to each other provided the structure of each of the protein domains is known, this is not the case for carbohydrates. The distribution of the CH vectors is by far not isotropic e.g. in the most abundant pyranose rings in order to determine a sufficiently accurate alignment tensor. Thus, measurement of only CH dipolar couplings is not sufficient to define the alignment tensor for each individual sugar moiety. If the alignment tensor can be identified with a high accuracy for each sugar moiety the relative orientation and mobility of the monosaccharides can be detected.

One needs to keep in mind that this would also be important for the structure refinement because the refinement against an inaccurate alignment tensor would lead to incorrect structures. Therefore, it is important to increase the number of measurable dipolar couplings in carbohydrates to extract a well defined alignment tensor. Different groups have shown different ways for doing this by increasing the number of independently measured dipolar couplings. $^2J_{CH}$ heteronuclear dipolar couplings, in addition to HH dipolar couplings (Tian, *et al.*, 2001) or different HH dipolar couplings, can be found in the literature (Martin-Pastor and Bush, 2001; Neubauer *et al.*, 2001). With at least five dipolar couplings from independent dipolar tensors the alignment tensor for each sugar moiety can be derived (Peti, *et al.*, 2002).

This approach is reviewed on the example of the trisaccharide raffinose (O-α-D-galactopyranosyl-1-6-α-D-glucopyranosyl-β-1-2-D-fructofuranoside, Figure 15) (Neubauer, *et al.*, 2001).

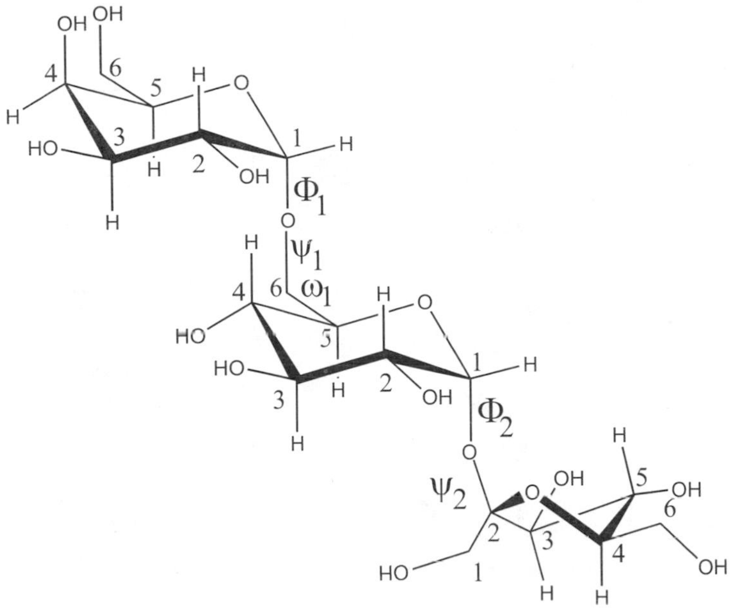

Figure 15. Constitution and configuration of raffinose. The definition of the torsional angles is given in the figure. Reproduced with permission from Neubauer *et al.*, (2001), *Helv. Chim. Acta.*, **84**, 243-258.

First the measurement of heteronuclear CH dipolar couplings for raffinose was performed: HSQC spectra (Bodenhausen and Ruben, 1980) without decoupling in the ^{13}C dimension and the SPITZE-HSQC (Carlomagno, *et al.*, 2000) sequence for the measurement of the CH and HH dipolar couplings in the CH$_2$-groups could be recorded. Additionally we used H,H-E.COSY (Griesinger, *et al.*, 1985; Griesinger, *et al.*, 1986; Griesinger, *et al.*, 1987; Mueller, 1987) for the measurement of the H,H dipolar couplings in carbohydrates. We want to focus on the new features in E.COSY spectra in anisotropic solutions and include some practical advice. This method can only be used if the alignment is not too strong. Therefore, different samples with different total lipid concentrations of the bicelles should be prepared and checked which one allows the most accurate measurement of the HH dipolar couplings in the E.COSY spectrum. In this case concentrations between 5% and 7.5% total lipid concentration were the optimum. It is more difficult to find the optimum for phages or other alignment media. These are just important for the measurement of HH dipolar couplings. There are three kinds of dipolar couplings that could be measured in raffinose in bicelles solution:

(a)

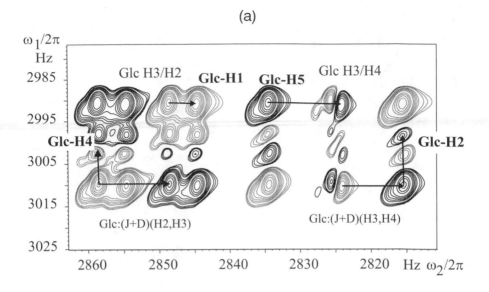

(b)

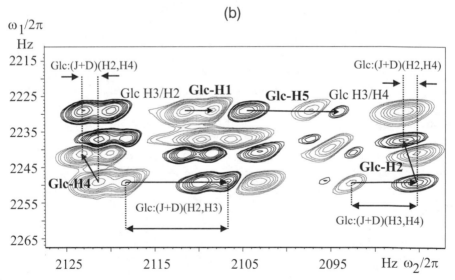

Figure 16. Isotropic (a, top) and anisotropic (b) solution spectra of raffinose showing the GlcH3/H2 and GlcH3/H4 cross peaks. While in isotropic solution, the passive spins show couplings only to one of the active spins, in anisotropic solution, E.COSY patterns arise due to the fact that H2 and H4 exhibit a dipolar coupling. The same is true for H1 and H3. The H2,H4 coupling can be readily read off from the cross peak. Reproduced with permission from Neubauer *et al.*, (2001), *Helv. Chim. Acta.*, **84**, 243-258.

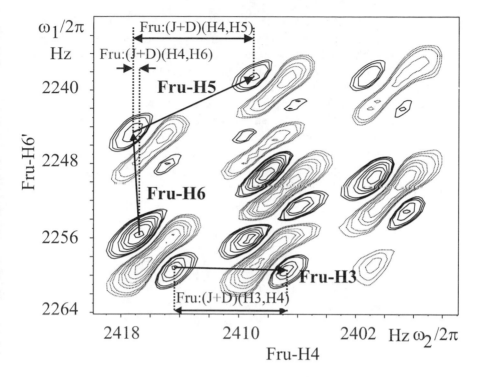

Figure 17. Cross peak between Fru-H6' and Fru-H4. The passive spins yielding E.COSY type shifts are indicated. The long range Fru(H6,H4) coupling can be extracted from this cross peak. Reproduced with permission from Neubauer *et al.*, (2001), *Helv. Chim. Acta.*, **84**, 243-258.

1.) Dipolar couplings having an isotropic counterpart: These dipolar coupling can be measured as heteronuclear backbone dipolar couplings in proteins. In the aligned phase the sum of the dipolar and the isotropic couplings is measurable. Therefore, the isotropic counterpart must be measured separately in isotropic solution (see Figure 16).

2.) Dipolar couplings having no isotropic counterpart: Since dipolar couplings may exist between spins A and B that are not scalar coupled (or the isotropic coupling is too small to be measured e.g. ^4J couplings), there will be new cross peaks in the E.COSY spectrum in anisotropic phase that are not observed in isotropic solution and thus easily identifiable. For the cross peaks in cases 1 and 2 the sign and the size of the dipolar couplings can be measured (see Figure 17).

3.)   Finally, the spectra also contains cross peaks that do not show
an E.COSY pattern, e.g. between protons belonging to different
monosaccharides in the trisaccharide. For these cross peaks, the sign of
the coupling constant cannot be determined with this method. However,
the size can be extracted by comparison with other cross peaks
according to the DISCO procedure (Oschkinat and Freeman, 1984;
Kessler *et al.*, 1985; Kessler and Oschkinat, 1985) procedure. These
dipolar couplings were not used in structure calculations (so called "non-
firm" dipolar couplings). They were used for cross checking after
structure calculation using all other dipolar couplings (see Figure 18).

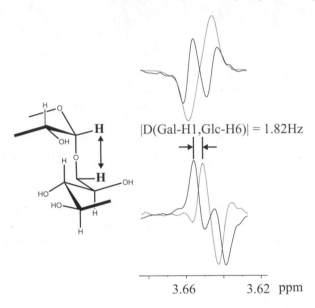

$|D(\text{Gal-H1},\text{Glc-H6})| = 1.82\text{Hz}$

3.66   3.62 ppm

Figure 18. Cross peak between Gal-H1 and Glc-H6. The dipolar coupling gives rise to the
cross peak. In this case, its size but not its sign can be determined by the so called DISCO
procedure using in addition the Glc-H6/H6' cross peak. Reproduced with permission from
Neubauer *et al.*, (2001), *Helv. Chim. Acta.*, **84**, 243-258.

In proteins where a large number of dipolar couplings can be easily
measured, mainly with an isotropic distribution of orientations, the
determination of the alignment tensor is unambiguous. In contrast, due to the
relatively low number of measured dipolar couplings per Monosaccharide
and the restricted orientation, the determination of the tensor size according
to Clore *et al.*, (1998) cannot be used in carbohydrates as explained above.
For a known structure such as the individual monosaccharides, one needs at
least five linearly independent equations to obtain the alignment tensor
(Ramirez and Bax, 1998; Cornilescu, *et al.*, 1999a; Meiler, *et al.*, 2000a;

Peti, et al., 2002). For glucose and galactose there are only two linearly independent CH orientations, which lead to two linearly independent equations. Thus, the tensor determination for each monosaccharide from CH couplings is impossible. However, this number can be increased well over five independent equations including H,H couplings. In order to gain insight into the error propagation of the estimation of the translation between the dipolar couplings in the alignment tensors, we performed the following test. We took the lowest energy structure of raffinose calculated (vide infra) (calculation method is shown below) and the firm experimental dipolar couplings (vide infra) and fit the alignment tensor individually to the monosaccharides (we named those dipolar couplings firm that are used for structure calculation). We obtained for galactose: D_{ax} = -3.09 ± 0.26 Hz, R = 0.45 ± 0.19, for glucose: D_{ax} = -5.6 ± 0.73 Hz, R = 0.56 ± 0.09, and for fructose: D_{ax} = 7.20 ± 0.93 Hz, R = 0.6 ± 0.08. The errors have been derived by randomly adding or subtracting ±0.2 Hz or 0 Hz to the experimental HH-dipolar couplings and ±0.4 Hz or 0 Hz to the experimental CH-dipolar couplings (these values are the errors of our experimental dipolar couplings). With this error propagation property of the dipolar couplings on the alignment tensor, for the structure calculations, we allowed for an error of approximately ±20% for the axial component and the rhombicity of the tensor.

A complete new procedure for structure calculation using dipolar coupling was used (Figure 19). The starting size of the tensor was taken from the experimental dipolar couplings and the neutron diffraction or X-ray structure. Thus, the methodology using a histogram was used knowing that the real tensor is different. For the structure calculation the tensor size and rhombicity was varied in a grid search to find the right tensor. The step size of the grid search was chosen to have 16 values for the axial component and 4 values for the rhombicity of the alignment tensor. The alignment tensor was back calculated from the best ten structures by Moore-Penrose-Inversion to check for self-consistency of the tensor. For those structures that were checked positive the "non-firm" dipolar couplings (these were not used in the structure calculations) were calculated and compared with the experimental ones.

For measurement of the dipolar couplings the carbohydrates were dissolved in liquid crystal solution (CHAPSO/DMPC 1:3; 5-15%) (Losonczi and Prestegard, 1998). In contrast to charged proteins where one can influence the orientation of the alignment tensor by adding CTAB (+) or SDS (-), we did not succeed when performing this in uncharged carbohydrates. Thus, uncharged carbohydrates like these used in this case can be seen as an ideal system where the orientation is completely sterically induced (Zweckstetter and Bax, 2000).

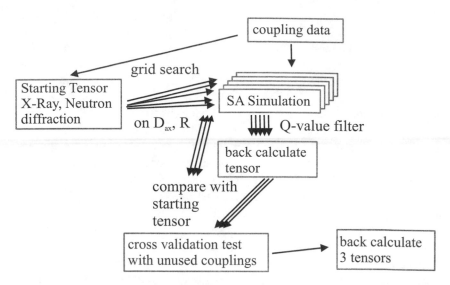

Figure 19. Flow diagram for the calculation of sugar structures using dipolar couplings. Using the experimental dipolar couplings and the known monosaccharidic structures, the alignment tensors for the individual sugars is calculated. With a grid search on the alignment tensor the conformation with the lowest Q value is calculated. The alignment tensor is back calculated from the new structure and compared with the original alignment tensor. After convergence of the alignment tensors, the unused dipolar couplings are used for cross validation.

Our protocol that used a optimized force field for carbohydrates (Homans, 1990) and XPLOR (Brünger, 1992), currently allows no ensemble calculation. Only one conformation can be calculated. Couplings that are subject to conformational averaging are excluded from the fitting for the determination of the alignment tensor and the structure calculation. The same holds for protons with overlapping chemical shifts. The glucose $C6H_2$ group dipolar couplings could be used for the structure calculation because the population of the C5, C6 torsion angle is mainly g (>80%) as is shown by J-coupling analysis. By contrast, the fructose $C6H_2$ group shows conformational averaging between 66% of g and 32% of g and could therefore not be used in the structure calculation protocol. The two protons at the galactose C6 are degenerate, as are the protons at fructose C1. The excluded couplings were used after calculation of the structure for cross validation against calculated theoretical dipolar couplings. Alignment tensors for each monosaccharide from firm heteronuclear D_{CH} dipolar couplings and the combination of firm heteronuclear and homonuclear dipolar $D_{CH} + D_{HH}$ dipolar couplings were extracted by fitting these firm dipolar coupling constants against the X-ray structure of raffinose (Jeffrey and Huang, 1990) individually for each monosaccharide and for the whole structure using *DipoCoup* (Meiler, *et al.*, 2000b).

Using the firm dipolar couplings (these couplings were used for structure calculation) we find the following tensor sizes for the monosaccharides from the CHAPSO/DMPC data: galactose: $D_{ax} = -3.11$ Hz, R = 0.35, glucose: $D_{ax} = -5.60$ Hz, R = 0.57, fructose: $D_{ax} = 7.01$ Hz, R = 0.56. The tensor sizes for glucose and fructose are similar assuming an error of 20%, whereas for galactose a reduction of the tensor size to about 60% is observed. We attribute this reduction of the monosaccharidic tensor size to dynamics. We therefore measured ^{13}C T_1, T_2 and NOE carbon relaxation data. A total correlation time of approximately $\tau_c = 150ps$ was calculated for raffinose. The average T_1 values of galactose are larger than in glucose and fructose, indicating a faster mobility. The mobility is anisotropic since the two axial CH vectors at C1 and C4 in galactose show a higher mobility than the equatorial vectors. This could be interpreted as mobility around both or either of the two parallel angles ϕ_1 or ω_1 angle of the glycosidic linkage between galactose and glucose in raffinose.

The results of the structure calculation was the following: For $D_{ax} = -4.50Hz$, R = 0.667 and $D_{ax} = 4.50Hz$, R = 0.583 the back calculated values from the ten lowest energy structures were $D_{ax} = -4.17 \pm 0.02Hz$, R = 0.58 $\pm$ 0.01 and $D_{ax} = -4.24 \pm 0.01Hz$, R = 0.58 $\pm$0.02, respectively. By adding or subtracting non dipolar couplings like $^3J_{HH}$ couplings between H_3H_4 and H_4H_5 protons in the fructose ring, no difference between the calculated structures could be seen. All calculations proved the $^4T_3$ configuration of the fructose ring.

It is interesting to note that fitting known substructures to the experimental dipolar coupling can easily derive guesses for conformational similarity. Fitting the neutron diffraction structure of saccharose (Brown and Levy, 1963; Brown and Levy, 1973) to the firm dipolar couplings for the saccharose part yields agreement with Q=0.26 and compares well with the Q-factor for the monosaccharide tensors. Thus it can be expected that the solution structure of the saccharose part in raffinose agrees well with the neutron diffraction structure of saccharose. Indeed, the so found values of ϕ_2 = 99.7 $\pm$ 3.8° and ψ_2 = -159.3 $\pm$ 3.5° agree very well with the neutron diffraction data ϕ_2 = 107.8° and ψ_2 = -159.8°. By contrast, fitting the X-ray structure (Jeffrey and Huang, 1990) of raffinose to the firm dipolar couplings yields a Q value of 0.54. Thus it is expected that the solution structure of raffinose deviates from its X-ray structure. Indeed, the Glc-Fru linkage has $\phi_2 = 107.8°$ and $\psi_2 = -159.8°$ whereas the X-ray structure of raffinose shows $\phi_2 = 81.6°$ and $\psi_2 = -105.5°$.

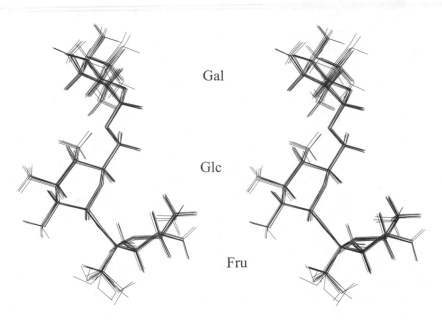

Figure 20. Stereo view of the average raffinose structure as derived in solution. The rmsd value of the structure is only 0.28 Å. The glycosidic angle between fructose and glucose is different in solution as compared to the X-ray structure. Reproduced with permission from Neubauer *et al.,* (2001), *Helv. Chim. Acta.,* **84**, 243-258.

To summarize our experiences for structure determination of small molecules we could show that by using dipolar couplings for structure determination the precision of the calculated structures is much higher than calculations without dipolar couplings. The rmsd of the heavy atoms is reduced from 1.08 Å (just NOE data) to 0.28 Å (including dipolar couplings) (Figure 20). The raffinose structure in solution is different from that in the crystal. This is especially true for the glycosidic linkage between the glucose and the fructose moiety. Future projects should also include the dynamical information of the dipolar couplings in ensemble calculation. This will allow a better description of the structure of molecules in solution.

5. THE IMPACT OF DIPOLAR COUPLINGS IN STRUCTURAL GENOMICS

To meet the demands of structural genomics, the process of fold prediction and structure elucidation need to be substantially accelerated. Dipolar couplings have the potential to play an important role. This is indicated in the fast increasing amount of publications on this topic (Fowler et al., 2000; Prestegard and Kishore, 2001; Prestegard et al., 2001; Zweckstetter and Bax, 2001). Nevertheless, new software for the automatic evaluation of NOESY spectra ARIA (Linge et al., 2001) and CANDID (Herrmann et al., 2002) will also accelerate the usage of the "classical" NOE parameter. X-ray crystallographers will provide high resolution structures very fast and in an increasing number. Integrated approaches that make clever use of easily obtainable experimental information and from the pdb database have therefore a great potential.

The number of known protein sequences and therefore the number of proteins for which a three dimensional structure determination becomes necessary for, increases rapidly. This is because projects such as the human genome project produce thousands of genes i.e. protein sequences.

One goal of post genomic research is to determine all protein folds, the number of which are expected to be limited. Sequence profile methods currently have a large impact in fold recognition. Ab initio structure prediction works up to 80-150 amino acids and may emerge as a powerful tool for structure prediction in the future. To obtain a complete coverage of folds most effectively it is important to focus on the elucidation of structures with novel folds rather than rediscovering known folds on new proteins. Blast threading and ab initio approaches rely on the analysis of primary and secondary structure in the context of a three dimensional structure database. Dipolar couplings are easy to measure and they restrain the conformation of proteins rather well. Therefore, it would be desirable to calculate structures just from dipolar couplings information only. Complete substitution is hampered by the fact that dipolar couplings do not contain translational but only angular information and, therefore, require measurements of extreme accuracy if no NOE's are used. An alternative is to use models to assist dipolar couplings to obtain the translational restraints. This was first proposed by the Annila group (Aitio et al., 1999; Annila et al., 1999). They showed that the most probable structure can be predicted by a comparison of measured dipolar couplings with dipolar couplings that are calculated from tentative 3D structures derived from known structures of homologous proteins. A similar approach was taken in DipoCoup (Meiler, et al., 2000b) as described below. Using this idea but also incorporating chemical shifts as additional easily measurable restraints, Bax and co-workers introduced a method called Molecular Fragment Replacement (MFR) for structure determination using dipolar couplings (Delaglio *et al.*, 2000). This method searches for 7-residue fragments of proteins in the PDB database that fulfill the experimental chemical shifts (this allows the prediction of the angles ϕ and ψ in the protein backbone (Cornilescu *et al.*, 1999b)) and the experimental dipolar couplings. This was shown for the example of ubiquitin where the difference in the backbone atom positions between the structure calculated based on NOE, J-coupling, H-bond and dipolar coupling data and the MFR calculated structure is less than 1Å. It could also be shown that based on dipolar couplings small conformational changes can be calculated (Chou *et al.*, 2000b).

In this Section, we will describe a method based on the program *DipoCoup* that allows a 3D homology search based on dipolar couplings. Since residual dipolar couplings give information about the relative orientations of two bond vectors with respect to one alignment tensor, the relative orientation of the two vectors with respect to each other is described. If these vectors belong to two secondary structure elements or to two domains for example, their relative orientation is defined and thus a set of dipolar couplings is capable to define the protein fold. This fact is used for establishing a three dimensional fold search on the basis of dipolar coupling information by comparing a experimental set of dipolar couplings with theoretically recalculated data for all known protein folds. This approach has the potential to predict folds of a new protein with little primary structure homology to proteins with known folds.

By the same token, structure elucidation of a new protein with a structure homologous to a known fold will be accelerated. After the method of fitting experimental dipolar couplings over a known three dimensional structure was established by Loszonski *et al.*, (1999), Annila, *et al.* (1999) discussed a first example of using these experimental NMR parameters in an early stage for 3D homology searches. The possibility for using protein fragments generated from PDB and chosen by aligning similar dipolar couplings and chemical shifts for structure determination was shown (Delaglio *et al.*, 2000). Meiler *et al.* (2000b) offer a free program *DipoCoup* for performing this homology search, as well as several other analyses dealing with dipolar coupling data on the internet (Meiler, 1996-2001). The approach is discussed at theoretically calculated and experimentally determined dipolar coupling data and pseudocontact shifts. All calculations where performed using the program *DipoCoup*.

Experimental dipolar couplings between nuclei i and j - D^{ij} and pseudocontact shifts between nucleus i and electron e - δ_{PC}^{ie} are related to the alignment tensor (principal components: A_{xx}, A_{yy}, A_{zz}) or to the magnetic susceptibility tensor (principal components: χ_{xx}, χ_{yy}, χ_{zz}) and to the orientation of a specific vector with respect to the alignment tensor expressed by the projection angles θ_z^{ij} and φ_x^{ij} according to Eq. 15. The vector is either the vector $\vec{r}_{ij}$ between the two coupled atoms i and j in the case of dipolar coupling (Eq. 15a), or the vector $\vec{r}_{ie}$ between the electron spin e (paramagnetic center) and the active nucleus i (Eq. 15b).

$$D^{ij}\left(\theta_z^{ij},\varphi_x^{ij}\right) = \frac{-\mu_0 h S \gamma_i \gamma_j}{8\pi^3 r_{ij}^3}\left[\begin{array}{c}\frac{1}{6}\left(2A_{zz}-A_{xx}-A_{yy}\right)\left(3\cos^2\theta_z^{ij}-1\right)\\ +\frac{1}{2}\left(A_{xx}-A_{yy}\right)\cos 2\varphi_x^{ij}\sin^2\theta_z^{ij}\end{array}\right]\quad [15a]$$

$$\delta_{PC}^{ie}\left(\theta_z^{ie},\varphi_x^{ie}\right) = \frac{10^6}{6\pi r_{ie}^3}\left[\begin{array}{c}\frac{1}{6}\left(2\chi_{zz}-\chi_{xx}-\chi_{yy}\right)\left(3\cos^2\theta_z^{ie}-1\right)\\ +\frac{1}{2}\left(\chi_{xx}-\chi_{yy}\right)\cos 2\varphi_x^{ie}\sin^2\theta_z^{ie}\end{array}\right]\quad [15b]$$

In the case of paramagnetic alignment, the susceptibility tensor $\hat{\chi}$ is related to the alignment tensor $\hat{A}$ by $\hat{\chi} = \hat{A}\dfrac{15\mu_0 kT}{4B_0\pi}$. Eq. 15 uses the alignment tensor as the frame of reference (Fig 21). The formal dependence of dipolar couplings and pseudo contact shifts is the same while the prefactors differ. The prefactor is constant for dipolar couplings (Eq. 15a) if the distance r_{ij} is constant. However, the prfactor varies for pseudocontact shifts and also for dipolar couplings between nuclei whose distance is not fixed a priori in the bonding network because the distance r_{ie} or r_{ij} cannot be regarded as constant (Ghose and Prestegard, 1997; Clore and Garrett, 1999).

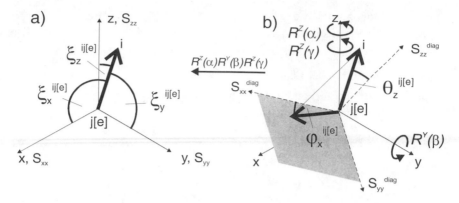

Figure 21. a) Coordinate system of the molecule *(x, y, z)* with a bond vector between the two nuclei *i* and *j* or a nucleus *i* and an electron *e*. The projection angle of the vector onto the *x, y, z* axes are ξ_x^{ij}, ξ_y^{ij} and ξ_z^{ij}, respectively. b) Representation of the vector in the frame of the tensor $S_{xx}^{diag}, S_{yy}^{diag}, S_{zz}^{diag}$. The Euler rotation transforms the tensor into the coordinate system of the molecule. The orientation of the bond vector r_{ij} is defined by the angles θ_z^{ij} and φ_x^{ij}. Reproduced from *J. Biomol. NMR.*, **17**:283 (2000) with kind permission of Kluwer Academic Publishers.

Figure 22 (overleaf): Schematic features of the program *DipoCoup*. Arrows (a) describe the analysis of experimental dipolar couplings and/or pseudocontact shifts from protein **A** without the knowledge of its three dimensional structure and without the use of the database. Arrows (c) describe the analysis of the three dimensional structure of protein **B** for calculating theoretical dipolar couplings or pseudo contact shifts. Arrows (b) indicate the fitting procedure of protein **A** to the known three dimensional structure of protein **B**. This is used to obtain the orientation of the alignment tensor derived from the experimental data for **A** in the molecular frame of protein **B**. The quality of the fit is measured by the *Q*-value. Alternatively a whole database can be searched finding homologous structures or structure fragments to a single molecule **B**. Reproduced from *J. Biomol. NMR.*, **17**:283 (2000) with kind permission of Kluwer Academic Publishers.

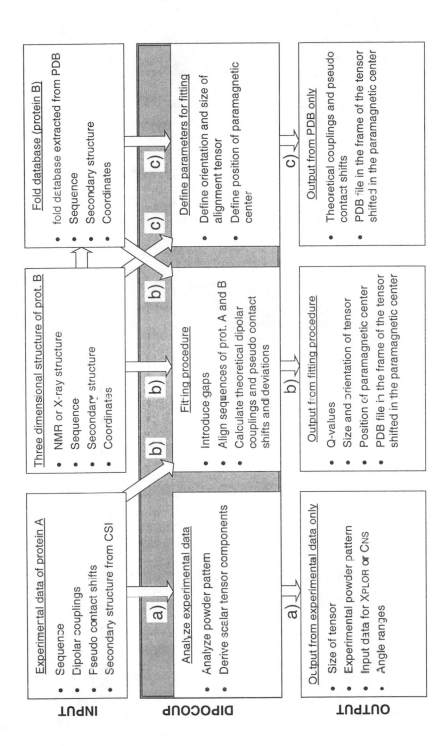

However, the measured orientation value cannot be translated directly in a combination of θ_z^{ij} and φ_x^{ij}. An infinite number of combinations of θ_z^{ij} and φ_x^{ij} exist that fulfill an experimental value. However if one pair of angles θ_z^{ij} and φ_x^{ij} can be found to be correct due to the alignment of a whole molecule, four orientations of the molecule fulfill all experimental values, since the signs of angles θ_z^{ij} and φ_x^{ij} can be reversed independently in Eq. 15 without the change of either dipolar couplings or pseudo contact shifts.

In the context of a 3D homology search, the coordinate system of a protein in the 3D structure data file is the natural frame of reference. Therefore, we express Eq. 15 in this coordinate system which is rotated by three Euler angles α, β, γ with respect to the alignment tensor (Figure 21). Eq. 16 expresses the dipolar couplings in the molecular frame (an identical equation is obtained for pseudocontact shifts δ_{PC}^{ie} by replacing j with e; all following equations are given only for dipolar couplings):

$$
D^{ij}\left(\xi_x^{ij},\xi_y^{ij},\xi_z^{ij}\right) = F_{ij}\begin{pmatrix}\cos\xi_x^{ij}\\\cos\xi_y^{ij}\\\cos\xi_z^{ij}\end{pmatrix}^T\begin{pmatrix}-S_{yy}-S_{zz} & S_{xy} & S_{xz}\\ S_{xy} & S_{yy} & S_{yz}\\ S_{xz} & S_{yz} & S_{zz}\end{pmatrix}\begin{pmatrix}\cos\xi_x^{ij}\\\cos\xi_y^{ij}\\\cos\xi_z^{ij}\end{pmatrix}
$$

$$
= F_{ij}\begin{pmatrix}\left(\cos^2\xi_y^{ij}-\cos^2\xi_x^{ii}\right)S_{yy}\\ +\left(\cos^2\xi_z^{ij}-\cos^2\xi_x^{ij}\right)S_{zz}\\ +\left(2\cos\xi_x^{ij}\cos\xi_y^{ij}\right)S_{xy}\\ +\left(2\cos\xi_x^{ij}\cos\xi_z^{ij}\right)S_{xz}\\ +\left(2\cos\xi_y^{ij}\cos\xi_z^{ij}\right)S_{yz}\end{pmatrix}
$$

[16]

$$
\textit{with } F_{ij} = \frac{-\mu_0 h S \gamma_i \gamma_j}{8\pi^3 r_{ij}^3}
$$

In this molecular frame the alignment tensor is no longer diagonal and can be expressed by a symmetric three by three traceless matrix holding five independent elements S_{yy}, S_{zz}, S_{xy}, S_{xz} and S_{yz}, which are the elements of the Saupe matrix (Saupe, 1968). The eigenvalues of this matrix S_{xx}^{diag}, S_{yy}^{diag}, S_{zz}^{diag} are identical to the principal components of the alignment tensor A_{xx}, A_{yy}, A_{zz}. The angles ξ_x^{ij}, ξ_y^{ij}, ξ_z^{ij} define the projection angles of the bond vector r_{ij} or the vector between the nucleus and the electron r_{ie} onto the molecular frame. For a given structure and experimental dipolar couplings $D_{exp.}^{ij}$, the five independent tensor contributions can be determined directly by solving the linear system of equations given from Eq. 17 for a set of experimental dipolar couplings for n pairs of nuclei i and j requiring $D^{ij}_{exp.} = D^{ij}_{theor.}$ (Losonczi, et al., 1999).

$$
\begin{pmatrix} D^{ij1}_{exp.}/F_{ij} \\ \vdots \\ D^{ijn}_{exp.}/F_{ij} \end{pmatrix} \overset{!}{=} \begin{pmatrix} D^{ij1}_{theor.}/F_{ij} \\ \vdots \\ D^{ijn}_{theor.}/F_{ij} \end{pmatrix} =
$$

$$
\begin{pmatrix}
\cos^2 \xi^{ij1}_y & \cos^2 \xi^{ij1}_y & 2\cos\xi^{ij1}_x & 2\cos\xi^{ij1}_x & 2\cos\xi^{ij1}_x \\
-\cos^2 \xi^{ij1}_x & -\cos^2 \xi^{ij1}_x & \bullet\cos\xi^{ij1}_y & \bullet\cos\xi^{ij1}_y & \bullet\cos\xi^{ij1}_y \\
\vdots & \vdots & \vdots & \vdots & \vdots \\
\cos^2 \xi^{ijn}_y & \cos^2 \xi^{ijn}_y & 2\cos\xi^{ijn}_x & 2\cos\xi^{ijn}_x & 2\cos\xi^{ijn}_x \\
-\cos^2 \xi^{ijn}_x & -\cos^2 \xi^{ijn}_x & \bullet\cos\xi^{ijn}_y & \bullet\cos\xi^{ijn}_y & \bullet\cos\xi^{ijn}_y
\end{pmatrix}
\begin{pmatrix} S_{yy} \\ S_{zz} \\ S_{xy} \\ S_{xz} \\ S_{yz} \end{pmatrix} = \mathbf{C}\vec{S} \tag{17}
$$

This system of equations can be solved by multiplication of the pseudo inverse of the rectangular matrix $\mathbf{C}$, i. e. by calculating the Moore-Penrose-Inverse of the matrix yielding the vector S. Rebuilding the Saupe matrix from these values and analyzing its eigensystem yields the eigenvalues of the tensor S_{xx}^{diag}, S_{yy}^{diag}, S_{zz}^{diag} as well as its orientation given by the eigenvectors. It can be expressed in terms of three Euler angles in α, β, and γ.

$$
\begin{pmatrix}
-S_{yy} - S_{zz} & S_{xy} & S_{xz} \\
S_{xy} & S_{yy} & S_{yz} \\
S_{xz} & S_{yz} & S_{zz}
\end{pmatrix} = \left(R^Z(\alpha) R^Y(\beta) R^Z(\gamma) \right)^T
$$

$$
\bullet \begin{pmatrix}
S_{xx}^{diag} & 0 & 0 \\
0 & S_{yy}^{diag} & 0 \\
0 & 0 & S_{zz}^{diag}
\end{pmatrix} \tag{18}
$$

$$
\bullet R^Z(\alpha) R^Y(\beta) R^Z(\gamma)
$$

The solution of the Moore-Penrose inversion problem is equivalent to finding a solution D^{ij}_{theor} with the least square deviation for a given experimental set of D^{ij}_{exp}. Experimental errors cannot be taken into consideration during this approach. Therefore, a careful analysis afterward is necessary according to (Losonczi, *et al.*, 1999).

The 3D homology search program *DipoCoup* offers three general means of analyzing dipolar couplings and pseudocontact shifts (Figure 22) of the protein **A** under investigation by comparing it to one or several selected proteins **B** from the database. In procedure (a) one can analyze the experimental dipolar couplings and/or pseudocontact shifts of protein **A**, as well as obtain secondary structure information from ^{13}C chemical shift index data, CSI, (Spera and Bax, 1991; Wishart *et al.*, 1992; Wishart and Sykes, 1994; Wishart *et al.*, 1995). The program can handle different sets of dipolar couplings in combination with pseudo contact shifts for one alignment tensor. Dipolar couplings for atom pairs with defined distances (e. g.: N-H^N, Cα-Hα, Cα-CO) in protein **A** scaled with F_{ij}^{-1} can be visualized in a histogram. The eigenvalues S_{xx}^{diag}, S_{yy}^{diag}, S_{zz}^{diag} of the tensor can then be determined from the histogram (Clore, *et al.*, 1998). With this information the program can generate input files for XPLOR- or CNS-simulated annealing protocols which use residual dipolar couplings as restraints. It also calculates the projection angle ranges that allow the use of dipolar couplings in XPLOR- or CNS-calculations without requiring definition of the orientation of the alignment tensor (compare to section 3 and (Meiler *et al.*, 2000a)).

3D homology searching is prepared in (c) by calculating NMR properties of a potentially homologous protein **B** which is extracted from a structure database. From the given three dimensional structure of protein **B** a set of dipolar couplings and pseudocontact shifts can be generated. To do this, the program first adds hydrogen atoms that may be missing in the structure and corrects the bond lengths between all heavy atoms and their bound hydrogen atoms according to (Ottiger and Bax, 1998). For a given alignment tensor and paramagnetic center theoretical dipolar couplings and pseudocontact shifts can be calculated, visualized and stored. Also the three dimensional structure can be stored and oriented in the frame of the alignment tensor and shifted to the appropriate position with respect to the paramagnetic center.

Finally, in procedure (b), both the experimental data from the protein **A** under investigation and the three dimensional structure from the protein **B** can be checked for matching 3D folds. One or several proteins **B** from the protein data bank (PDB) can be used, allowing one to compare the experimental data to a database of proteins. Hydrogen atoms are added or corrected for proteins **B** as previously described in (c). The secondary structure elements of proteins **B** are calculated from the coordinates by analyzing hydrogen bonds and φ- and ψ-angles. Then the alignment of residues a of **A** and b of **B** is performed such that $D_{exp}(a)$ is assigned to the respective atoms of residue b of protein **B**. This set of "experimental" dipolar couplings is used to calculate the alignment tensor and its orientation according to Eq. 17. In this case no analysis of the histogram is required. As a quality measure, the Q-value of the dipolar couplings (analogous for pseudocontact shifts) is used (Cornilescu *et al.*, 1999a):

$$Q = \sqrt{\sum_{ij} \left(D^{ij}_{exp} - D^{ij}_{theor}\right)^2 \bigg/ \sum_{ij} \left(D^{ij}_{exp}\right)^2}$$

Q is a normalized square deviation and is equivalent to $\sqrt{2}$ times the R-factor (Clore and Garrett, 1999). Moreover the program calculates the correlation coefficient R (not to be confused up with the R-factor) and thus offers a second quality value. Taking the specially conserved structure of the secondary structure elements into consideration, both quality values are calculated not only for the overall protein but also for regions of well defined secondary structure and for loop regions separately.

The alignment of the residues a of protein **A** and b of protein **B** is not based on primary sequence homology. Rather, the sequences are aligned to have a minimum Q-value. The program aligns first all amino acids of protein **A** over the amino acids of protein **B** starting with the first for both proteins, respectively. After calculation of the Q factor for this alignment the sequence of protein **A** is shifted by one residue and the procedure is repeated until the last amino acid of **A** is aligned with the last amino acid of **B**. This ensures that the terminal secondary structure elements of protein **A** are fully used in the alignment process. This procedure avoids discovering false positive hits due to a changing number of dipolar couplings and CSI data used. A check for matches of the secondary structure elements is performed. Secondary structure elements are derived from the CSI for protein **A** and by analysis of H-bonds and ϕ and ψ (procedure (c) in Figure 22) for protein **B**. To achieve optimal alignment, the secondary structure elements of **A** can be disconnected and aligned individually with the matching secondary structure elements of **B**. By default, disconnection of the secondary structure elements in **A** occurs at boundaries of secondary structure elements, e.g., from β-sheet to random coil. However, the user may also suggest other positions for disconnecting the sequence if additional information is required or other ideas need to be verified. If no secondary structure alignment is possible the alignment with minimal Q without the use of secondary structural information is presented. The program allows an overall or partial search of the PDB database as will be described subsequently.

If pseudocontact shifts are given the position of the paramagnetic center can be either explicitly defined in the three dimensional structure or can be optimized by an interactive grid search protocol. For the optimization, a starting position, a step size, and the size of the cube to be searched must be supplied. The program searches this given cube using the starting step size and restarts the search with the best point of the previous search and a decreased step size and a decreased size of the cube until the step size is smaller than a predefined target value (e.g. 0.1Å).

The program is applied to three different protein structures: For Rhodniin, a theoretical set of dipolar couplings and pseudo contact shifts using a NMR structure was calculated (Maurer and Griesinger, 1998). A 3D homology search was performed on a restricted database of proteins according to procedure (b) of Figure 22. For cyclophilin A we recorded experimental dipolar couplings and procedure (a) of Figure 2 is used to analyze the experimental data. The dipolar couplings are fit against the known NMR and X-ray structures, and the orientation of the alignment tensor is determined. The third example is the protein *Hgi*CIC. This protein contains a helix-turn-helix motive. Using experimentally derived dipolar couplings, a 3D homology search was performed on a restricted set of the PDB.

Rhodniin consists of 103 amino acid and contains two similarly folded domains of 45 amino acids connected by a flexible linker of 10 amino acids. A set of $^1D_{NH}$ dipolar couplings and pseudocontact shifts for amide hydrogens was calculated from the known NMR structure of the protein for the N-terminal domain, assuming a specific size and orientation of the alignment tensor and a specific position of a paramagnetic center. Only 36 couplings of the domain were used for the following calculations. The eigenvalues were set to be $S_{zz}^{diag}=4,58 \cdot 10^{-4}$, $S_{yy}^{diag} = -2,96 \cdot 10^{-4}$ and, $S_{xx}^{diag} = -1,62 \cdot 10^{-4}$ resulting in a rhombicity of 0.2. This set of dipolar couplings and pseudocontact shifts is used as an *"experimental"* test set.

Measured dipolar couplings were fit to the NMR structure of Rhodniin by omitting and including pseudocontact shifts. As expected, the dipolar couplings are reproduced in the first case (Figure 23) when pseudocontact shifts were omitted. With a normalized square deviation of $Q = 0.00$ the tensor size and orientation reproduce the predefined values exactly. The Q value was 0.08 in the second case, when the tensor and the position of the paramagnetic center were recalculated. The paramagnetic center had a deviation of $0.786Å$ to its original position, which was caused by the final grid search step size of $0.5Å$ yielding a maximum deviation of $\frac{1}{2}\sqrt{3} \, Å \approx 0.866Å$. This deviation is also the reason for Q being larger than 0.00. Deletion of one, two or three amino acids after residues 15 and 39, as well as the addition of amino acids at the same positions, does not influence the result of the calculation. The sequence alignment is always correct, irrespective of the usage of pseudo contact shifts.

The "experimental" set of dipolar couplings was fit to the X-ray structure of ovomucoid (a homologous protein to the N-terminal domain of

Rhodniin). The "experimental" values, as well as the values calculated for the best fit, are given in Figure 23, together with the visualization of both structures in the frame of the resulting alignment tensor. The program finds an eight amino acid shift in the sequence alignment (Table 8) which agrees with the primary sequence alignment for Rhodniin and ovomucoid. In this case, the normalized square deviation was $Q = 0.30$.

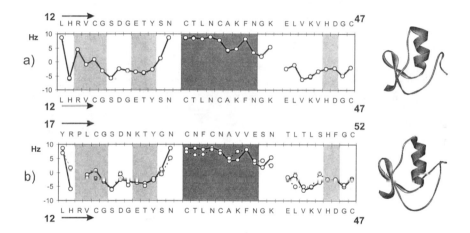

Figure 23.   Results for fitting a theoretical set of dipolar couplings for the N-terminal domain of Rhodniin to the Rhodniin structure (protein **A**) itself (a) and to ovomucoid (protein **B**, a Kazal inhibitor), which is homologous in sequence and structure (b). The black lines indicate the theoretical calculated coupling values, the dotted lines indicate dipolar couplings calculated for the final fit. On the upper x-axis the amino acid number of protein **B**, on the lower x-axis the amino acid number of protein **A** is found. Secondary structure elements are shown by light gray areas (β-sheet) and dark gray areas (α-helix). The three dimensional structures are given in the coordinate system of the tensor (y- and z- axis are in the paper plain, x- axis is perpendicular to the paper plane). Reproduced from *J. Biomol. NMR.*, **17**:283 (2000) with kind permission of Kluwer Academic Publishers.

Table 8. Results of fitting procedure for the experimental set of dipolar couplings of N-terminal domain of Rhodniin over ovomucoid. Identical amino acids in both sequences are labeled by | and similar amino acids are labeled by *.

```
Rhodniin   : 12 L H R V C G S D G E T Y S N P C
                *   * | | | |       | |   | | |
ovomucoid  : 20 T R P L C G S D N K T Y G N P C


Rhodniin   :    T L N C A K F N G K P E L V L V H D G C 47
                    |           * |   |   |   |   | |
ovomucoid  :    N F C N A V V E S N P T L T L S H F G C 55
```

To speed up the process of a three dimensional homology search, a subset of 125 folds was extracted from the PDB with a diverse set of folds according to Rost and Sander (Rost and Sander, 1994). Loading the data and calculating secondary structure elements for all protein in the fold database takes approximately five minutes on a 450MHz Pentium II Processor. The search itself takes under 1 second for the entire database if no gaps are introduced. This time increases to 48 seconds if disconnecting of protein parts, as explained above, if a gap size of up to 5 amino acids is allowed.

This database search using the earlier mentioned theoretical set of dipolar couplings for the N terminal domain of Rhodniin (a typical Kazal inhibitor) yields ovomucoid (1ovo_a) as the 2nd best hit with a Q-value of *0.45* and porcine pancreatic secretory trypsin inhibitor (1tgs_i) as the 16th best hit with a Q-value of *0.53*. Both proteins are known as Kazal inhibitors and are homologous to Rhodniin. In nine of the best 16 examples the α-helix of the Rhodniin domain is fit over a β-strand of the protein from PDB. This observation can be explained by the parallel orientation of N-H^N bond vectors in both secondary structure elements. Dipolar couplings are therefore of the same size in both secondary structure elements, which makes a distinction difficult.

Figure 24 (overleaf)   Results of a search in a database of 125 folds extracted from PDB for the theoretical set of dipolar couplings for N-terminal domain of Rhodniin. The black lines indicate the experimental coupling values (protein **A**, Rhodniin), dotted lines indicate dipolar couplings calculated for protein **B** from the database. The upper and lower x-axis shows the amino acid number of protein **B** and protein **A**, respectively. Secondary structure elements are represented similar to Figure 3. The results are ordered by increasing normalized square deviations (Q-values). (a) ovomucoid (1ovo_a) with a Q-value of *0.30*, (b) fragment of an oxidoreductase (6fdr residues 164-99) with a Q-value of *0.31*, (c) is again a proteinase inhibitor (1tgs_i) with a Q-value of *0.32* and (d) is a part of an intramolecular oxidoreductase (4xia_a residues 181-216) with a Q-value of *0.35*. Subsequent hits have considerably worse matches with Q-values above *0.40*. Reproduced from *J. Biomol. NMR.*, **17**:283 (2000) with kind permission of Kluwer Academic Publishers.

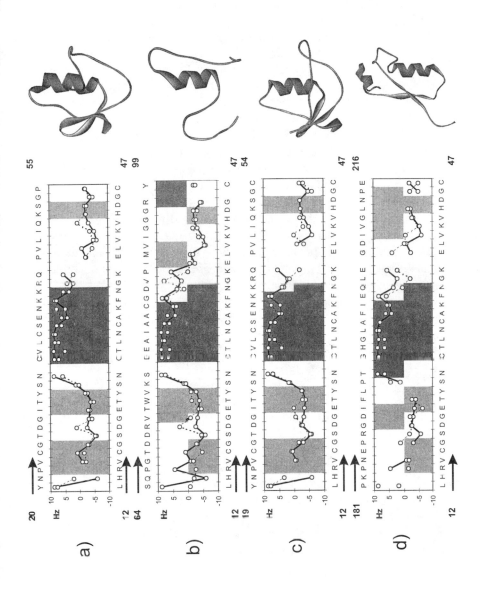

More significant results with fewer false positive answers and lower Q-values are obtained when secondary structure information from CSI is utilized by two simple rules: First, the alignment of β-strands over α-helices is excluded and second, only residues in well defined secondary structure regions are used for the calculation of Q-values. Using these rules, the two Kazal inhibitors of our database are ranked 1st (ovomucoid, 1ovo_a, Q = 0.30) and 3rd (porcine pancreatic secretory trypsin inhibitor, 1tgs_i, Q = 0.32). Figure 24 presents the first four hits of this search for which structures are displayed in the coordinate system of the tensor. The second result is part of dihydrofolate reductase (6dfr) with a Q-value of 0.31 and the fourth result is part of D-xylose isomerase (4xia_a) with a Q-value of 0.35. Results (b) and (d) have a similarly oriented α-helix, and at least one of the three β-strands present in Rhodniin with a similar orientation with respect to each other. Since the three β-strands are very short (three residues per strand) and nearly parallel, all dipolar couplings within them are of the same size. Therefore, this matches very well with one larger β-strand or an extended region when all N-H^N bonds are parallel (d). Matches b) and d) have a primary sequence homology of 11% and 5%. Thus the program finds 3D homology irrespective of sequence homology.

The result of the first homology search suggests Rhodniin to be homologous to other Kazal inhibitors. Therefore, a more thorough search for Kazal type inhibitors was performed in the PDB and a subset of such inhibitors was extracted. The Q-values of all eight structures range from 0.27 to 0.49 (Table 9).

The second example is cylclophilin, for which only 69 fast and easily determinable dipolar couplings were extracted and fit to the NMR structure and X-ray structures. Results are given in Figure 25, together with the three dimensional structures in the alignment tensor frame of reference. Q-values are 0.28 and 0.21 for NMR- and X-Ray- structure, respectively. The good agreement of both structures with the experimental data proves that it is not necessary to determine all couplings for fitting. Moreover, the possibility of calculating dipolar couplings for other residues allows the acceleration of further interpretation of spectra. While we detect 3D homology to other known cyclophilins, searching in a data bank of 125 shows finds small parts of the whole sequence, in particular helix-strand-strand motives.

Table 9. Results of fitting procedure for the experimental set of dipolar couplings of N-terminal domain of Rhodniin and over Rhodniin itself and an ensemble of eight Kazal inhibitors in complex with serine proteases or alone. For 1tbq the data of the N-terminal domain are fit over the homologue C-terminal domain of Rhodniin.

protein name	pdb code	fit range	Q
Rhodniin	---	12-47	0.00
Rhodniin in complex with thrombin (Res.: 2.6Å)	1tbr	12-47	0.27
Rhodniin in complex with thrombin (Res.: 3.1Å)	1tbq	65-101	0.27
ovomucoid	1ovo	20-55	0.30
human pancreatic secretory inhibitor in complex with trypsin	1cgi	20-55	0.30
procine pancreatic secretory inhibitor in complex with trypsin	1tgs	19-54	0.32
human pancreatic secretory trypsin inhibitor	1hpt	20-55	0.36
pig proteinase inhibitor (Kazal type)	1pce	24-59	0.38
leech-derived inhibitor with procine in complex with trypsin	1ldt	10-45	0.49

Figure 25 (overleaf). Results of fitting an experimental set of dipolar couplings for cylophilin A (protein **A**) to the NMR-structure (a) and to the X-ray structure (b) (protein **B**, respectively). Definition of lines and shaded areas are the like in Figure 24. *Q*-values are *0.28* and *0.21* for (a) and (b) respectively. The three dimensional structure is given in the coordinate frame of the tensor extracted from the fitting procedure (c). Reproduced from *J. Biomol. NMR.*, **17**:283 (2000) with kind permission of Kluwer Academic Publishers.

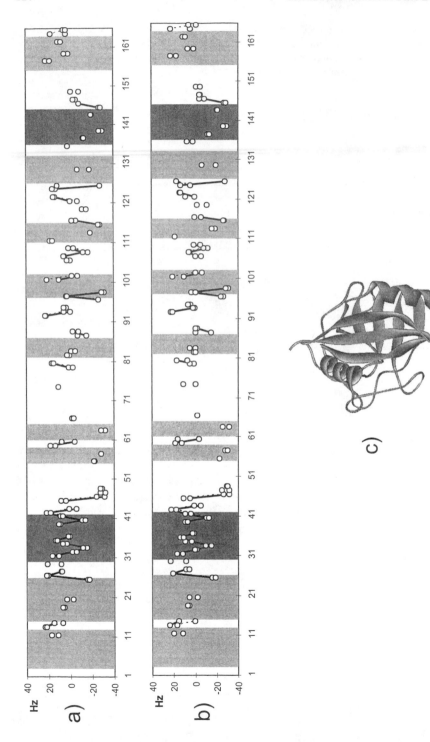

5.1   Acknowledgements

This work was supported by the Fonds der Chemischen Industrie. It was further supported by the DFG and the MPG. We are grateful to M. Maurer for the NMR structure of Rhodniin and for extracting the dipolar couplings of cyclophilin. We are thankful for enlightening discussions and for the trigger factor structure with Dr. K. Fiebig and Dr. M. Vogtherr. Moreover, we thank A. Zeidler, M. Müller, B. Coligaev and F. Schmäschke for downloading, analyzing and preparing the subset of the PDB. We thank Prof. Dr. G. Fischer and Dr. H. Bang (MPI, Halle) for the preparation of cyclophilin A and Prof. Dr. M. Kröger and Dr. K. Meyer-Rogge (University Gießen) for the preparation of *Hgi*CIC (His6) [C46S]. All spectra were recorded at the Large Scale Facility for Biomolecular NMR at the University of Frankfurt. W.P. and J.M. were both supported by a Kekulé stipend of the Fonds der Chemischen Industrie. We would like to thank all the members of the Griesinger group in Frankfurt and Göttingen for their support and help.

6.   REFERENCES

Aitio, H., Annila, A., Heikkinen, S., Thulin, E., Drakenberg, T., Kilpeläinen, I., (1999), *Protein Science*, **8**: 2580.
Al-Hashimi, H. M., Bolon, P. J., Prestegard, J. H., (2000), *J. Magn. Reson.*, **142**: 153.
Annila, A., Aitio, H., Thulin, E., Drakenberg, T., (1999), *J. Biomol. NMR*, **14**: 223.
Aue, W. P., Bartholdi, E., Ernst, R. R., (1976), *J. Chem. Phys.*, **64**: 2229.
Barrientos, L. G., Dolan, C., Gronenborn, A. M., (2000), *J. Biomol. NMR*, **16**: 329.
Barrientos, L. G., Louis, J. M., Gronenborn, A. M., (2001), *J. Magn. Reson.*, **149**: 154.
Bax, A., Tjandra, N., (1997), *J. Biomol. NMR*, **10**: 289.
Bayer, P., Varani, L., Varani, G., (1999), *J. Biomol. NMR*, **14**: 149.
Beger, R. D., Marathias, V. M., Volkman, B. F., Bolton, P. H., (1998), *J. Magn. Reson.*, **135**: 256.
Bertini, I., Felli, I. C., Luchinat, C., (2000), *J. Biomol. NMR*, **18**: 247.
Bewley, C. A., Gustafson, K. R., Boyd, M. R., Covell, D. G., Bax, A., Clore, G. M., Gronenborn, A. M., (1998), *Nat. Struct. Biol.*, **5**: 571.
Bodenhausen, G., Ruben, D. J., (1980), *Chem. Phys. Lett.*, **69**: 185.
Bolon, P. J., Al-Hashimi, H. M., Prestegard, J. H., (1999), *J. Mol. Biol.*, **293**: 107.
Bolon, P. J., Prestegard, J. H., (1998), *J. Am. Chem. Soc.*, **120**: 9366.
Brown, G. M., Levy, H. A., (1973), *Acta Cryst.*, **B29**: 790.
Brown, G. M., Levy, H. A., (1963), *Science*, **141**: 921.
Brunner, E., Arnold, M. R., Kremer, W., Kalbitzer, H. R., (2001), *J. Biomol. NMR*, **21**: 173.
Brunner, E., Kalbitzer, H. R., (2000), unpublished.
Brunner, E., Ogle, J., Wenzler, M., Kalbitzer, H. R., (2000), *Biochem. Biophys. Res. Com.*, **272**: 694.
Cai, M., Huang, Y., Zheng, R., Wei, S.-Q., Ghirlando, R., Lee, M. S., Craigie, R., Gronenborn, A. M., Clore, G. M., (1998), *Nat. Struct. Biol.*, **5**: 903.
Cai, M., Wang, H., Olejniczak, E. T., Meadows, R. P., Gunasekera, A. H., Xu, N., Fesik, S. W., (1999), *J. Magn. Reson.*, **139**: 451.

Carlomagno, T., Peti, W., Griesinger, C., (2000), *J. Biomol. NMR*, **17**: 99.

Cavagnero, S., Dyson, J. H., Wright, P. E., (1999), *J. Biomol. NMR*, **13**: 387.

Chou, J. J., Delaglio, F., Bax, A., (2000a), *J. Biomol. NMR*, **18**: 101.

Chou, J. J., Li, S., Bax, A., (2000b), *J. Biomol. NMR*, **18**: 217.

Choy, W. Y., Tollinger, M., Mueller, G. A., Kay, L. E., (2001), *J. Biomol. NMR*, **21**: 31.

Clore, G. M., Garrett, D. S., (1999), *J. Am. Chem. Soc.*, **121**: 9008.

Clore, G. M., Gronenborn, A. M., (1998), *PNAS*, **95**: 5891.

Clore, G. M., Gronenborn, A. M., Bax, A., (1998), *J. Magn. Reson.*, **133**: 216.

Clore, G. M., Starich, M. R., Gronenborn, A. M., (1998), *J. Am.Chem. Soc.*, **120**: 10571.

Clore, G. M., Straich, M. R., Bewely, C. A., Cai, M., Kuszewski, J., (1999), *J. Am. Chem. Soc.*, **121**: 6513.

Cornilescu, G., Marquardt, J. L., Ottiger, M., Bax, A., (1999a), *J.Am.Chem.Soc.*, **120**: 6836.

Cornilescu, G., Delaglio, F., Bax, A., (1999b), *J. Biomol. NMR*, **13**: 2889.

Delaglio, F., Kontaxis, G., Bax, A., (2000), *J. Am. Chem. Soc.*, **122**: 2142.

Delaglio, F., Wu, Z. R., Bax, A., (2001), *J. Magn. Reson.*, **149**: 276.

Desvaux, H., Gabriel, J.-C. P., Berthault, P., Camerel, F., (2001), *Angew. Chem.*, **113**: 387.

Fischer, M. W., Losonczi, J. A., Weaver, J. L., Prestegard, J. H., (1999), *Biochemistry*, **38**: 9013.

Flemming, K., Gray, D., Prasannan, S., Matthews, S., (2000), *J. Am. Chem. Soc.*, **122**: 5224.

Fowler, C. A., Tian, F., Al-Hashimi, H. M., Prestegard, J. H., (2000), *J. Mol. Biol.*, **304**: 447.

Gardner, K. H., Kay, L. E., (1998), *Annu. Rev. Biophys. Biomol. Struct.*, **27**: 357.

Gemmecker, G., Fesik, W. S., (1991), *J. Magn. Reson.*, **95**: 208.

Ghose, R., Prestegard, J. H., (1997), *J. Magn. Res.*, **128**: 138.

Girvin, M. E. (1994), *J. Magn. Reson. Ser. A*, **108**: 99.

Griesinger, C., Søerensen, O. W., Ernst, R. R., (1985), *J. Am. Chem. Soc.*, **107**: 6394.

Griesinger, C., Søerensen, O. W., Ernst, R. R., (1986), *J. Chem. Phys.*, **85**: 6837.

Griesinger, C., Søerensen, O. W., Ernst, R. R., (1987), *J. Magn. Reson.*, **75**: 474.

Halbeek, H. v., (1994), *Curr. Opin. Str. Biol.*, **4**: 697.

Hansen, M. R., Mueller, L., Pardi, A., (1998a), *Nat. Sruct. Biol.*, **5**: 1065.

Hansen, M. R., Rance, M., Pardi, A., (1998b), *J. Am. Chem. Soc.*, **120**: 11210.

Herrmann, T., Güntert, P., Wüthrich, K., (2002), *J. Mol. Biol.*, sumitted.

Hilbers, C. W., MacLean, C., (1968), *Chem. Phys. Lett.*, **2**: 445.

Hilbers, C. W., MacLean, C., (1970), *Chem. Phys. Lett.*, **7**: 587.

Hilbers, C. W., MacLean, C. ,(1971), *Ber. der Bunsenges.*, **75**: 277.

Homans, S. W., (1990), *Biochemistry*, **29**: 9110.

Huang, X., Moy, F., Powers, R., (2000), *Biochemistry*, **39**: 13365.

Ishii, Y., Markus, M. A., Tycko, R., (2001), *J. Biomol. NMR*, **21**: 141.

Jeffrey, G. A., Huang, D. B., (1990), *Carbohydr. Res.*, **206**: 173.

Kaikkonen, A., Otting, G., (2001), *J. Am. Chem. Soc.*, **123**: 1770.

Kay, L. E., (1998), *Nat. Struct. Biol.*, NMR supplement: 513.

Kessler, H., Müller, A., Oschkinat, H., (1985), *Magn. Reson. Chem.*, **23**: 844.

Kessler, H., Oschkinat, H., (1985), *Angew. Chem.*, **97**: 689.

Kiddle, G. R., Homans, S. W., (1998), *FEBS Lett.*, **436**: 128.

Koenig, B. W., Hu, J.-S., Ottiger, M., Bose, S., Hendler, R. W., Bax, A., (1999), *J. Am. Chem. Soc.*, **121**: 1385.

Kontaxis, G., Bax, A., (2001), *J. Biomol. NMR*, **20**: 77.

Kontaxis, G., Clore, G. M., Bax, A., (2000), *J. Magn. Reson.*, **143**: 184.

Kramer, F., Peti, W., Griesinger, C., Glaser, S. J., (2001), *J. Magn. Reson.*, **149**: 58.

Kung, H. C., Wang, K. Y., Goljer, I., Bolton, P. H., (1995), *J. Magn. Reson.*, Series **B109**: 323.

Landersjo, C., Hoog, C., Maliniak, A., Widmalm, G., (2000), *J. of Phys. Chem. B*, **104**: 5618.

Linge, J. P., O'Donoghue, S. I., Nilges, M., (2001), *Meth. Enzymol.*, **339**: 71.

Losonczi, J. A., Andrec, M., Fischer, M. W. F., Prestegard, J. H., (1999), *J. Magn. Res.*, **138**: 334.

Losonczi, J. A., Prestegard, J. H., (1998), *J. Biomol. NMR*, **12**: 447.

Martin-Pastor, M., Bush, C. A., (2000a), *Biochemistry*, **39**: 4674.

Martin-Pastor, M., Bush, C. A., (2000b), *Carb. Res.*, **323**: 147.

Martin-Pastor, M., Bush, C. A., (2001), *J. Biomol. NMR*, **19**: 125.

Maurer, M., Griesinger, C., (1998), *personal communication*,

Medek, A., Hajduk, P. J., Mack, J., Fesik, S. W., (2000), *J. Am. Chem. Soc.*, **122**: 1241.

Meiler, J., (1996-2001), *www.jens-meiler.de*,

Meiler, J., Blomberg, N., Nilges, M., Griesinger, C., (2000a), *J. Biomol. NMR*, **16**: 245.

Meiler, J., Peti, W., Griesinger, C., (2000b), *J. Biomol. NMR*, **17**: 283.

Meiler, J., Prompers, J., Peti, W., Griesinger, C., Brüschweiler, R., (2001), *J. Am. Chem. Soc.*, **123**: 6098.

Meissner, A., Duus, J. O., O.W., S., (1997), *J. Biomol. NMR*, **10**: 89.

Montelione, T. G., Wagner, G., (1989), *J. Am. Chem. Soc.*, **111**: 5474.

Mueller, G. A., Choy, W. Y., Skrynnikov, N. R., Kay, L. E., (2000a), *J. Biomol. NMR*, **18**: 183–188.

Mueller, G. A., Choy, W. Y., Yang, D. W., Forman-Kay, J. D., Venters, R. A., Kay, L. E., (2000b), *J. Mol. Biol.* **300**: 197.

Mueller, L., (1987), *J. Magn. Reson.*, **72**: 191.

Neubauer, H., Meiler, J., Peti, W., Griesinger, C., (2001), *Helv. Chim. Acta*, **84**: 243.

Ojennus, D. D., Mitton-Fry, R. M., Wuttke, D. S., (1999), *J. Biomol. NMR*, **14**: 175.

Olejniczak, E. T., Meadows, R. P., Wang, H., Cai, M., Nettesheim, D. G., Fesik, S. W., (1999), *J. Am. Chem. Soc.*, **121**: 9249.

Oschkinat, H., Freeman, J., (1984), *J. Magn. Reson.*, **60**: 164.

Ottiger, M., Bax, A., (1999), *J. Biom. NMR*, **13**: 187.

Ottiger, M., Bax, A., (1998), *J. Biomol. NMR*, **12**: 361.

Ottiger, M., Delaglio, F., Bax, A., (1998a), *J. Magn. Reson.*, **131**: 373.

Ottiger, M., Delaglio, F., Marquardt, J. L., Tjandra, N., Bax, A., (1998b), *J. Magn. Reson.*, **134**: 365.

Otting, G., Rückert, M., Levitt, M. H., Moshref, A., (2000), *J. Biomol. NMR*, **16**: 343.

Palmer, A. G., Williams, J., Mcdermott, A., (1996), *J. Phys. Chem.*, **100**: 13293.

Parac, T. M., Vogtherr, M., Maurer, M., Pahl, A., Rüterjans, H., Griesinger, C., Fiebig, K. M., (2001), *J. Biomol. NMR*, **20**: 193.

Pellecchia, M., Vander Kooi, C. W., Keliikuli, K., Zuiderweg, E. P. R., (2000), *J. Magn. Reson.*, **143**: 435.

Permi, P., Annila, A., (2000), *J. Biomol. NMR*, **16**: 221.

Pervushin, K., Riek, R., Wider, G., Wüthrich, K., (1997), *Proc Natl Acad Sci U S A*, **94**: 12366.

Peshkovsky, A., McDermott, A. E., (1999), *J. Phys. Chem. A*, **103**: 8604

Peti, W., Griesinger, C., (2000), *J. Am. Chem. Soc.*, **122**: 3975.

Peti, W., Meiler, J., Brüschweiler, R., Griesinger, C., (2002), *J. Am. Chem. Soc.*, in press.

Prestegard, J. H., (1998), *Nature Struct. Biol.*, 517.

Prestegard, J. H., Kishore, A. I., (2001), *Current Opinion in Chemical Biology*, **5**: 584.

Prestegard, J. H., Valafar, H., Glushka, J., Tian, F., (2001), *Biochemistry*, **40**: 8677.

Prosser, S. R., Losonczi, J. A., Shiyanovskaya, I. V., (1998), *J. Am. Chem. Soc.*, **120**: 11010.

Ramirez, B. E., Bax, A., (1998), *J. Am. Chem. Soc.*, **120**: 9106.

Riley, S. A., Augustine, M. P., (2000), *J. Phys. Chem. A*, **104**: 3326.

Rost, B., Sander, C., (1994), *Proteins Struct. Funct. Genet.*, **19**: 55.

Ruckert, M., Otting, G., (2000), *J. Am. Chem. Soc.*, **122**: 7793.

Ruessink, B. H., MacLean, C., (1987), *Mol. Phys.*, **60**: 1059.

Rundlöf, T., Landersjö, C., Lycknert, K., Maliniak, A., Widmalm, G., (1998), *Magn. Reson. Chem.*, **36**: 773.

Sanders, C. R., Schwonek, J. P., (1992), *Biochemistry*, **31**: 8898.

Sanders II, C. R., Hare, B. J., Howard, K. P., Prestegard, J. H., (1994), *Prog. NMR Spec.*, **26**: 421.

Sass, J., Cordier, F., Hoffmann, A., Rogowski, M., Cousin, A., Omichinski, J. G., Lowen, H., Grzesiek, S., (1999), *J. Am. Chem. Soc.*, **121**: 2047.

Sass, J., Musco, G., Stahl, S. J., Wingfield, P. T., Grzesiek, S., (2000), *J. Biomol. NMR*, **18**: 303.

Sattler, J., Schleucher, J., Griesinger, C., (1999), *Prog. NMR Spec.*, **34**: 93.

Saupe, A., (1968), *Angew. Chem. Int. Ed. Engl.*, **7**: 97.

Schleucher, J., Sattler, M., Griesinger, C., (1993), *Angew. Chem.*, **105**: 1518.

Schulte-Herbrüggen, T., Madi, Z. L., Søerensen, O. W., Ernst, R. R., (1991), *Mol. Phys.*, **72**: 847.

Spera, S., Bax, A., (1991), *J. Am. Chem. Soc.*, **113**: 5490.

Thompson, G. S., Shimizu, H., Homans, S. W., Donohue-Rolfe, A., (2000), *Biochemistry*, **39**: 13153.

Tian, F., Al-Hashimi, H. M., Craighead, J., Prestegard, J. H., (2001), *J. Am. Chem. Soc.*, **123**: 485.

Tian, F., Bolon, P. J., Prestegard, J. H., (1999), *J. Am. Chem. Soc.*, **121**: 7712.

Tian, F., Fowler, C. A., Zartler, E. R., Jenney Jr., F. A., Adams, M. W., Prestegard, J. H., (2000), *J. Biomol. NMR*, **18**: 23.

Tjandra, N., (1999), *Structure*, **7**: R205.

Tjandra, N., Bax, A., (1997), *Science*, **278**: 1111.

Tjandra, N., Grzesiek, S., Bax, A., (1996), *J. Am. Chem. Soc.*, **118**: 6264.

Tjandra, N., Marquardt, J., Clore, G. M., (2000), *J. Magn. Reson.*, **142**: 393.

Tjandra, N., Omichinski, J. G., Gronenborn, A. M., Clore, G. M., Bax, A., (1997), *Nat. Struct. Biol.*, **4**: 732.

Tolman, J. R., Al-Hashimi, H. M., Kay, L. E., Prestegard, J. H., (2001), *J. Am. Chem. Soc.*, **123**: 1416.

Tolman, J. R., Flanagan, J. M., Kennedy, M. A., Prestegard, J. H., (1995), *Proc. Natl. Acad. Sci.*, **92**: 9279.

Tycko, R., Blanco, F. J., Ishii , Y., (2000), *J. Am. Chem. Soc.*, **122**: 9340 .

van Zijl, P. C. M., Ruessink, B. H., Bulthuis, J., MacLean, C., (1984), *Acc. Chem. Res.*, **17**: 172.

Vijay-Kumar, S., Bugg, C. E., Cook, W. J., (1987), *J. Mol. Biol.*, **194**: 531.

Vuister, G. W., Bax, A., (1993), *J. Am. Chem. Soc.*, **115**: 7772.

Wang, H., Eberstadt, M., Olejniczak, T., Meadows, R. P., Fesik, S. W., (1998a), *J. Biomol. NMR*, **12**: 443.

Wang, Y.-X., Marquardt, J. L., Wingfield, P., Stahl, S. J., Lee-Huang, S., Torchia, D., Bax, A., (1998b), *J. Am. Chem. Soc.*, **120**: 7385.

Weisemann, R., Rüterjans, H., Schwalbe, H., Schleucher, J., Bermel, W., Griesinger, C., (1994), *J. Biomol. NMR*, **4**: 231.

Willker, W., Leibfritz, D. (1992), *J. Magn. Reson.*, **99**: 421.

Wishart, D. S., Bigam, C. G., Holm, A., Hodges, R. S., Sykes, B. D., (1995), *J. Biomol. NMR*, **5**: 67.

Wishart, D. S., Sykes, B. D., (1994), *J. Biomol. NMR*, **4**: 171.

Wishart, D. S., Sykes, B. D., Richards, F. M., (1992), *Biochemistry*, **31**: 1647.
Wu, Z. R., Bax, A., (2001), *J. Magn. Reson.*, **151**: 242.
Yang, D., Tolman, J. R., Goto, N. K., Kay, L. E., (1998), *J. Biomol. NMR*, **12**: 325.
Yang, D. W., Venters, R. A., Mueller, G. A., Choy, W. Y., Kay, L. E., (1999), *J. Biomol. NMR.*, **14**: 333.
Zweckstetter, M., Bax, A., (2000), *J. Am. Chem. Soc.*, **122**: 3791.
Zweckstetter, M., Bax, A., (2001), *J. Am. Chem. Soc.*, **123**: 9490.

Chapter 8

Protein Structure Refinement using Residual Dipolar Couplings

Angela M. Gronenborn
Laboratory of Chemical Physics, NIDDK, NIH, Bethesda, MD 20892

Abstract   Residual dipolar couplings arise from small degrees of alignment of molecules in a magnetic field. Media employed for imparting alignment onto biomolecules lacking sufficient magnetic susceptibility anisotropies themselves comprise dilute aqueous phospholipid bicelles, colloidal suspensions of rod-shaped viruses and lamellar phases of quasiternary surfactant systems. The magnitude of the residual dipolar couplings depends upon the degree of ordering and allows the determination of the corresponding internuclear vectors with respect to the molecule's alignment frame. Inclusion of dipolar constraints into NMR structure calculations leads to improved accuracy of the resulting structures, especially in cases where the information content provided by traditional NOE constraints is limited. This chapter describes the different media used for alignment and the application of dipolar coupling information for protein structure refinement and fold determination.

1.   INTRODUCTION

Traditionally, the determination of three-dimensional structures of biological macromolecules employs NOE derived distance restraints and torsion angle restraints extracted from J-couplings (Wüthrich, 1986; Clore and Gronenborn, 1989). A key limitation inherent to this approach concerns the strictly local nature of these parameters, since they solely define distances and angles between atoms close in space within the structure. Despite this limitation, protein structure determination by NMR has been extremely successful, primarily because the large number of short interproton distances between amino acids far apart in sequence renders these distances conformationally highly restrictive.

Nevertheless, the use of only short distance information may limit the accuracy of NMR derived structures, especially for elongated structures where the cumulative error may be significant or in cases where only few contacts are available between structural elements. Examples of such systems include modular and multi-domain proteins and linear nucleic acids. Recently, NMR methods have been developed which allow the extraction of structural restraints characterizing long-range order. In particular, residual dipolar couplings (Tolman et al., 1995; Kung et al., 1995; Tjandra et al., 1996, 1997; Tolman and Prestegard, 1996; Tjandra and Bax, 1997a,b; Hansen et al., 1998, Wang et al., 1998b), which contain information about the orientation of the internuclear vector relative to the molecular susceptibility tensor, have been used to derive angular restraints and employed in NMR based structure calculations (Tjandra et al., 1997; Cai et al., 1998; Bewley at al., 1998; Garrett et al., 1999; Bayer et. al, 1999).

The existence of magnetic field induced residual dipolar couplings in proteins in isotropic solution was known for a number of years, but their utility for structural studies was only realized with the advent of high field magnets and heteronuclear methods which allow their precise determination. Increased field strength was important, since the size of the residual one bond dipolar coupling scales with the square of the magnetic field and novel experiments had to be devised to extract these extremely small couplings from J-modulated HSQC experiments with a precision of a fraction of a Hertz (Tjandra et al., 1996) or use of Selective Coupling Enhanced HSQC experiments (Tolman and Prestegard, 1996a). Given the small degree of alignment of diamagnetic proteins in the magnetic field, resulting in minute residual dipolar couplings of typically < 0.2 Hz, the practicality of extracting such couplings reliably seemed limited. More promising, on the other hand, appeared the magnetic field induced alignment for paramagnetic proteins, nucleic acids and protein/nucleic acid complexes which exhibit magnetic susceptibility anisotropies of greater -10×10^{-34} m^3 per molecule resulting in residual dipolar couplings of ~0.5 Hz for N-H vectors and ~0.9 Hz for C-H vectors.

The major breakthrough with respect to any potential routine use of dipolar coupling derived structural restraints was the demonstration that tunable degrees of molecular alignment can be achieved by placing the molecule under investigation into a dilute, aqueous liquid crystalline phase of dihexanoyl phosphatidylcholine (DHPC) and dimyristoyl phosphatidylcholine (DMPC) bicelles (Tjandra and Bax, 1997a; Bax and Tjandra, 1997; reviewed by Prestegard, 1998; Bax, 2000). Sufficient high degrees of alignment can be achieved in this manner resulting in one-bond dipolar couplings of 5-40 Hz which are easily detectable by simply measuring the splittings in 2D and 3D coupled HSQC spectra. This opened the door for developing experiments to extract other types of residual dipolar couplings (Ottiger *et al.*, 1998a,b) and resulted in a flurry of activity studying alignment media with the aim of improving the initial bicelle system (Ottiger and Bax, 1998, 1999; Wang *et al.*, 1998; Losonczi and Prestegard, 1998; Cavagnero *et al.*, 1999) as well as the discovery of novel ones (Clore *et al.*, 1998c; Prosser *et al.*, 1998; Hansen *et al.*, 1998a; Barrientos *et al.*, 2000; Rückert *et al.*, 2000; Fleming *et al.*, 2000, Tycko *et al.*, 2000).

The degree of alignment in any given liquid crystalline medium can easily be assessed by monitoring the deuterium quadrupolar splitting of the HDO signal in the sample prepared as an aqueous solution containing 90% H_2O/10% D_2O. This signal arises from exchange between bulk water and water bound to the oriented liquid crystal. Quadrupole splittings ranging from 10 Hz to 30 Hz are generally observed for the different media of variable composition. Interestingly, the degree of alignment of the solute macromolecule under investigation is not directly correlated to the deuterium quadrupole splitting across different liquid crystalline media, although a linear dependence on concentration is frequently observed for each individual medium. This is most likely due to the fact that the distribution and location of bound water molecules on the surfaces of the different aligning media (bicelles of varying composition, phages and viruses, and surfactant phases) are very distinct and strongly dependent on the local surface structure. Nevertheless, measuring the deuterium quadrupole splitting of the HDO resonance is generally the best and most efficient way to assess whether a medium is suitable for alignment purposes.

2.   LIQUID CRYSTALLINE MEDIA BASED ON PHOSPHOLIPIDS

The most common medium used for purposes of partial alignment of solute macromolecules with the magnetic field is a lyotropic liquid crystal consisting of a binary mixture of DMPC and DHPC. Binary mixtures of these phospholipids form disc shaped particles (Sanders and Schwonek, 1992), commonly referred to as bicelles (Sanders and Landis, 1995). These particles adopt a nematic liquid crystalline phase in the magnetic field, aligning with their normal orthogonally to the field direction. Using dilute mixtures of bicelles in aqueous solution, biological macromolecules such as proteins, nucleic acids or complexes thereof can be dissolved in the aqueous portion, rendering their rotational diffusion rates essentially unaffected by the bicelles. In this manner it is possible to take advantage of high resolution NMR methodology while simultaneously extracting residual dipolar couplings arising from the small degree of molecular alignment imparted onto the solute molecules by the oriented bicelles. For most applications, the concentration of the bicelles must be kept small ($\leq$ 5%) to avoid line broadening caused by unresolved dipolar couplings, possibly caused by transient binding to the bicelles. However, under these dilute conditions the liquid crystal becomes less stable and the temperature range over which a stable nematic phase is observed is limited. In addition, the samples occasionally phase separate within the time required to collect the NMR data. A careful study on the preparation of dilute bicelle mixtures was carried out by Ottiger and Bax (1998). The authors showed that 3% or 5% aqueous w/v DMPC/DHPC mixtures at ratios of DMPC/DHPC of 3/1 form stable liquid crystal phases over a temperature interval of 29-42°C and 33-45°C, respectively. This somewhat limited temperature range can be extended by using a ternary mixture of DHPC, DMPC and charged amphiphiles such as hexadecyl(cetyl)trimethyl ammonium bromide (CTAB) or SDS (Losonczi and Prestegard, 1998) or by using shorter chain phospholipids instead of DMPC (Wang *et al.*, 1998a). In addition, the long-term stability of the phospholipid based liquid crystal phase exhibits a strong pH dependence. Acid- and base catalyzed hydrolosis of the carboxyl ester bonds are well known to occur, limiting the solute pH to a narrow range around pH 7. Replacement of the diacyl phospholipids by non-hydrolyzable dialkyl ananlogs overcomes this limitation (Ottiger and Bax, 1999; Cavagnero *et al.*, 1999). Despite all these improvements one has to keep in mind, that the stability of the liquid crystal phase is also affected by the solute, namely the protein solution under investigation. Solubility and stability of the protein naturally influence the choice of solvent conditions for the aqueous component, i.e. concentration, buffer choice and pH, and ionic strength. Each of these parameters affects the stability of the liquid crystalline phase in a complex manner.

3.   NEMATIC PHASES OF ROD-SHAPED VIRUSES AND FILAMENTOUS PHAGES

Suspensions of charged, rod-shaped viruses, such as tobacco mosaic virus (TMV) and filamentous bacteriophages fd/M13 and Pfl were known to undergo a magnetic field induced isotropic-nematic phase transition at moderate concentrations (Torbet and Maret, 1979; Nakamura and Okano, 1983; Fraden *et al.*, 1989) and indeed, structural studies on TMV, Pfl and M13 using X-ray diffraction were carried out on magnetic field oriented specimen (Stubbs *et al.*, 1977; Makowski *et al.*, 1980; Glucksman *et al.*, 1986; Glucksman *et al.*, 1992). Solutions of magnetically aligned virus or phage therefore seemed like an attractive alternative to the above described phospholipid bicelles. Within a short time two independent reports of successfully using virus/phage solutions as a medium to measure residual dipolar couplings appeared (Clore *et al.*, 1998c; Hansen *et al.*, 1998a). We used TMV and fd solutions whereas Pardi's group used Pf1. The molecular structures of the virus and phages are very similar. They are long, negatively charged rods, in which a cylinder of coat proteins is arranged in a helical fashion around either an RNA or DNA single stranded genome. TMV is approximately 15 nm wide and ~3000 nm long, while the bacteriophages are both ~6.6 nm in diameter and ~880 nm (fd/M13) or ~1900 nm (Pf1) long. This particle size is considerably larger than the one observed for the phospholipid bicelles which possess a thickness of ~4 nm and length of ~50 nm. The large viral particle size leads to high macroscopic viscosity, however the microscopic tumbling rates of the dissolved macromolecules are not affected. Measuring T_2 relaxation times revealed no significant changes for most solute molecules in the absence or presence of phage solution, similar to findings for the bicelle systems. Concentrations of viruses of ~25-30 mg/ml, ~50 mg/ml and 17-46 mg/ml were used for fd, TMV and Pf1, respectively, and residual dipolar couplings on dissolved proteins, nucleic acids and complexes thereof were reported. Similar to the bicelle systems, partial molecular alignment of the dissolved macromolecules arises from collisions with the aligned virus particles, thus imposing a preferred direction of diffusion rather than alignment via transient binding. It appears, that it is possible to align macromolecules over a wide range of temperatures and buffer conditions using dilute colloidal phage suspensions (Barrientos *et al.*, 2001). Indeed, the degree of alignment and the size of the residual dipolar coupling is proportional to the length of the phage, as expected from Onsager theory for semiflexible charged rods (Barrientos *et al.*, 2001). Residual dipolar couplings have been measured and used in structure refinement for hairpin RNAs (Hansen *et al.*, 1998) and the enzyme I-HPr complex (Garrett *et al.*, 1999).

4.   LIQUID CRYSTAL PHASES OF SURFACTANTS

In the search for alternative and robust liquid crystalline media suitable for partially aligning biomolecules for structural studies, dilute quasitemary systems of surfactant/salt/alcohol forming Helfrich lamellar phases were investigated (Prosser *et al.*, 1998; Barrientos *et al.*, 2000). In the early 1970s Helfrich predicted that lamellar phases should exist in a form which is stabilized principally by repulsive entropic forces. These phases consist of bilayers, which can be swelled by solvent such that the spacing between the bilayers is much larger than the thickness of the bilayer itself, which undergoes large amplitude fluctuations. The morphology of this lamellar liquid crystalline (LLC) phase consists of ~3 nm thick bilayers with large interlamellar spacings (McGrath, 1997). These Helfrich lamellar phases will readily take up water soluble polymers of considerable size ($r_g \sim$ 200A) whereas classical lamellar phases are unable to do so or will take up only small amounts. Thus Helfrich lamellar phases were thought to constitute another medium for studying biological macromolecules under partial alignment. Prosser *et al.* used a 2% aqueous solution of CPCl/hexanol (1/1) in 200 mM NaCl and demonstrated that residual dipolar couplings up to 15 Hz could be measured on ubiquitin. We prepared surfactant phases from CPBr/hexanol and NaBr and found that solutions of 3 - 6% CPBr/hexanol in 20-30 mM NaBr gave excellent results. Dipolar couplings were measured for the GB1 domain as well as a DNA oligonucleotide. Although initially expected to form lamellar phases, further characterization of the CPBr/hexanol/NaBr liquid crystal phases revealed that different particle morphologies exist for slightly varying compositions.

5.   ALIGNMENT USING ANISOTROPIC MATERIALS

5.1   Transient Binding to Anisotropic Materials

In contrast to alignment caused by steric interactions with liquid crystalline media as described above, transient binding to oriented particles can give rise to residual dipolar couplings as well. Such data has been reported for proteins interacting with purple membrane fragments (Koenig *et al.*, 1999; Sass *et al.*, 1999) or Pf1 phage (Ojennus *et al.*, 1999). Binding is assumed to occur as evidenced by significant line broadening and a large decrease in ^{15}N $T_{1\rho}$ and T_2 relaxation times of the solute proteins. In this case the degree of alignment is assumed to depend on the binding properties of the biomolecule under investigation, and any information derived from residual couplings reflect the conformation in the bound state. The weak binding and fast exchange interaction between the medium and solute molecule is therefore reminiscent of the transferred NOE effect.

5.2   Using Cellulose Crystallites or Strained Gels

Other media and methods for creating weakly aligned states for measuring residual dipolar couplings are still being searched for. The use of a suspension of cellulose crystallites has been reported (Fleming *et al.*, 2000). Due to the large negative diamagnetic anisotropy of individual cellulose crystallites, which generally have a length of several 100 nm and a width of ca. 10 nm, alignment in the magnetic field occurs, and proteins can be dissolved in such suspensions. Another approach exploits the anisotropy of strained polymeric gels. Either compression (Tycko *et al.*, 2000) or both vertical or radial squeezing of polyacrylamide gels (Sass *et al.*, 2000) was employed, and the proteins were introduced into the gels simply by diffusion. There is no doubt that other materials will be found and exploited for imparting anisotropy onto biomolecules, such as giant rubber like micelles made from blocked copolymers (Bates *et al.*, 1999) or stretched polymer hydrogels (Liu *et al.*, 1999).

There are several reasons why it is important to have different alignment media available. First, not every medium is compatible with the properties of the molecules or systems under investigation. Proteins that interact with membranes are clearly not compatible with bicelle based media, and negatively charged molecules, such as nucleic acids tend to bind to positively charged bicelles. In addition, very flexible or partially folded proteins have a tendency to strongly interact with bicelles. Likewise, positively charged proteins can potentially interact with the negatively charged phage particles at neutral pH values, thereby increasing the electrostatic component of the alignment, resulting in large linewidths or collapse of the liquid crystalline phase. As an example, the protein ubiquitin with a pI of ~ 6.5 interacts strongly with PF1, unless high ionic strength is used to screen the charges an the surface of the phage. Second, different alignment media frequently result in different orientations of the solute molecule with respect to the magnetic field, because the alignment tensors in two different alignment media will exhibit different orientations. This property allows to lift the degeneracies in the orientation of a given interatomic bond, inherent in the relationship between dipolar coupling and internuclear vector orientation. A dipolar coupling measured in a given liquid crystalline medium positions the vector between the two coupled partners on one of the two possible, oppositely oriented cones. If the alignment tensor in the second medium has a different orientation relative to the molecular frame of the molecule, the same vector will now reside on two different cones. Thus the true orientation of this particular interatomic vector has to lie at the intersections between the two cones. However, it should be pointed out that one can never distinguish between the true direction and the reverse one from a single dipolar coupling interaction, thus a twofold degeneracy in the orientation will always exist.

6.   REFINEMENT USING RESIDUAL DIPOLAR COUPLINGS

A key aspect of any NMR structure determination is that the ensemble of calculated structures satisfies all of the experimental NMR constraints, exhibits only very small deviations from idealized covalent geometries, such as bond lengths, bond angles, and planarity, and displays good non bonded contacts. It therefore is of utmost importance for devising any calculational strategy, that the global minimum of the target function is reliably and efficiently located. The use of residual dipolar couplings, which impose a tight restriction on the orientation of a bond (if measured for directly bonded nuclei) should therefore greatly improve the quality of traditional NMR structures, calculated based on NOE distance restraints, coupling constants and chemical shifts. The simplest way to incorporate the geometric content of the dipolar couplings into a structure calculation is by means of an error function. For each measured dipolar contribution a term

$$E_{dipolar} = k_{dipolar} \left(D_{calc} - D_{obs} \right)^2 \qquad [1]$$

(where k_{dipol} is a force constant and D_{calc} and D_{obs} are the observed and calculated values of the dipolar couplings) is added to the various other terms, like distance and torsion angle restraints, covalent geometry and non-bonded contacts in the simulated annealing or minimizing protocols. $E_{dipolar}$ is evaluated by calculating the θ and ϕ angles between the appropriate bond vectors (e.g., N-H, Cα-H, Cα-C' etc) and an external arbitrary axis system. Since the orientation of the axis system is not known *a priori*, it is defined by an artificial tetra-atomic molecule comprising atoms X, Y, Z and O, with three mutually perpendicular bonds, X-O, Y-O, and Z-O, representing the x, y, and z axes of the anisotropy tensor which is allowed to float during the calculation (Tjandra *et al.*, 1997; Clore *et al.*, 1998a). The expression for the residual dipolar coupling between two directly coupled nuclei can be simplified to the form

$$D(\theta, \phi) = D_a \left(3 \cos^2 \theta - 1 \right) + 3/2\, R \left(\sin^2 \theta \cos 2 \phi \right) \qquad [2]$$

where D_a and D_r are the axial and rhombic components of the traceless second rank diagonal tensor D in units of Hz. R is the rhombicity defined by D_r / D_a ; θ is the angle between the inter-atomic vector and the z axis of the tensor and ϕ is the angle which describes the position of the projection of the inter-atomic vector on the x-y plane of the tensor. D_a and D_r subsume a number of constants, including the gyromagnetic ratio of the two nuclei involved, the distance between them, the generalized order parameter S for internal motion of the internuclear vector, the magnetic field strength and the medium permeability.

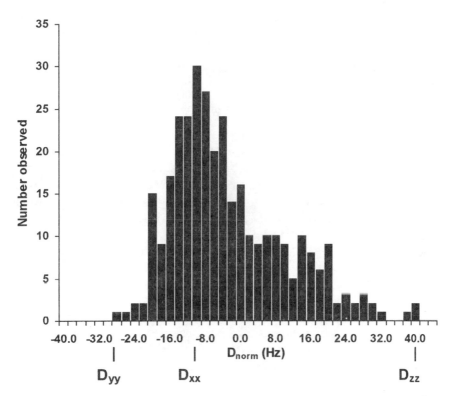

Figure 1. Histogram depicting the distribution of normalized residual dipolar couplings measured for the protein CV-N.

Naturally, using dipolar couplings in NMR structure determination and refinement is predicated on the assumption that motional averaging will not compromise the data. The magnitude of dipolar coupling depends on the generalized order parameter S for internal motions of the interatomic vector (Lipari and Szabo, 1982), thus different contributions have to be considered, at least in principle. Rather than using individual, residue specific S values, it seems reasonable to assume uniform S values for all those residues for which heteronuclear relaxation measurements indicate a well ordered conformation as evidenced by experimental S^2 values of 0.7 – 0.9 (corresponding to S values of 0.85 to 0.95). Dipolar coupling restraints for residues which experience either slow conformational exchange or low order parameters ($S^2 < 0.6$) need to be excluded from the data set (Tjandra *et al.*, 1997). As an aside, it should be pointed out, that D_a and D_r scale with S, rather than S^2, thus the assumption of an overall S value introduces at most an error of a few percent in the dipolar couplings for the ordered regions of the molecule, well within the error of the experimental measurements.

In order to use Eqs. [1] and [2] for structure refinement the values of D_a and R have to be determined. They are obtained by either iteratively best fitting the alignment tensor for a given structure during the course of refinement as described above, or directly from the experimental data by examining the distribution of dipolar couplings (Clore *et al.*, 1998b). A histogram of the ensemble of normalized residual dipolar couplings for the protein cyanovirin-N is illustrated in Fig.1. Extracting D_a and R from this distribution exploits the fact, that different, fixed-distance inter-nuclear vector types in a molecule are approximately uniformly and isotropically distributed in space relative to the alignment tensor of the molecule. The magnitude of the axial and rhombic components of the molecular alignment tensor are related to the extrema and mode of the coupling histograms, which in the absence of random errors look almost like perfect powder patterns. The highest probability dipolar coupling value, therefore, coincides with the magnitude of the bond vector aligned along the x axis of the alignment tensor (Clore *et al.*, 1998b). Using this approach, the accuracy with which D_a and R are determined, clearly depends on the accuracy in the estimates for the two extrema and the maximum of the distribution, which in turn depends on the number of dipolar couplings observed and the degree of aniosotropy in the orientation of the internuclear vectors. In cases where it is not possible to measure a large set of dipolar couplings, it is possible to use a maximum likelihood method method for extracting D_a and R (Warren and Moore, 2001). Alternatively, singular value decomposition for calculation of the Saupe order matrix allows the determination of the axial and rhombic components of the alignment tensor (Losonczi *et al.*, 1999). Still another way to obtain the alignment tensor exists, if the alignment is purely steric, such as frequently observed in neutral bicelles. In this case, the alignment tensor is predicted based on the shape of the solute molecule using an obstruction model (Zweckstetter *et al.*, 2000). This approach is very useful if an initial low resolution structure is available, either based on a set of traditional NMR restraints or a model of a related molecule.

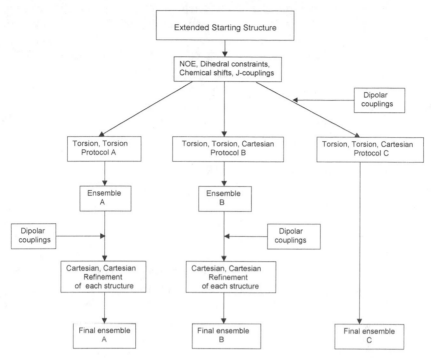

Figure 2. Flowchart illustrating several variants of the structure calculation protocol incorporating residual dipolar couplings using the program CNS

Below I outline a procedure for the refinement of protein structures using residual dipolar couplings. The approach is demonstrated for the potent HIV-1 inactivating protein cyanovirin-N (CV-N). Five types of dipolar couplings were measured: $^1D_{NH}$, $^1D_{C\alpha H\alpha}$, $^1D_{C\alpha C'}$, $^1D_{NC'}$, $^2D_{HNC'}$ and used for refinement. All structure calculations were performed using the program CNS (Brünger *et al.*, 1998). A flowchart for three versions of the protocol is outlined in Figure 2. In all cases, the starting structure comprises an extended strand, that is folded into a final structure by simulated annealing, subject to experimental constraints.

Refinement against dipolar couplings represents a difficult optimization problem, since each dipolar coupling is compatible with two orientations of the associated bond vectors, pointing in opposite directions. In addition, dipolar coupling constraints are of a very different qualitative nature, compared to NOE based distance constraints when used in a simulated annealing protocol. The success of the latter for structure determination based on distance constraints is based on the fact, that the experimentally determined interatomic distances are highly correlated. For example, if amino acid 20 is close to amino acid 90 within a folded protein structure, then the distance between atom X of amino acid 20 and atom Y of amino acid 90 restrains the distance between X and any other atom of amino acid 90, such as atom Z. Furthermore, it is highly probable, that any atom on residue 21 is also close to one ore more atoms of residue 90. In this manner, the potential surface of the optimization in three dimensional space resembles a funnel with a relatively smooth surface. For the case of dipolar couplings, no such straightforward correlation esists. On the contrary, dipolar couplings (or bondvector orientations) tend to compete with each other and with the distances. If, for example, during the simulated annealing a NH bond becomes oriented such, that it's dipolar coupling is satisfied, this does not necessarily lead to a better agreement for the dipolar coupling of the adjacent NC' bond. Thus, each interatomic vector orientation represents an independent orientational parameter. As a result, calculational strategies based on dipolar couplings alone are fraught with difficulties. But even the simple addition of a constraint term for dipolar couplings to proven methodologies of NMR structure calculations can result in structures becoming trapped in deep local minima. As a consequence, the convergence properties of the procedure can be severely curtailed and careful adjustment of the protocols is necessary.

In order to avoid such trapping in the determination of a NMR structure, it is advantageous to employ a two stage simulated annealing protocol (protocol A and B). In the first stage, all conventional experimental constraints, such as NOE and H-bond based distance constraints, dihedral angle constraints, coupling constants and carbon chemical shifts are employed to calculate an ensemble of models. This first stage employs either a torsion, torsion – (protocol A) or a torsion, torsion, Cartesian space (protocol B) algorithm. Details of the implementation for both algorithms are provided in Table 1. During the second stage, each structure of the ensemble is refined against the dipolar couplings. These are added as constraints in a simulated annealing procedure, performed in Cartesian space, using both a high temperature step and a low temperature slow cooling step (details in Table 1). In this case, it is possible to improve the position of adjacent bonds, since only small adjustments in orientation are

Table 1   Simulated Annealing Protocols

	STAGE 1			STAGE 2		
	Torsion (T), Torsion (T) Dynamics			**Cartesian (C), Cartesian (C) Dynamics**		
	High temperature (T)	Slow-cooling (T)	Powell minimization	High temperature (C)	Slow-cooling (C)	Powell minimization
A and B	Temp=50000 K N_{steps}=1000 Time step=0.015 ps k_{vdW}=0.1 k =k_{fin}	Temp=50000-0 K Temp step=250 K N_{steps}=4000 Time step=0.010 ps k_{vdW}=0.1-1.0 k =k_{fin}	N_{steps}=200 N_{ovdw}=10 k_{NOE}=75 k_{DIH}=400 k_{vdW}=1.0 $k_{CS H}$=7.5 k =k_{fin}	Temp=1000 K Time step=0.005 ps N_{steps}=666 k_{vdW}=0.1 k_{ANGL}=0.4 k_{fin} k_{IMP}=0.1 k_{fin} k D_X=0.01 k =k_{fin} N_{steps}=334 k_{vdW}=0.1 k =k_{fin}	Temp=1000-1 K Temp step=25 K N_{steps}=20000 Time step=0.002 ps k_{vdW}=0.1-4.0 k D_X=0.01- k_{fin} k =k_{fin}	N_{steps}=200 N_{ovdw}=10 k_{NOE}=75 k_{DIH}=400 k_{vdW}=1.0 $k_{CS H}$=7.5 k D_X=k_{fin} k =k_{fin}
B only		Additional Slow-cooling (C) Temp=1000-1 K Temp step=25 K N_{steps}=20000 Time step=0.002 ps k_{vdW}=1.0-4.0 k =k_{fin}				

		STAGE 2	
	Torsion (T), Torsion (T), Cartesian (C) Dynamics		Powell minimization
	High temperature (T)	Slow-cooling (T) Additional Slow-cooling (C)	
C	Temp=50000 K N_{steps}=1000 Time step=0.015 ps k_{vdW}=0.1 k D_X=0.01 k =k_{fin}	Temp=50000-0 K Temp=1000-1 K Temp step=250 K N_{steps}=20000 N_{steps}=4000 Time step=0.002 ps Time step=0.010 ps Time step=0.002 ps k_{vdW}=0.1-1.0 k_{vdW}=1.0-4.0 k D_X=0.01- 0.1 k_{fin} k D_X=(0.1-1.0) k_{fin} k =k_{fin} k =k_{fin}	N_{steps}=200 N_{ovdw}=1) k_{NOE}=7½ k_{DIH}=400 k_{vdW}=1.) $k_{CS H}$=7.5 k D_X=k_{fin} k =k_{fin}

Final values of force constants (k_{fin}):
k_{BOND}=1000 kcal mol⁻¹ Å⁻²; k_{ANGL}=500 kcal mol⁻¹ rad⁻²; k_{IMP}=500 kcal mol⁻¹ rad⁻²; k_{NOE}=150 kcal mol⁻¹ Å⁻²; k_{DIH}=200 kcal mol⁻¹ rad⁻²; k_{vdW}=4.0 kcal mol⁻¹ Å⁻⁴;
k $^3J_{HNα}$=1.0 kcal mol⁻¹ Hz⁻²; $k_{CS C}$=0.5 kcal mol⁻¹ ppm⁻²; $k_{CS H}$=7.5 kcal mol⁻¹ ppm⁻²; k D_{NN}=1.0 kcal mol⁻¹ Hz⁻²; k D_{CaHa}=1.0 kcal mol⁻¹ Hz⁻²;
k D_{NC}=0.108 kcal mol⁻¹ Hz⁻²; k D_{CaC}=0.035 kcal mol⁻¹ Hz⁻²; k D_{NC}=0.05 kcal mol⁻¹ Hz⁻²

necessary. For example, considering the above mentioned NH and NC' bonds, now the positioning of the NC' bond can be improved by the NC' dipolar coupling, since the starting structure is already very close to the true structure.

Particular individual features in the implementation of the simulated annealing protocols as shown here for protocol A and protocol B do not result in significant differences in the quality of the final families of structures, as evidenced by the data summarized in Table 2. Note, that the slightly higher rate of convergence in the initial stage of protocol B is offset by the additional time required for the added slow cooling Cartesian space step compared to protocol A. The latter improves the quality of the structures in the first stage of the calculation. This observation underscores the known fact that good quality "conventional" NMR structures, calculated by simulated annealing, usually comprise a slow cooling/annealing module in their simulation protocol. Protocol C is a one-stage procedure, which was designed to include the dipolar couplings as constraints from the start. In order to circumvent the above outlined problems associated with false minima traps in this implementation, the dipolar couplings are introduced gradually. During the first slow cooling stage in torsion angle space, the force constants for all dipolar couplings are ramped to only 1/10 their usual value. Only during the last annealing step in Cartesian space are the dipolar coupling force constants increased to their final values, which are identical to those employed in the second stage module of protocols A and B.

A summary of all experimental constraints, pertinent structural statistics as well as parameters describing the precision and quality of the final ensembles of structures is provided in Table 2.

As noted previously (Bewley et al., 1998), inclusion of the additional dipolar constraints improves the precision of the structures. The coordinate precision increases from 0.48 Å to 0.21 Å for backbone atoms and from 0.71 Å to 0.48 Å for all non hydrogen atoms (protocol A). The improvement is not as dramatic for protocol B, where increases from 0.32 Å to 0.20 Å for backbone atoms and from 0.56 Å to 0.48 Å for all non hydrogen atoms are observed. For all three sets of final ensembles, the coordinate precision is very similar, namely 0.21 Å, 0.20 Å and 0.22 Å for backbone atoms (or 0.48 Å, 0. 48 Å, and 0.50 Å for all heavy atoms), calculated with protocols A, B, and C, respectively. Ensembles of structures calculated with the different protocols are shown in Fig.3 for the backbone atoms, and for a set of structures including sidechains in Fig.4. It may appear surprising, that no increase in coordinate precision is observed between structures calculated with NH dipolar couplings only (Table 2B, column 2) and those calculated including all other dipolar couplings (Table 2B, column 3). This is related to the fact that inclusion of only residual NH dipolar couplings mainly

Table 2(A) Structural Statistics for CV-N

Convergance [a]	#	Protocol A			Protocol B		Protocol C
		30%	+NH DC 100%	+all DC 100%	45%	+all DC 100%	30%
		rmsd +/-sd	rmsd +/-sd	rmsd +/-sd	rmsd +/-sd	rmsd +/-sd	rmsd +/-sd
Covalent & geometric restraints							
Bonds (Å)		0.0020 +/- 0.0008	0.0019 +/- 0.0002	0.0020 +/- 0.0001	0.0020 +/- 0.0002	0.0020 +/- 0.0003	0.0020 +/- 0.0001
Angles (deg)		0.367 +/- 0.016	0.367 +/- 0.010	0.500 +/- 0.009	0.350 +/- 0.010	0.493 +/- 0.013	0.505 +/- 0.012
Impropers (deg)		0.285 +/- 0.023	0.308 +/- 0.016	0.430 +/- 0.013	0.261 +/- 0.013	0.427 +/- 0.016	0.437 +/- 0.019
Number of contacts with distance <1.6 Å		0.9 +/- 0.7	1.8 +/- 1.4	2.7 +/- 2.1	0.9 +/- 0.7	3.7 +/- 2.5	2.0 +/- 1.8
Experimental restraints							
NOE (Å)	1161	0.0042 +/- 0.0019	0.0040 +/- 0.0008	0.0037 +/- 0.0009	0.0037 +/- 0.0008	0.0037 +/- 0.0007	0.0041 +/- 0.0006
H-bonds (Å)	109	0.0061 +/- 0.0038	0.0036 +/- 0.0018	0.0046 +/- 0.0017	0.0032 +/- 0.0015	0.0046 +/- 0.0012	0.0063 +/- 0.0015
Dihedral constraints (deg)	339	0.142 +/- 0.047	0.080 +/- 0.001	0.141 +/- 0.020	0.129 +/- 0.038	0.138 +/- 0.028	0.170 +/- 0.036
$^{3}J_{HN\text{-}H\alpha}$-couplings (Hz)	82	0.479 +/- 0.056	0.433 +/- 0.012	0.601 +/- 0.012	0.368 +/- 0.022	0.594 +/- 0.008	0.593 +/- 0.011
Chemical shifts (ppm)							
$^{13}C_\alpha$	81	0.889 +/- 0.025	0.841 +/- 0.012	0.856 +/- 0.013	0.839 +/- 0.019	0.852 +/- 0.010	0.856 +/- 0.013
$^{13}C_\beta$	74	1.119 +/- 0.029	1.073 +/- 0.009	1.081 +/- 0.012	1.034 +/- 0.019	1.114 +/- 0.011	1.125 +/- 0.012
^{1}H	362	0.256 +/- 0.016	0.289 +/- 0.011	0.292 +/- 0.008	0.261 +/- 0.015	0.293 +/- 0.008	0.289 +/- 0.009
Dipolar couplings (Hz)							
D_{HN}(exp.err.= 0.5-1.0)	83		0.23 +/- 0.02	0.44 +/- 0.01		0.44 +/- 0.01	0.44 +/- 0.01
$D_{C\alpha H\alpha}$ (1-1.5)	76			1.03 +/- 0.04		1.02 +/- 0.04	1.05 +/- 0.05
$D_{C\alpha C'}$ (1-1.5)	43			1.270 +/- 0.02		1.27 +/- 0.01	1.30 +/- 0.01
D_{NC} (0.5-1.0)	65			0.550 +/- 0.005		0.550 +/- 0.001	0.570 +/- 0.008
$^{2}D_{HNC}$ (1-1.5 Hz)	62			1.21 +/- 0.01		1.21 +/- 0.01	1.22 +/- 0.01

a) Acceptance criteria: RMSD for bond lengths < 0.01 Å; RMSD for impropers < 1.0 deg; RMSD for angles < 1.0 deg; no dihedral angle violations > 5 deg; no NOE viol. > 0.5 Å; no angle viol. > 5 deg; no improper viol. > 5 deg; no bond length viol. > 0.05 Å

Table 2(B) Structural Statistics for CV-N

Structure Quality	Protocol A			Protocol B		Protocol C
		+ NH DC	+ all DC		+ all DC	
Atomic RMSD vs mean back bone (N, C$_\alpha$, C')	0.48 +/- 0.07	0.18 +/- 0.02	0.21 +/- 0.03	0.32 +/- 0.05	0.20 +/- 0.03	0.22 +/- 0.03
non-hydrogen	0.71 +/- 0.07	0.45 +/- 0.03	0.48 +/- 0.04	0.56 +/- 0.05	0.48 +/- 0.04	0.50 +/- 0.04
Ramachandran plot (average structure)	#res (%)	#res (%)	#res (%)	#res (%)	#res (%)	#res (%)
most allowed	74 (81.3)	72 (79.1)	79 (86.8)	71 (78.0)	79 (86.8)	79 (86.8)
add. allowed	16 (17.6)	18 (19.8)	10 (11.0)	19 (20.9)	10 (11.0)	10 (11.0)
gen. allowed	1 (1.1)	1 (1.1)	2 (2.2)	1 (1.1)	2 (2.2)	2 (2.2)
residue identity	Ser 52	Ser 52	Ser 52 Lys 3	Ser 52	Ser 52 Lys 3	Ser 52 Lys 3
non allowed	0 (0.0)	0 (0.0)	0 (0.0)	0 (0.0)	0 (0.0)	0 (0.0)
no of bad contacts	2	2	1	2	1	1
identity of bad contacts	(64-68, 12-19)	(64-68, 74-78)	(64-68)	(64-68, 74-78)	(64-68)	(64-68)

increases the precision of structures, without significantly improving their accuracy. Indeed, all the structures may simply move to the next local minimum. The atomic r.m.s. shifts for the backbone atoms of the average structures between the ensembles calculated with protocol A is displayed in Figure 5A while the differences between the average structures calculated with protocols A, B, and C is illustrated in Figure 5B.

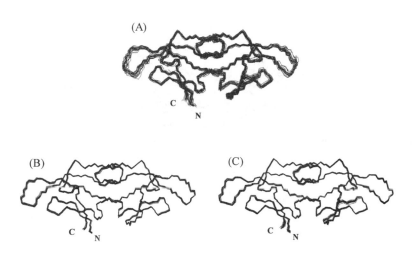

Figure 3.   Best fit superpositions of backbone (N, Cα, C') atoms of 20 CV-N structures calculated with different protocols. (A) Protocol A without dipolar couplings; (B) Protocol A with dipolar couplings; and (C) Protocol C with dipolar couplings.

The most significant improvements in the quality of the structures upon inclusion of all residual dipolar couplings relate to the Ramachandran statistics. As is evident from the data summarized in Table 2B, the percentage of residues found in the most allowed regions of the Ramachandran map increases from ca. 80% to ca. 90% and the number of bad contacts is reduced by 50% in the average structure calculated with dipolar couplings. Therefore, structures calculated with one-bond residual dipolar couplings (as listed in Table 2), exhibit superior packing characteristics, even without the inclusion of a conformational database potential (Kuszewski *et al.*, 1996, 1997).

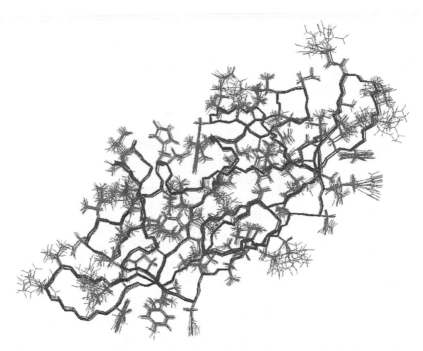

Figure 4. Best fit superpositions of backbone (N, Cα, C') atoms and of selected sidechain heavy atoms of CV-N. Nine structures calculated with protocol C are superimposed.

7.   VALIDATION OF PROTEIN FOLDS

The most powerful and attractive use of residual dipolar couplings lies in their application for validation of structures. In this context, structures may be derived from modelling, either *ab initio* or homology, or from low resolution experimental data. If the model structure is accurate, the dipolar couplings will exhibit excellent agreement between those calculated based on the structure and the experimentally measured ones. Exploitation of this fact may lead to the most direct and important contribution of NMR based methodology in 'structural genomics'. Although large efforts are underway

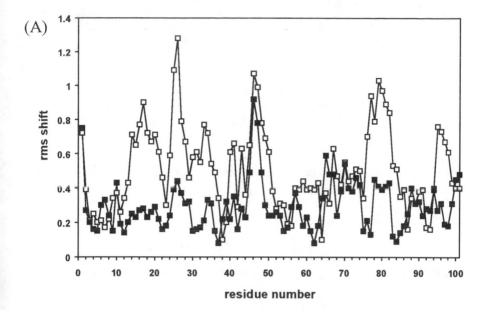

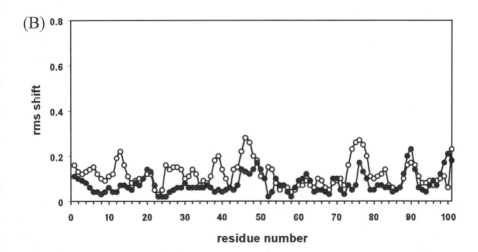

Figure 5.   Atomic r.m.s. shift of backbone (N, Cα, C') coordinates between the average structures calculated with different protocols. (A) Differences between the structures calculated with protocol A; □ without dipolar couplings versus with all dipolar couplings; ■ with NH dipolar couplings only versus with all dipolar couplings. (B) Differences between the structures calculated with protocol B (•) or protocol C (o) and those calculated with protocol A using the complete set of dipolar couplings.

to determine as many protein structures by NMR in a high throughput manner, traditional methodology for NMR structure determination for proteins of intermediate size (~ 30-45 kDa) is still relatively slow. For instance, to solve the structure of the 40 kDa complex between the N-terminal domain of enzyme I and HPr, the NMR measurement time alone extended over almost 5 months (Garrett *et al.*, 1999), with an additional time of at least 6 months for data interpretation and structure calculation. It therefore seems imperative to explore alternative avenues, if NMR is envisaged as a viable tool for structure determination in the post-genomic era. Exploitation of strategies based on residual dipolar couplings may overcome these shortcomings and lead to new conceptual applications of NMR.

One possible avenue in this direction is based on the combination of structure prediction and experimental validation or selection. Based on our increasing understanding of the important factors that govern folding and stability of proteins, it has been proposed that only a finite set of protein folds exist (Wolf *et al.*, 2000). With improving force fields in structure prediction, it is conceivable, that for a particular amino acid sequence a set of several thousand different possible folds can be generated, one of which may be the correct one. The crucial question in this scenario then remains, how to identify it. Using NMR may provide the answer. It is relatively fast and straightforward to obtain backbone resonance assignments for isotopically labelled proteins and the measurement of residual dipolar couplings in dilute liquid crystalline media is also fast and straightforward. It therefore is possible to experimentally obtain sets of residual dipolar couplings for the backbone, which in turn can be compared with those calculated for the different theoretically predicted structures. Thus NMR becomes an experimental filter for these theoretically predicted structures.

8.   CONCLUDING REMARKS

The accessibility of anisotropic NMR parameters in solution for biological macromolecules opens the door for future imaginative exploitation of this diverse wealth of physical information. Applications with respect to improving the accuracy of protein structures (Clore and Gronenborn, 1998; Clore *et al.*, 1999) in defining the long range orientation of the RNA in a protein-RNA complex (Bayer et. al., 1999) and domain orientation in multidomain proteins (Fischer *et al.*, 1999; Skrynnikov *et al.*, 2000; Chou *et al.*, 2000) as well as recognition of protein folds (Annila *et al.*, 1999; Delaglio *et al.*, 2000) have already been reported. No doubt further developments will occur. In the era of structural genomics, NMR is poised to make a major contribution. Since dipolar constraints can be readily measured on partially aligned proteins, they can be incorporated into powerful methodologies for validation of theoretical models. Not every protein structure of the completed human genome and those of model organisms will be solved experimentally by NMR or X-ray crystallography to high resolution. However, all of them are amenable to structure prediction, and validation of these predicted structures and folds can be achieved rapidly by NMR using residual dipolar couplings.

8.1   Acknowledgments

I am indebted to all my collaborators and colleagues mentioned in the text and references who have developed the technologies described in this article and who have been a constant source of stimulating discussions. Particularly, Drs. Ad Bax, Nico Tjandra and Anatoliy Dobrodumov provided invaluable intellectual contributions. Special thanks go to Dr. Anatoliy Dobrodumov for preparing the tables and figures. The work in the author's laboratory was in part supported by the Intramural AIDS Targeted Antiviral Program of the Office of the Director of the National Institutes of Health.

9.   REFERENCES

Annila, A., Aittio, H., Thulin, E., and Drakenberg, T., (1999), *J. Biomol. NMR.* **14**, 223-230.
Barrientos, L. G., Dolan, C., and Gronenborn, A. M., (2000), *J. Biomol. NMR.* **16**, 329-337.
Barrientos, L. G., Louis, J. M., and Gronenborn, A. M., (2001), *J. Magn. Reson.* **148**, 159-162.
Bax, A., and Tjandra, N., (1997), *J. Biomol. NMR.* **10**, 289-292.
Bayer, P., Varani, L., and Varani, G., (1999), *J. Biomol. NMR.* **14**, 149-155.
Bewley, C. A., Gustafson, K. R., Boyd, M. R., Covell, D. G., Bax, A., Clore, G. M., and Gronenborn, A. M. (1998), *Nat. Struct. Biol.* **5**, 571-578.
Brünger, A. T., Adams, G. M., Clore, G. M., DeLano, W. L., Gros, P., Grosse-Kunstleve, R. W., Jiang, J. S., Kuszewski, J., Nilges, M., Pannu, N. S., Read, R. J., Rice, L. M., Simonson, T. and Warren, G. L., (1998), *Acta Cryst. D. Biol. Crystallogr.* **54**, 905-921.
Cai, M., Huang, Y., Zheng, R., Wei, S.,Q., Ghirlando, R., Lee, M.,S., Craigie, R., Gronenborn, A.,M., and Clore, G.,M., (1998), *Nat.,Struct.,Biol.* **5**, 903-909.
Cavagnero, S., Dyson, J. H., and Wright, P. E., (1999), *J. Biomol. NMR* . **13**, 387-391
Clore, G. M., and Gronenborn, A. M., (1989), *CRC Critical Rev. Biochem. and Molec. Biol.* **24**, 479-564.
Clore, G. M., and Gronenborn, A. M., (1998), *Proc. Natl. Acad. Sci.U.S.A.* **95**, 5891-5898.
Clore, G. M., Gronenborn, A. M., and Tjandra, N. (1998a), *J. Magn. Reson.* **131**, 159-162.
Clore, G. M., Gronenborn, A. M., and Bax, A. (1998b), *J. Magn. Reson.* **133**, 216-221.
Clore, G. M., Starich, M. R., and Gronenborn, A. M. (1998c), *J. Amer. Chem. Soc.* **120**, 10571-10572.
Clore, G. M., Starich, M. R., Bewley, C. A., Cai, M., and Kuszewski, J., (1999), *J. Amer. Chem. Soc.* **121**, 6513-6514.

Chou, J. J., Li. S. and Bax, A., (2000), *J. Biomol. NMR.* **18**, 217-227.

Delaglio, F., Kontaxis, G., and Bax, A., (2000), *J. Amer. Chem. Soc.* **122**, 2142-2143.

Fischer, M. W. F., Losonczi, J. A., Weaver, J. L., and Prestegard, J. H., (1999), *Biochemistry* **38**, 9013-9022.

Fleming, K., Gray, D., Prasannan, S., and Matthews, S., (2000), *J. Amer. Chem. Soc.* **122**, 5224-5225.

Fraden, S., Maret, G., Caspar, D. L. D., and Meyer, R. B., (1989), *Phys. Rev. Lett.* **63**, 2068-2071.

Garrett, D. S., Seok, Y. J., Peterkofsky, A., Gronenborn, A. M., and Clore, G. M., (1999), *Nat. Struct. Biol.* **6**, 166-173.

Glucksman, M. J., Hay, R. D., and Makowski, L., (1986), *Science* **231**, 1273-1276.

Glucksman, M. J., Bhattacharjee, S., and Makowski, L., (1992), *J. Mol. Biol.* **226**, 455-470.

Gomati, R., Appell, J., Bassereau, P., Marignan, J., and Porte, G., (1987), *J. Phys. Chem.* **91**, 6203-6210.

Hansen, M. R., Rance, M., and Pardi, A., (1998a), *J. Amer. Chem. Soc.* **120**, 11210-11211.

Hansen, M. R., Mueller, L., and Pardi, A., (1998b), *Nat. Struct. Biol.* **5**, 1065-1074.

Hus, J.-C., Marion, D., and Blackledge, M., (2000), *J. Mol. Biol.* **298**, 927-936.

Koenig, B. W., Hu, J.-S., Ottiger, M., Bose, S., Hendler, R. W., and Bax, A., (1999), *J. Amer. Chem. Soc.* **121**, 1385-1386.

Kung, H. C., Wang, K. Y., Golier, I., and Bolton, P. H., (1995), *J. Magn. Reson. Ser. B.* **109**, 323-325.

Kuszewski, J., Gronenborn, A. M., and Clore, G. M., (1996), *Protein Science* **5**, 1067-1080.

Kuszewski, J., Gronenborn, A. M., and Clore, G. M., (1997), *J. Magn. Reson.* **125**, 171-177.

Lei, L., Li, P., and Asher, S. A., (1999), *J. Amer. Chem. Soc.* **121**, 4040-4046.

Lipari, G., and Szabo, A., (1982), *J. Amer. Chem. Soc.* **104**, 4546-4559.

Losonczi, J. A. and Prestegard, J. H., (1998), *J. Biomol. NMR* **12**, 447-451.

Losonczi, J. A., Andrec, M., Fischer, M. W., and Prestegard, J. H., (1999), *J. Magn. Reson.* **138**, 334-342.

Makowski, L., Caspar, D. L., and Marvin, D. A., (1980), *J. Mol. Biol.* **140**, 149-181.

McGrath, K. M., (1997), *Langmuir* **13**, 1987-1995.

Nakamura, H., and Okano, K., (1983), *Phys. Rev. Lett.* **50**, 186-189.

Ojennus, D. D., Mitton-Fry, R. M., and Wuttke, D. S., (1999), *J. Biomol. NMR.* **14**, 175-179.

Ottiger, M. and Bax, A., (1998), *J. Biomol. NMR.* **12**, 361-372.

Ottiger, M., Delaglio, F., and Bax, A., (1998a), *J. Magn. Reson.* **131**, 373-378.

Ottiger, M., Delaglio, F., Marquardt, J. L., Tjandra, N., and Bax, A., (1998b), *J. Magn. Reson.* **134**, 365-369.

Ottiger, M., and Bax, A., (1999), *J. Biomol. NMR.* **13**, 187-191.

Prestegard, J. H., (1998), *Nat. Struct. Biol.* **5** Suppl, 517-522.

Prestegard, J. H., Tolman, J. R., Al-Hashimi, H. M. and Andrec, M., (1999) in "Structure Computation and Dynamics in Protein NMR", Biological Magnetic Resonance, Volume 17, N. R. Krishna and L. J. Berliner, Editors, Kluwer Academic/Plenum Publishers, pp. 311-356.

Prosser, R. S., Losonczi, J. A., and Shiyanovskaya, I. V., (1998), *J. Amer. Chem. Soc.* **120**, 11010-11011.

Rückert, M., and Otting, G., (2000), *J. Amer. Chem. Soc.* **122**, 7793-7797.

Sanders II, C. R., and Landis, G. C., (1995), *Biochemistry* **34**, 4030-4040.

Sanders II, C. R., and Schwonek, J. P., (1992), *Biochemistry* **31**, 8898-8905.

Sass, H. -J., Cordier, F., Hoffmann, A., Rogowski, M., Cousin, A., Omichinski, J. G., Lowen, H., and Grzesiek, S., (1999), *J. Amer. Chem. Soc.* **121**, 2047-2055.

Sass, H. -J., Musco, G., Stahl, S. J., Wingfield, P. T. and Grzesiek, S., (2000), *J. Biomol. NMR.* **18**, 303-309.

Skrynnikov, N. R., Goto, N. K., Yang, D., Choy, W. -Y., Tolman, J. R., Mueller, G. A. and Kay, L. E., (2000), *J. Mol. Biol.* **295**, 1265-1273.

Stubbs, G., Warren, S., and Holmes, K. (1977), *Nature* **267**, 216-221.

Tian, F., Bolon, P. J., and Prestegard, J. H. (1999), *J. Amer. Chem. Soc.* **121**, 7712-7713.

Tjandra, N., Grzesiek, S., and Bax, A., (1996), *J. Amer. Chem. Soc.* **118**, 6264-6272.

Tjandra, N., and Bax, A., (1997a), *Science* **278**, 1111-1114.[erratum in Science (1997) 278, 1697].

Tjandra, N., and Bax, A., (1997b), *J. Magn. Reson.* **124**, 512-515.

Tjandra, N., Omichinski, J. G., Gronenborn, A. M., Clore, G. M., and Bax, A., (1997), *Nat. Struct. Biol.* **4**, 732-738.

Tolman, J. R., Flanagan, J. M., Kennedy, M. A., and Prestegard, J. H., (1995), *Proc. Natl. Acad. Sci. U.S.A.* **92**, 9279-9283.

Tolman, J. R., Flanagan, J. M., Kennedy, M. A., and Prestegard, J. H., (1997), *Nat. Struct. Biol.* **4**, 292-297.

Tolman, J. R., and Prestegard, J. H., (1996a), *J. Magn. Reson. Series B* **112**, 245-252.

Tolman, J. R., and Prestegard, J. H., (1996b), *J. Magn. Reson. Series B* **112**, 269-274.

Torbert, J. and Maret, G., (1979), *J. Mol. Biol.* **134**, 843-845.

Tycko, R., Blanco, F. J., and Ishii, Y., (2000), *J. Amer. Chem. Soc.* **122**, 9340-9341.

Wang, H., Eberstadt, M., Olejniczak, E. T., Meadows, R. P., and Fesik, S.W., (1998a), *J. Biomol. NMR* **12**, 443-446.

Wang, Y. -X., Marquardt, J. I.., Wingfield, P., Stahl, S. J., Lee-Huang, S., Torchia, D., and Bax, A., (1998b), *J. Amer. Chem. Soc.* **120**, 7385-7386.

Warren, J. J., and Moore, P. B., (2001), *J. Magn. Reson.,* **149**, 271-275.

Won, Y.-Y., Davis, H. T., and Bates, F. S., (1999), *Science* **283**, 960-963.

Wolf, Y. I., Grishin, N. V., and Koonin. E.V., (2000), *J. Mol. Biol.* **299**, 897-905.

Wüthrich, K., (1986), NMR of Proteins and Nucleic Acids (Wiley, New York),

Zweckstetter, M., and Bax, A., (2000), *J. Am. Chem. Soc.* **122**, 3791-3792.

Chapter 9

Hydrogen Bond Scalar Couplings – A New Tool In Biomolecular NMR[$]

Stephan Grzesiek[1], Florence Cordier[1], and Andrew J. Dingley[2,3]
[1]*Department of Structural Biology, Biozentrum, University of Basel, 4056 Basel, Switzerland,*
[2]*Institute of Physical Biology, Heinrich-Heine-Universität, 40225 Düsseldorf, Germany,*
[3]*Institute of Structural Biology, IBI-2, Forschungszentrum Jülich, 52425 Jülich, Germany*

Abstract: This chapter gives a summary of the recent findings on electron-mediated scalar couplings across hydrogen bonds. The main emphasis is put on a discussion of the various types and sizes of such couplings which have been detected in biological macromolecules. Various experimental schemes for the detection of H-bond couplings are summarized and possible applications are suggested.

1.   INTRODUCTION

The concept of a hydrogen bond (H-bond) has long been recognized (Huggins, 1919; Latimer and Rodebush, 1920; Pauling, 1960) as a very intuitive explanation for the weak attractive forces between an electronegative acceptor atom and a hydrogen atom attached to a second electronegative atom. Compared to covalent bonds, the activation energies for the formation of an H-bond and the bond energies themselves are small, such that H-bonds are readily formed and broken between various partners under conditions where donor and acceptor groups undergo diffusive processes under ambient temperatures in many common solvents. This property explains the key role that H-bonds play in the formation and

---

[$] This chapter is an updated version of an article previously published as "Scalar couplings across hydrogen bonds" by S. Grzesiek, F. Cordier, and A.J. Dingley, *Methods in Enzymology* (2001), **338**, 111-133. Reprinted by permission of the publisher.

stabilization of biomacromolecular structures and as participants in many chemical reactions (Fersht, 1985; Jeffrey and Saenger, 1991).

In biological systems, H-bond donors and acceptors are predominantly nitrogen and oxygen atoms. However, also sulfur, metallic cofactors and sometimes π-electron rich aromatic groups are involved in hydrogen bonding. In most cases, evidence for individual hydrogen bonds in biomacromolecular structures is derived from the spatial proximity of donor and acceptor groups once the structure of a molecule is determined by diffractive or NMR techniques. More detailed information about H-bonds from x-ray diffraction is particularly hard to obtain since the weak scattering density of the hydrogen atom can only be detected in structures solved at highest resolutions, i.e. better than ~ 1.0 Å. Such structures currently (year 2000) comprise about 0.6 % of all crystallographic structures entered in the Brookhaven protein data bank. Only in these cases, it is possible to ascribe individual spatial positions to the hydrogen atoms which are independent of the use of standard covalent geometries. Neutron diffraction has yielded high resolution information on the position of either proton or deuterium nuclei in H-bonds for a small number of biomacromolecules for which large crystals can be grown. Recent developments in this field (Niimura, 1999) have reduced the required size of crystals to about 1 mm^3.

Numerous NMR observables for individual nuclei in hydrogen-bonded moieties have been shown to correlate with the stability and relative geometry of individual H-bonds (Becker, 1996). Such parameters include: the reduced hydrogen exchange rates with the solvent (Hvidt and Nielsen, 1966; Wagner, 1983), fractionation factors (Bowers and Klevitt, 2000; Hvidt and Nielsen, 1966; LiWang and Bax, 1996; Loh and Markley, 1994), isotope shifts for the substitution of the hydrogen bonded proton by ^{2}H and ^{3}H (Altman et al., 1978; Gunnarsson et al., 1976), the isotropic and anisotropic chemical shifts of the hydrogen-bonded proton and of other nuclei within the donor and acceptor groups (Asakawa et al., 1992; Markowski et al., 1977; McDiarmott and Ridenour, 1996; Shoup et al., 1966; Takahashi et al., 1998; Tessari et al., 1997; Tjandra and Bax, 1997; Wagner et al., 1983), the size of the electric field gradient at the position of the proton within the H-bond as observed by the deuterium quadrupolar coupling constant (Boyd et al., 1997; LiWang and Bax, 1997), the sequential one-bond $^1J_{C'N}$ coupling constants in proteins (Juranic et al., 1995; Juranic et al., 1996).

Recently, electron-mediated, scalar couplings which are active between magnetic nuclei on both sides of the hydrogen bridge have been discovered in biological macromolecules and their complexes (Tables 1–3). These couplings are closely related to similar inter- and intramolecular couplings across hydrogen bridges in smaller chemical compounds (Crabtree et al., 1996; Golubev et al., 1999; Kwon and Danishefsky, 1998; Shenderovich et *al.*, 1998). The scalar coupling effect can be used to "see" individual H-bonds within a biological macromolecule. This means that the frequencies of all three H-bond partners, i.e. of the donor, the acceptor and the proton itself, can be correlated by a single two- or three-dimensional NMR experiment. Thus in favorable cases, the hydrogen bond connectivity pattern of nucleic acids and proteins (and therefore the secondary structure) can be established directly via a COSY experiment. In addition to this direct structural use, the size of the scalar coupling has been shown to be influenced by the relative geometry of the H-bond partners. Therefore valuable information about the "strength" of individual H-bonds can be derived from an analysis of the size of the coupling constants.

Since scalar couplings had been thought to be associated with normal covalent bonds, the initial observations of H-bond couplings came as a surprise. However, it seems now well established (Bagno, 2000; Barfield *et al.*, 2001; Benedict *et al.*, 2000; Case, 2000; Dannenberg *et al.*, 1999; Del Bene, 2000; Del Bene and Bartlett, 2000; Del Bene and Jordan, 2000; Del Bene *et al.*, 2000; Dingley *et al.*, 1999; Pecul *et al.*, 2000; Perera and Bartlett, 2000; Scheurer and Brüschweiler, 1999; Shenderovich *et al.*, 1998) that the H-bond couplings are the result of the same nucleus → electron → nucleus polarization mechanism as any other covalent scalar coupling. Therefore the H-bond couplings behave experimentally in exactly the same way as their more usual covalent counterparts.

It is the purpose of this Chapter (1) to give an overview over the range of groups for which such couplings have been observed so far and to summarize their sizes (2) to discuss experiments for their detection and to point at experimental problems and (3) to suggest possible applications.

Table 1: Observed trans hydrogen bond couplings in nucleic acids

	Size [Hz]	N_d	N_a	Type	References
$^{h2}J_{NN}$	5-11	G-N1, U/T-N3, C-N3$^+$	C-N3, A-N1, A-N7, G-N7	Watson-Crick, Hoogsteen, reverse Hoogsteen, GA mismatch	(Dingley and Grzesiek, 1998; Dingley et al., 1999; Luy and Marino, 2000; Pervushin et al., 2000; Pervushin et al., 1998; Wöhnert et al., 1999)
$^{h1}J_{HN}$	0.5-4	G-N1, U/T-N3, C-N3$^+$	C-N3, A-N1, A-N7, G-N7	Watson-Crick, Hoogsteen	(Dingley et al., 1999; Pervushin et al., 2000; Pervushin et al., 1998; Yan et al., 2000)
$^{h2}J_{NN}$	2.5-8	A-N6, G-N2	A-N7, G-N7	A-A mismatch, G$_4$-tetrad, GCGC-tetrad, A$_2$G$_4$-hexad	(Dingley et al., 2000; Majumdar et al., 1999a; Majumdar et al., 1999b)
$^{h3}J_{NC}$	~0.2[1)	G-N1	G-C6	G$_4$-tetrad	(Dingley et al., 2000)
$^{h4}J_{NN}$	~0.14[1)	G-N1	G-N1	G$_4$-tetrad	(Liu et al., 2000b)

1) sign not determined experimentally

Table 2: Observed trans hydrogen bond couplings in proteins

	Size [Hz]	Donor	Acceptor	References
C^α—H^α ... N—H•••• O=C, O=C, $^{h3}J_{NC}$	-0.2 – -0.9	backbone amide	backbone carbonyl or side chain carboxylate	(Cordier and Grzesiek, 1999; Cornilescu et al., 1999a; Cornilescu et al., 1999b; Wang et al., 1999)
N—H•••• O=C, $^{h2}J_{HC}$	-0.6 – 1.3	backbone amide	backbone carbonyl	(Cordier et al., 1999; Meissner and Sørensen, 2000b)
N—H•••• O=C—C^α, $^{h3}J_{HC\alpha}$	0 – 1.4	backbone amide	backbone carbonyl	(Meissner and Sørensen, 2000a)
N—H•••• N, $^{h2}J_{NN}$	8 – 11	Histidine-Nε2	Histidine-Nε2	(Hennig and Geierstanger, 1999)
N—H•••• O=P, $^{h3}J_{NP}$	4.6[1]	backbone amide	GDP-phosphate	(Mishima et al., 2000)
N—H•••• O=P, $^{h2}J_{HP}$	3.4[1]	backbone amide	GDP-phosphate	(Mishima et al., 2000)
N—H•••S ••••Me, $^{h2}J_{HMe}$	0.3 – 4[1]	backbone amide	Cysteine-S coordinating ^{113}Cd or ^{199}Hg	(Blake et al., 1992a; Blake et al., 1992b)

1) sign not determined experimentally

Table 3: Observed trans hydrogen bond couplings in protein - nucleic acid complexes

	Size [Hz]	Donor	Acceptor	References
N—H•••O=P $^{h3}J_{NP}$	$1.7 - 4.6^{1)}$	backbone amide	phosphate	(Löhr et al., 2000; Mishima et al., 2000)
N—H•••O=P $^{h2}J_{HP}$	$0.5 - 3.4^{1)}$	backbone amide	phosphate	(Löhr et al., 2000; Mishima et al., 2000)
O—H•••O=P $^{h2}J_{HP}$	$0.6 - 1.7^{1)}$	serine/threonine OH	phosphate	(Löhr et al., 2000)
—N$_d$—H, H•••N$_a$ $^{h2}J_{NN}$	6	arginine $^{\eta}$N-H$_2$	G-N7	(Liu et al., 2000c)

1) sign not determined experimentally

2. H-BOND SCALAR COUPLINGS IN BIOMOLECULES

2.1 Nucleic Acids

Imino-N-H•••N-aromatic H-bonds

Scalar couplings across H-bonds are expected to be strongest when the number of bonds between the two interacting nuclei is small. It is therefore no surprise that the strongest H-bond couplings in biomolecules are observed within nucleic acid base pairs when both donor and acceptor are ^{15}N-labelled nitrogen atoms. Figure 1A shows as an example the Watson-Crick base pairing scheme for U-A (T-A) and G-C pairs in RNA (DNA) where the central hydrogen bond connects the imino nitrogen N3 of U (T) to the aromatic acceptor nitrogen N1 in A and N1 of G is hydrogen bonded to the acceptor N3 in C. In such a situation, trans H-bond couplings only need to bridge one ($^{h1}J_{HN}$) or two ($^{h2}J_{NN}$) bonds (including the H-bond) in order to be observable (see also Table 1). The symbol $^{hn}J_{AB}$ was introduced by Wüthrich and coworkers as a notation (Pervushin et al., 1998) for trans H-bond scalar couplings between nuclei A and B in order to emphasize that one of the n bonds connecting the two nuclei in the chemical structure is actually an H-bond.

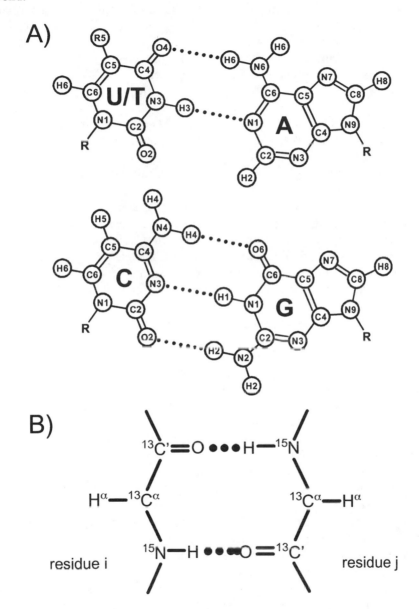

Figure 1. Typical H-bond patterns in nucleic acid base pairs and in the backbone of proteins. A. Watson-Crick U-A (T-A) and G-C nucleotide base pairs in RNA (DNA). R5 = H (RNA) or CH₃ (DNA). B. antiparallel β-sheet conformation of two amino acids in a protein

Figure 2 shows as an example an E.COSY type detection of both $^{h1}J_{NN}$ and $^{h2}J_{NN}$ couplings in the U-A base pairs of the 69 nucleotide left terminal domain of the potato spindle tuber viroid. The experiment used is a normal, two-dimensional ^{1}H-^{15}N TROSY scheme (Pervushin et al., 1997) detecting the narrow components of the uridine imino ^{1}H3-^{15}N3 doublets in both the ^{1}H and ^{15}N dimension. Besides additions for a water flip-back WATERGATE (Grzesiek and Bax, 1993; Piotto et al., 1992), the only further modification of the TROSY scheme is that all the ^{15}N pulses are applied as band selective pulses which only affect the chemical shift region of the uridine ^{15}N3 donor nuclei (~160 ppm). Therefore the spin states of the adenosine ^{15}N1 nuclei resonating around 230 ppm are not disturbed by radio frequency pulses during the entire experiment. Since the U-^{15}N3 and the A-^{15}N1 chemical shift regions are well separated, this can easily be achieved with most proposed selective pulses; in this case the 90° and 180° ^{15}N pulses were applied as sinc pulses with a duration of 300 and 600 µs, respectively. As a result, the uridine imino^{15}N3 and ^{1}H3 spins evolve during the t_1 and t_2 evolution periods of the experiment under the influence of the scalar coupling to the unperturbed adenosine ^{15}N1 nucleus. Clearly visible in Figure 2 are E.COSY type splittings of the uridine ^{15}N3-^{1}H3 correlations which are the result of the $^{h2}J_{NN}$ and the $^{h1}J_{HN}$ coupling to the adenosine ^{15}N1 nucleus which is either in the up or in the down state during the evolution periods of the ^{15}N3 and ^{1}H3 nuclei. A quantification of these splittings yields values of approximately 7 Hz ($^{h2}J_{NN}$) and 2-3 Hz ($^{h1}J_{HN}$), respectively. In contrast to the HNN-COSY detection scheme (Dingley and Grzesiek, 1998) described below, the E.COSY TROSY used for this illustration does not provide the resonance frequencies of the acceptor nuclei. However, E.COSY provides the relative signs of the $^{h2}J_{NN}$ and the $^{h1}J_{HN}$ couplings. The negative slope of the E.COSY pattern in Figure 2 indicates that $^{h2}J_{NN}$ and $^{h1}J_{HN}$ have the same sign since the gyromagnetic moments of ^{1}H and ^{15}N have opposite signs. As the $^{h1}J_{HN}$ couplings have been determined to be positive (Pervushin et al., 1998), this indicates that in accordance with theoretical predictions (Dingley et al., 1999; Scheurer and Brüschweiler, 1999) the $^{h2}J_{NN}$ coupling is also positive.

In general, the size of $^{h2}J_{NN}$ couplings between imino ^{15}N nuclei and aromatic ^{15}N nuclei in nucleic acid base pairs ranges between 5 and 11 Hz whereas $^{h1}J_{HN}$ couplings between the imino proton and aromatic ^{15}N acceptor nuclei have been observed in the range of 0.5 to 4 Hz. Table 1 lists ^{15}N-imino donor/aromatic ^{15}N-acceptor combinations in various base pair types for which trans H-bond couplings have been observed so far. The results of several density functional theory simulations (Barfield et al., 2001; Dingley et al., 1999) show that with the exception of charged Hoogsteen C$^+$•G base pairs, the size of the $^{h1}J_{HN}$ and $^{h2}J_{NN}$ couplings does not depend strongly on the base pair type. The range of $^{h2}J_{NN}$-values listed in Table 1 is rather the result of slight variations of the H-bond geometry in the different base pair types (see below).

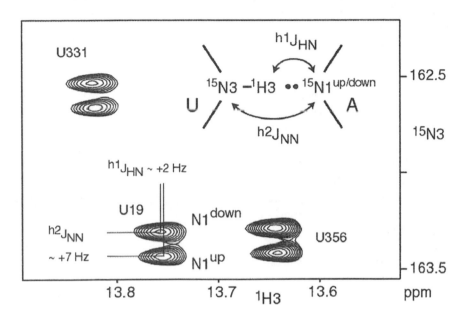

Figure 2.   Simultaneous observation of $^{h2}J_{NN}$ and $^{h1}J_{HN}$ couplings in the U-A base pairs of the left terminal domain of the potato spindle tuber viroid. (Barfield *et al.*, 2001; Dingley and Grzesiek, 1998) A ^{1}H-^{15}N E.COSY TROSY was recorded on a sample of the 1.6 mM uniformly ^{13}C/^{15}N-enriched 69-nucleotide RNA domain. Excitation of the ^{15}N-resonances was restricted to the uridine ^{15}N3 region by means of selective sinc pulses (300 μs for 90°) centered at 163 ppm. The data matrix consisted of 150*(t$_1$) × 1024*(t$_2$) data points with acquisition times of 150 ms (t$_1$) and 71 ms (t$_2$). Data were acquired at 25°C on a 600-MHz Bruker DMX instrument, total experimental time 18 h.

Amino-N-H•••N-aromatic H-bonds

Due to intermediate-fast rotations around the C-N bond, the two amino protons of the nucleic bases in DNA and RNA are often only observed as a single broad resonance. In cases where this rotation is sufficiently slowed down, observations of $^{h2}J_{NN}$ and $^{h1}J_{HN}$ couplings between the amino ^{15}N or ^{1}H nuclei and an aromatic ^{15}N acceptor are possible by detection via the amino protons (Table 1). Such observations of $^{h2}J_{NN}$ couplings have been made for amino groups in A-A mismatch pairs (Majumdar *et al.*, 1999a), GGGG (Dingley *et al.*, 2000) and GCGC (Majumdar *et al.*, 1999b) tetrads as well as A$_2$G$_4$ hexads (Majumdar *et al.*, 1999b). Whereas $^{h2}J_{NN}$ couplings of ~7 Hz have been measured for the guanosine tetrads, the A-A mismatch pairs showed $^{h2}J_{NN}$ couplings of only 2.5 Hz. Very likely, these differences in the coupling constant values reflect differences in the H-bond geometries of the different base pair types. In cases where the amino proton is too broad to be detected or exchanged against deuterium, an alternative detection of $^{h2}J_{NN}$ couplings has been proposed recently (Hennig and Williamson, 2000; Majumdar et al., 1999b).

$^{h1}J_{HN}$ couplings between the hydrogen-bonded amino proton and its nitrogen acceptor have also been observed in the G_4-quartets [A. Dingley, unpublished results]. With values of 2-3 Hz for these couplings, there is no indication of any fundamental difference in the mechanism of magnetization transfer across H-bonds from either imino or amino groups to aromatic nitrogen acceptors.

Imino-N-H•••carbonyl H-bonds

Besides the aromatic nitrogen acceptor atoms, many nucleic acid base pairs involve also the oxygen atoms of carbonyl groups as H-bond acceptors (e.g. Figure 1A). Although the size of scalar couplings across the H-bridge to the magnetic oxygen isotope ^{17}O is likely to be on the same order as the couplings to a nitrogen ^{15}N acceptor, the fast relaxation of this quadrupolar nucleus would prevent such an observation in solution. Instead H-bond couplings to the next possible nucleus of the carbonyl acceptor group, the carbonyl ^{13}C carbon have been observed in proteins (see below) and more recently also in the nucleic acid base pairs of guanosine quartets (Table 1) (Dingley et al., 2000). The observed three-bond couplings ($^{h3}J_{NC}$) from the imino ^{15}N donor to the carbonyl ^{13}C acceptor nucleus are only 0.2 Hz in size and thus smaller than the average of the analogous couplings in proteins (Table 2, Figure 4). Nevertheless, due to the favorable relaxation properties of the G-quartet system, all the expected $^{h3}J_{NC}$ correlations could be detected across H-bridges. The information obtained from this experiment is sufficient to establish the G-quartet structure and the number of non-equivalent G-quartets within the entire molecule as well as to derive all the H-bond partners for the single guanine bases within each quartet.

Very recently four-bond couplings $^{h4}J_{NN}$ of about 0.14 Hz have also been observed in a similar G-quartet system (Liu et al., 2000b) where the scalar couplings connect the imino $^{15}N1$ nuclei in the guanosine N1-H1•••O6=C6-N1 hydrogen bonds (Table 1).

2.2   Proteins

Amide-N-H•••carbonyl H-bonds

The predominant H-bond in proteins is the bridge between the backbone amide proton of one amino acid and the backbone carbonyl oxygen atom of a second amino acid. The canonical patterns of hydrogen bonds in protein secondary structures are given by $H_i \bullet\bullet\bullet O_k$, $O_i \bullet\bullet\bullet H_k$, $H_{i-2} \bullet\bullet\bullet O_{k+2}$, $O_{i-2} \bullet\bullet\bullet H_{k+2}$ for anti-parallel β-sheets, by $H_i \bullet\bullet\bullet O_k$, $O_i \bullet\bullet\bullet H_{k+2}$, $H_{i+2} \bullet\bullet\bullet O_{k+2}$, $O_{i+2} \bullet\bullet\bullet H_{k+4}$ for parallel β-sheets, by $O_i \bullet\bullet\bullet H_{i+4}$ for α-helices, and by $O_i \bullet\bullet\bullet H_{i+3}$ for 3_{10}-helices where H_i and O_k stand for the backbone amide proton and oxygen atoms of the hydrogen bonded residues i and k, respectively. Figure 1B shows as an example the two H-bonds between two amino acids which are part of an anti-parallel β-sheet. As in the case of the nucleic acid H-bonds to carbonyl, on the acceptor side the oxygen nucleus is not accessible for the detection of trans H-bond couplings. Interactions which have been detected so far are $^{h3}J_{NiC'k}$, $^{h2}J_{HiC'k}$, and $^{h3}J_{HiC\alpha k}$ couplings which are active between the amide ^{15}N or ^1H nucleus of residue i and the carbonyl or alpha ^{13}C nucleus of residue k (Table 2). The values determined for the $^{h3}J_{NiC'k}$ coupling constants cover a range of -0.2 to -0.9 Hz (Table 2). Although the absolute size of these couplings is quite small and their use is therefore limited to small to midsize proteins, meanwhile a number of studies (Cordier and Grzesiek, 1999; Cornilescu et al., 1999a; Cornilescu et al., 1999b; Meissner and Sørensen, 2000b; Wang et al., 1999) have been carried out, showing that complete backbone hydrogen bond networks can be detected and that the size of the couplings correlates with donor acceptor distances (Cordier and Grzesiek, 1999; Cornilescu et al., 1999b). Ranges of - 0.6 – 1.3 Hz have been observed for $^{h2}J_{HiC'k}$ couplings (Cordier et al., 1999; Meissner and Sørensen, 2000b). Thus, these values are similar to the size of $^{h3}J_{NiC'k}$ couplings. Surprisingly, even three-bond couplings of the $^{h3}J_{HiC\alpha k}$ type have now been observed (Meissner and Sørensen, 2000a) with values of up to 1.4 Hz (Table 2).

Amide-N-H•••carboxylate H-bonds

H-bond $^{h3}J_{NC'}$ couplings of -0.1 to -0.4 Hz have also been detected for backbone amides H-bonded to the carboxylate side chains of glutamic and aspartic acids (Cornilescu et al., 1999b) (Table 2). Apparently, the magnetization transfer across these H-bonds is very similar to the backbone amide to carbonyl H-bonds.

Histidine-imidazole-N-H•••N H-bonds

For a pair of H-bonded histidine side chains, $^{h2}J_{NN}$ couplings in the range of 8 to 11 Hz have been observed (Hennig and Geierstanger, 1999) between a protonated $^{15}N\epsilon2$-nucleus and an unprotonated $^{15}N\epsilon2$ (Table 2). Thus, these values are close to the values for $^{h2}J_{NN}$ couplings found for the imino and amino N-H•••N$_{aromatic}$ H-bonds of nucleic acid base pairs.

Amide-N-H•••S•••Metal Ion H-bonds

Probably the earliest observations of H-bond couplings in proteins were made by Summers and coworkers (Blake *et al.*, 1992b) in ^{113}Cd- or ^{199}Hg-substituted rubredoxin (Table 2). In this protein, J interactions of up to 4 Hz connected backbone amide protons and the cysteine-coordinated metal atom. Apparently these J-couplings were mediated via the hydrogen bonds from the backbone amides to the S atoms of the cysteine residues.

2.3   Protein – Nucleic Acid Complexes

N-H•••phosphate and O-H•••phosphate H-bonds

Recently, H-bond couplings involving protein amide or hydroxyl groups as donors and phosphate groups as acceptors have been described in molecular complexes of Ras p21 and GDP (Mishima *et al.*, 2000) as well as flavodoxin and riboflavin 5'-monophosphate (Löhr *et al.*, 2000) (Table 3). It was found that depending on the geometry both $^{h2}J_{HP}$ and $^{h3}J_{NP}$ can be as large as 3–5 Hz. Given the larger extent of phosphorous d-orbitals participating in the phosphate oxygen bonds, as compared to the more restricted carbon p-orbitals participating in the carbonylic oxygen bonds, it is perhaps not surprising that the H-bond couplings to ^{31}P in phosphates are larger than the analogous couplings to ^{13}C in the carbonyl groups. The detection of such couplings is particularly important because long-range information on the position of phosphorus in biomolecular complexes is very hard to obtain by traditional high resolution NMR methods.

Arginine-$^{\eta}$N-H$_2$•••N-nucleic acid H-bonds

Another important development is the direct detection of H-bonds between amino acid side chains and nucleic acids by scalar couplings. Very recently, $^{h2}J_{NN}$-couplings between arginine side-chain guanidinium moieties and guanine base ^{15}N7 nuclei have been observed in a complex of the HTLV-1 Rex peptide in complex with an RNA aptamer (Table 3) (Liu *et al.*, 2000c). With values of about 6 Hz these couplings are again an indication that the chemical nature of the donor or acceptor group has a very limited influence on the size of the $^{h2}J_{NN}$-couplings.

3.   H-BOND SCALAR COUPLINGS IN OTHER SYSTEMS

H-bond couplings have also been detected as intra- and intermolecular interactions in smaller compounds. A number of observations indicates that *intramolecular* H-bonds involving hydroxyl groups can yield couplings between the OH protons and protons on the other side of the H-bridge on the order of a few Hertz (Crabtree *et al.*, 1996; Kwon and Danishefsky, 1998). The detection of *intermolecular* H-bond scalar couplings in smaller compounds than biomacromolecules had long been hampered by the very limited lifetime of such small complexes. Due to the pioneering work of the Limbach group (Benedict *et al.*, 2000; Dunger *et al.*, 2000; Golubev *et al.*, 1999; Shenderovich *et al.*, 1998), the detection of such *intermolecular* H-bond scalar couplings has been made possible by the use of cryogenic solvents which extent the lifetime of these complexes. For solutions of hydrogen bonded clusters of F^- and $(HF)_n$, $^{h2}J_{FF}$ couplings of up to 146 Hz and $^{h1}J_{FH}$ couplings of up to 41 Hz have been reported (Shenderovich *et al.*, 1998). Similarly, $^{h2}J_{FN}$ couplings of 96 Hz and one-bond couplings of 20-80 Hz from the proton to donor and acceptor atoms have been observed in almost symmetric H-bridges of the type F•••H•••N in a binary system consisting of ^{15}N-collodine and HF (Golubev *et al.*, 1999).

4. "THROUGH SPACE" COUPLINGS

The observation of the H-bond scalar couplings closely ties in with a large number of earlier reports of scalar couplings that were attributed to a "through space" electronic interaction because the number of intervening covalent bonds would have been too large to explain the effect (Kainosho and Nakamura, 1969; Petrakis and Sederholm, 1961). It is particularly instructive that some of these "through space" couplings were observed in cases (Blake *et al.*, 1992b; Kainosho and Nakamura, 1969) where the forces between the interacting atoms are repulsive. Clearly, repulsive forces can also lead to electronic correlations which mediate scalar magnetization transfer. Therefore, the observation of the H-bond couplings alone does not implicate that the electronic interactions across the H-bridge are attractive and covalent in nature (Ghanty *et al.*, 2000).

5. EXPERIMENTS

Since the H-bond scalar couplings are based on the same forces as their more usual covalent counterparts, their quantum-mechanical description and in particular the spin-spin coupling Hamiltonian used for the description of NMR experiments is identical. Therefore all the methods of detection, quantification, and transfer of magnetization described for other scalar couplings (Bax *et al.*, 1994; Vuister *et al.*, 1998) can be applied in analogous ways to the H-bond scalar couplings.

5.1 $^{h2}J_{NN}$-Couplings

The conceptually simplest scheme for the detection of coupling partners and the quantification of $^{h2}J_{NN}$ correlations is the non-selective HNN-COSY experiment (Dingley and Grzesiek, 1998; Pervushin et al., 1998). In this scheme, the ^{15}N-evolution period of a ^1H-^{15}N-TROSY has been replaced by a homonuclear ^{15}N-^{15}N COSY scheme. As described in detail in the original publications, this scheme transfers magnetization from the proton (H) within the hydrogen bridge via the covalent $^1J_{HN}$ coupling to the ^{15}N nucleus of the H-bond donor (N_d) and then via the $^{h2}J_{NN}$ coupling to the ^{15}N acceptor nucleus (N_a). After a ^{15}N frequency labeling period (t_1), the magnetization is transferred back to the original proton and its oscillations are detected in the receiver during the t_2-period. As in similar quantitative J-correlation schemes, this leads to resonances in the 2-dimensional spectrum at frequency positions ω_{Na}-ω_H (cross peaks) and ω_{Nd}-ω_H (diagonal peaks). The value of the $^{h2}J_{NN}$ coupling constant is then derived from the intensity ratio of the cross and diagonal peaks by the formula $|^{h2}J_{NN}| = \text{atan}[(-I_{Na}/I_{Nd})^{1/2}]/(\pi T)$ where T is the COSY transfer time.

Figure 3 shows as an example the region of the Hoogsten-Watson-Crick T•A-T triplets of an intramolecular DNA-triplex of 32 nucleotides (Dingley et al., 1999). Panel A shows the diagonal resonances corresponding to the thymidine $^1H3$-$^{15}N3$ correlations, whereas panel B depicts the cross resonances corresponding to the Watson-Crick $^1H3(T)$-$^{15}N1(A)$ and the Hoogsteen $^1H3(T)$•$^{15}N7(A)$ correlations. Due to the considerable size of the $^{h2}J_{NN}$ couplings, the sensitivity of the HNN-COSY experiment is sufficient that all H-N•••N type hydrogen bonds could be observed for this intramolecular triplex (Dingley et al., 1999) and even for an RNA domain of 69 nucleotides (Dingley and Grzesiek, 1998) in overnight experiments at ~1.5 mM concentrations. Note that for large molecules with short transverse relaxation times of the ^{15}N-donor (πT_2 $^{h2}J_{NN}$ << 1), the sensitivity for the observation of cross-peaks is highest when the total COSY transfer time is set to a value of approximately T_2.

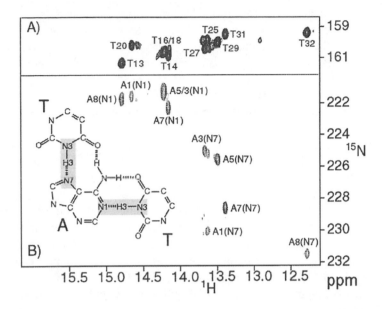

Figure 3. Part of a quantitative-J_{NN} HNN-COSY spectrum of an uniformly $^{13}C/^{15}N$ labeled intramolecular DNA triplex (Dingley *et al.*, 1999) This triplex consists of 5 Hoogsteen-Watson–Crick T•A-T and 3 Hoogsteen C$^+$•G-C base triplets. The spectral region corresponds to the 10 imino resonances of the Hoogsteen-Watson–Crick T•A-T triplets. Positive contours depict diagonal resonances (A), negative contours (B) correspond to cross peaks resulting from H-bond scalar $^{15}N-^{15}N$ magnetization transfer. Resonances are labeled with assignment information. The insert in panel B shows the chemical structure of the T•A–T triplet.

Two experimental difficulties should be pointed out for the HNN-COSY. The first is that the quantification of the coupling constants from the intensity ratio of cross and diagonal peaks is made difficult by the different line widths of the ^{15}N donor and acceptor resonances (Figure 3A/B). The different line widths are the result of the different relaxation mechanisms for both nuclei. Whereas the width of donor ^{15}N resonances is narrowed by the partial cancellation of the ^1H-$^{15}N_d$ dipole and $^{15}N_d$ CSA relaxation in the TROSY scheme, the very large (300-400 ppm) CSA (Hu *et al.*, 1998) of the ^{15}N acceptor nuclei in the nucleic bases dominates the relaxation for the acceptor resonances. As a practical consequence, it is not sufficient to approximate the intensity ratios of the cross and diagonal peaks by their amplitude ratios. Instead the intensities of both donor and acceptor resonances need to be determined by an appropriate peak integration scheme in order to derive accurate values for the coupling constants. Since the amplitude of the oscillations in the ^{15}N time domain is proportional to the magnetic excitation and to the peak integral, time domain fitting routines are a practical alternative to the peak integration in the frequency domain (Delaglio et al., 1995; Dingley and Grzesiek, 1998).

The second problem stems from the frequency separation of ^{15}N donor and acceptor resonances and the limited strength of the available ^{15}N radio frequency pulses. In Watson-Crick and Hoogsteen imino-aromatic-N hydrogen bridges, frequency separations of the ^{15}N donor and acceptor resonances are in the range of 50 to 70 ppm. For typical ^{15}N radio frequency pulse strengths of about 6 kHz and a magnetic field strength of 14 Tesla, errors in $^{h2}J_{NN}$ values resulting from the finite excitation bandwidth have been estimated to be on the order of 10 % (Dingley and Grzesiek, 1998). Smaller errors can be achieved by the use of stronger radio frequency pulses, lower magnetic field strengths, or by ^{15}N excitation schemes which cover only selected regions of the donor and acceptor resonances in the different base pair types. Frequency separations between amino and aromatic ^{15}N resonances are typically in the range of 100 to 130 ppm. As has been pointed out by Majumdar and coworkers (Majumdar et al., 1999a), such ^{15}N frequency separations are too large to be covered effectively by the non-selective ^{15}N pulses of the homonuclear HNN-COSY. They therefore designed a pseudo-heteronuclear H(N)N-COSY experiment where selective ^{15}N pulses excite the amino and aromatic ^{15}N resonances separately and which yields excellent sensitivity (Majumdar et al., 1999a). An inconvenience of the latter experiment is that the resonances corresponding to the amino ^{15}N nuclei are not detected and a separate spin-echo difference experiment was used to quantify the $^{h2}J_{NN}$ values. A slightly improved version of this pseudo-heteronuclear H(N)N-COSY (Dingley et al., 2000) remedies this problem by the use of phase-coherent ^{15}N pulses such that both amino and aromatic ^{15}N resonances can be detected in a single experiment.

A further modification of the HNN-COSY scheme has been proposed for the observation of $^{h2}J_{NN}$ correlations where the hydrogen nucleus in the H-bond is unobservable (Hennig and Williamson, 2000; Majumdar et al., 1999b). This situation is often found due to intermediate conformational exchange of hydrogen bonding amino groups or due to exchange of the proton with the solvent. In some cases, the HNN-COSY can then be started and detected on a carbon-bound proton in the vicinity of the acceptor. This is possible for adenosine and guanosine ^{15}N7 acceptors as well as for adenosine ^{15}N1 acceptors which can be connected to the H8 or H2 proton by means of the $^{2}J_{H8N7}$ or $^{2}J_{H2N1}$ couplings of approximately 11–15 Hz (Ippel et al., 1996), respectively.

Another ingenious modification of the HNN-COSY scheme involves the replacement of the homonuclear ^{15}N COSY transfer by a ^{15}N-TOCSY transfer (Liu et al., 2000b). As the homonuclear TOCSY transfer is twice as fast as the COSY transfer, a significant sensitivity increase is achieved. The application is, however, limited to cases where the ^{15}N donor and acceptor resonances are at similar frequencies such that the power of the ^{15}N-TOCSY radio frequency pulses need not to be too strong. A very interesting application for this scheme was presented for the sensitive detection of very small (0.14 Hz, Table 1) $^{h4}J_{N1N1}$ couplings in imino-carbonyl hydrogen bridges of guanosine tetrads (Liu et al., 2000b).

5.2 $^{h1}J_{HN}$-Couplings

Due to their smaller size as compared to $^{h2}J_{NN}$ and due to the faster relaxation times of the proton resonances, the measurement of $^{h1}J_{HN}$-couplings is more challenging. A number of methods have been described for the quantification of $^{h1}J_{HN}$-couplings: an E.COSY experiment (Pervushin et al., 1998), a quantitative J-correlation TROSY scheme based on the more favorable relaxation properties of ^{1}H-^{15}N zero-quantum coherences (Pervushin et al., 2000), a quantitative HN-COSY (Dingley et al., 1999) and a modified HNN-TROSY (Yan et al., 2000).

5.3 $^{h3}J_{NC'}$-Couplings

The detection of these weak ($|^{h3}J_{NC'}| < 0.9$ Hz) correlations in proteins can be achieved with a conventional HNCO experiment where the nitrogen to carbonyl dephasing and rephasing delays are set to longer values than for the detection of the sequential one-bond $J_{NiC'i-1}$ correlations. In a standard HNCO experiment, the time for the two INEPT transfers between in-phase N^y_i and anti-phase $2N^x_iC'^z_{i-1}$ magnetization is usually set to values of about 25–30 ms. In an HNCO experiment suitable for the transfer by $^{h3}J_{NC'}$, this time is set to 133 ms $\approx 2/^1J_{NC'}$ such that the one-bond transfer from N^y_i to $2N^x_iC'^z_{i-1}$ is approximately refocused (Cordier and Grzesiek, 1999, Cornilescu et al., 1999a). In such a situation, the resulting HNCO spectrum contains mostly correlations which result from smaller, long-range $J_{NC'}$ couplings.

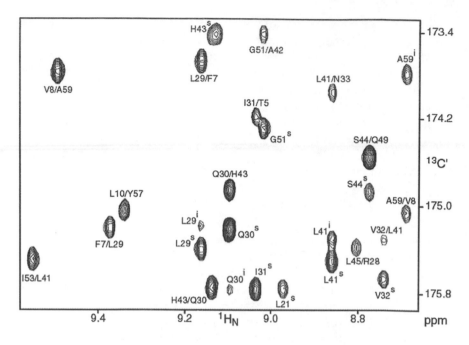

Figure 4. Selected region of the two-dimensional, long-range, quantitative-$J_{NC'}$ H(N)CO-TROSY spectrum recorded on a 3.5 mM sample of $^2H/^{15}N/^{13}C$ labeled c-Src SH3 in complex with the polyproline peptide RLP2 at 95% H_2O/5% D_2O, 25°C (Cordier *et al.*, 2000). The data matrix consisted of 65^* (t_1) × 512^* (t_2) data points (where n^* refers to complex points) with acquisition times of 39 ms (t_1) and 53 ms (t_2). Total measuring time was 17 h on a Bruker DMX600 instrument. Cross peaks marked as Res_i/Res_j are due to $^{3h}J_{NiC'j}$ trans-hydrogen bond scalar couplings between the ^{15}N nucleus of residue i and $^{13}C'$ nucleus of residue j. Residue names marked by the superscript 's' denote not completely suppressed, sequential, one-bond correlations between the ^{15}N nucleus of residue i and $^{13}C'$ nucleus of residue i-1. The superscript 'i' denotes intraresidue two-bond $^{15}N_i$-$^{13}C'_i$ correlations.

Figure 4 shows the results of this experiment on the c-Src SH3 domain (Cordier *et al.*, 2000) carried out as a two-dimensional H(N)CO where only amide proton and carbonyl ^{13}C frequencies were detected. Clearly visible for example are the correlations from the amide ^1H-(^{15}N) of residue F7 to the carbonyl ^{13}C nucleus of residue L29 and from the amide ^1H-(^{15}N) of residue L29 to the carbonyl ^{13}C of residue F7 which correspond to the anti-parallel beta sheet arrangement of these two amino acids (Figure 1B). Quantification of the long-range $J_{NC'}$ coupling constants can be achieved by comparison of the cross-peak intensities with the intensities of sequential amide-carbonyl correlations measured in a second reference experiment (Cordier and Grzesiek, 1999; Cornilescu *et al.*, 1999a). In the latter experiment transfer is tuned to the sequential $^1J_{NC'}$ coupling while the relaxation losses are kept identical to the long-range experiment. A slightly modified version of this long-range HNCO using highly selective ^{13}C-carbonyl pulses can also be used for the detection of $^{h3}J_{NC'}$ correlations in nucleic acids (Dingley et al., 2000). The quantitative long-range HNCO experiment only yields the absolute value of $^{h3}J_{NC'}$. In agreement with density functional theory simulations (Scheurer and Brüschweiler, 1999), the sign of the $^{h3}J_{NC'}$ coupling in proteins was determined as negative by a zero-quantum/double-quantum technique (Cornilescu et al., 1999b).

Due to the small size of the $^{h3}J_{NC'}$ couplings, the sensitivity of the long-range HNCO experiment is rather low. Whereas about 83 % of all expected backbone to backbone hydrogen bonds could be observed in a 12 h experiment on a 1.6 mM sample of uniformly $^{13}C/^{15}N$ labeled human ubiquitin (MWT 8.6 kDa), for larger systems, the shorter ^{15}N transverse relaxation times together with the long ^{15}N-^{13}C dephasing and rephasing times of 266 ms clearly limit the detection of $^{h3}J_{NC'}$ correlations. Several sensitivity improvements have been proposed.

An obvious enhancement in sensitivity can be obtained by the use of an HNCO-TROSY (Salzmann et al., 1998) at higher magnetic field strengths (> 14 T) which makes use of the cancellation of dipolar and CSA relaxation mechanisms for the down-field component of the amide ^{15}N doublet. Such an approach was successful to detect $^{h3}J_{NC'}$ correlations in the 30 kDa ribosome inactivating protein MAP30 (Wang et al., 1999) using uniform ^{2}H, ^{13}C, ^{15}N enrichment and a protein concentration of only 0.7 mM. Experience from this experiment shows that the number of detected $^{h3}J_{NC'}$ correlations (65) was not limited by the transverse ^{15}N relaxation time, but by the incomplete back-exchange of amide deuterons against protons and by the long amide proton T_1 relaxation times. Deuteration is essential for sensitivity gains by the HNCO-TROSY approach, because in protonated samples proton-proton spin-flips strongly reduce the line narrowing gain of TROSY (Kontaxis et al., 2000). For example, for protonated ubiquitin at 25°C and 14 T magnetic field strength, the "TROSY-T_2" and the T_2 of the amide ^{15}N singlet (using ^{1}H-decoupling) are almost identical (170–180 ms). In contrast, in perdeuterated (> 85%, non-exchangeable hydrogen positions), but amide protonated ^{13}C, ^{15}N-labeled ubiquitin, the "TROSY-^{15}N-T_2" increases to 340-400 ms. Clearly larger enhancements can be achieved at higher magnetic field strengths than 14 T (Kontaxis et al., 2000; Pervushin et al., 1997).

A second sensitivity improvement of the long-range HNCO has been proposed recently for non-deuterated protein samples (Liu et al., 2000a). In such samples, the transverse relaxation time of the amide ^{15}N nucleus is decreased to some extent by the scalar coupling to the $^{13}C_\alpha$ nucleus and its relatively short longitudinal relaxation time. This relaxation mechanism is commonly referred to as "scalar relaxation of the second kind" (Abragam, 1961). A similar decrease in imino ^{15}N transverse relaxation times due to the scalar relaxation effect has been observed in nucleic acids (Dingley et al., 2000). The effect is particularly pronounced for smaller macromolecules where ^{13}C T_1 relaxation times are short. In the case of proteins, the scalar contribution to the ^{15}N relaxation can be removed with relative ease by continuous decoupling of the $^{13}C_\alpha$ nucleus. Using this approach for the 12 kDa FK506 binding protein, a sensitivity increase by about 50% was observed as compared to the conventional long-range HNCO experiment (Liu et al., 2000a).

5.4 $^{h2}J_{HC'}$- and $^{h3}J_{HC\alpha}$-Couplings

At present, two schemes have been proposed to detect $^{h2}J_{HC'}$ correlations in proteins. The first uses very selective carbonyl pulses to separate the $^{h2}J_{HC'}$ correlations from interfering covalent two- and three-bond H_N-^{13}C-carbonyl scalar couplings (Cordier et al., 1999). The second scheme utilizes a non-selective E.COSY-HNCO-TROSY approach (Meissner and Sørensen, 2000b) which also yields the sign of the coupling constants. A very similar experiment was used to detect $^{h3}J_{HC\alpha}$ couplings (Meissner and Sørensen, 2000a).

5.5 $^{h3}J_{NP}$- and $^{h2}J_{HP}$-Couplings

Direct observations of $^{h3}J_{NP}$ correlations can be obtained by an HNPO or HNPO-TROSY experiment (Mishima et al., 2000) which is analogous to the HNCO or HNCO-TROSY experiment, the only difference being that the ^{13}C carbonyl pulses are replaced by ^{31}P pulses. For the quantification of the $^{h3}J_{NP}$ and $^{h2}J_{HP}$ couplings appropriate spin-echo difference HSQC schemes were proposed (Mishima et al., 2000).

6.   RELATION TO CHEMICAL SHIFT

Clear correlations of the chemical shift of the protons within the H-bond and the trans-H-bond coupling constants have been observed both for nucleic acids (Dingley et al., 1999) and for proteins (Cordier and Grzesiek, 1999). In nucleic acids N-H•••N H-bonds, $^{h1}J_{HN}$ is found to decrease whereas both $^{h2}J_{NN}$ and $^{1}J_{HN}$ increase linearly with increasing imino proton chemical shifts (Figure 5). Similarly, in protein N-H•••O=C H-bonds, values of $|^{h2}J_{HC'}|$ and $|^{h3}J_{NC'}|$ increase linearly with increasing amide proton chemical shifts (Figure 6). For proteins containing a large number of aromatic side chains, these correlations are considerably improved, when the proton chemical shifts are corrected for ring current effects (F. Cordier, unpublished results). Therefore these correlations could make it possible to separate the proton chemical shift into different contributions which can be understood in a quantitative way.

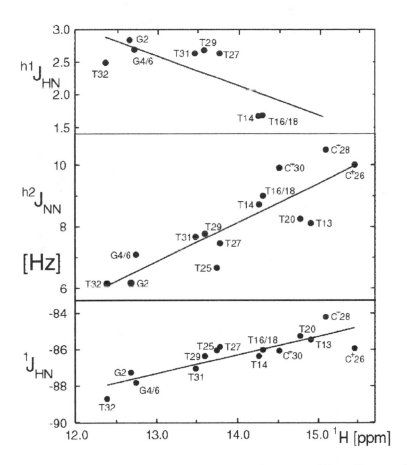

Figure 5. Correlations between the imino proton chemical shift and $^{h1}J_{HN}$, $^{h2}J_{NN}$, and $^{1}J_{HN}$ couplings in the intramolecular DNA triplex (Dingley *et al.*, 1999)..

The isotropic (Kuntz *et al.*, 1991; Wagner, 1983; Wagner *et al.*, 1983; Wishart *et al.*, 1991; Zhou *et al.*, 1992) and also the anisotropic chemical shift (Tessari *et al.*, 1997; Tjandra and Bax, 1997) of amide protons have been used as an indicator for the strength of the hydrogen bond in proteins.

In particular, there is a strong correlation between the hydrogen bond length and the isotropic shift of the amide proton (Kuntz *et al.*, 1991; Wagner, 1983; Wagner *et al.*, 1983; Wishart *et al.*, 1991; Zhou *et al.*, 1992). Short hydrogen bond lengths correspond to larger values for the isotropic chemical shift. Thus, the observed correlations between H-bond coupling constants and the isotropic proton chemical shifts indicate that both parameters are mainly determined by the relative geometry of the partners in the H-bond.

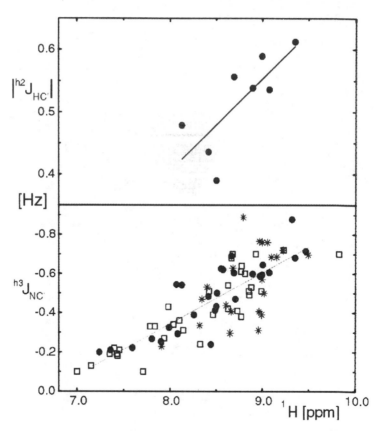

Figure 6. Correlations between the amide proton chemical shift and $^{h2}J_{HC'}$ and $^{h3}J_{NC'}$ couplings in proteins. Filled circles, open asterisks, and rectangles correspond to data from human ubiquitin, the c-Src SH3 domain, and from protein G, respectively. Proton chemical shifts are corrected for ring current effects.

7.   DEPENDENCE ON GEOMETRY

7.1   H-bond Lengths

Comparisons of the measured coupling constants to the geometry of the H-bond are hampered by the limited availability of very high resolution diffraction data. Especially, no crystallographic data are available for most of the nucleic acids for which H-bond couplings have been measured. In the available high resolution structures of nucleic acids, the variation of N1-N3 distances in Watson-Crick base pairs is clearly very limited. A survey of a number of such structures showed (Dingley et al., 1999) that typical N1-N3 distances are 2.92 ± 0.05 Å and 2.82 ± 0.05 Å for G-C and A-T (A-U) base pairs in DNA and RNA. The shorter crystallographic donor acceptor distances in A-T (A-U) base pairs as compared to G-C base pairs coincide with an increase in the value of the $^{h2}J_{NN}$ coupling constants from about 6–7 Hz (G-C) to about 7–8 Hz (A-T, A-U) in the NMR investigations (Dingley and Grzesiek, 1998; Dingley et al., 1999; Pervushin et al., 1998). Density functional theory simulations (Barfield et al., 2001) indicate that the different chemical nature of the donor and acceptor groups in Watson-Crick G-C and A-T (A-U) and in Hoogsteen A•T base pairs have a very limited influence on the $^{h2}J_{NN}$ coupling constants (≤ 0.2 Hz, for N-N distances from 2.7 to 4.0 Å). Therefore, the observed differences in $^{h2}J_{NN}$ values for G-C, A-T, A-U, and A•T base pairs should be largely due to the differences in donor-acceptor distances. Neglecting angular variations, it follows that at N-N distances of 2.8 to 2.9 Å, a change in the value of $^{h2}J_{NN}$ by 1 Hz corresponds to a change in donor-acceptor distance of 0.07 ± 0.01 Å. Decreases of $^{h2}J_{NN}$ coupling constants by about 1 Hz have been observed at the ends of helical stems and have been interpreted as a corresponding increase in the ensemble average of the donor-acceptor distance (Dingley et al., 1999).

In contrast to N-H•••N H-bonds in Watson-Crick base pairs, the geometry of N-H•••O=C H-bonds in proteins is more varied. A survey of a number of crystallographic structures (Baker and Hubbard, 1984) showed that typical values for N•••O and H•••O distances and for N-H•••O and H•••O=C angles are 2.99 ± 0.14 Å, 2.06 ± 0.16 Å, $157 \pm 11°$, $147 + 9°$ in the case of α-helices and 2.91 ± 0.14 Å, 1.96 ± 0.16 Å, $160 \pm 10°$, $151 \pm 12°$ in the case of β-sheet conformations, respectively. Thus, on average the N•••O and H•••O distances are 0.08–0.10 Å shorter in β-sheets as compared to α-helices. This coincides with an increase in the average strength of the observed $^{h3}J_{NC'}$ coupling from -0.38 ± 0.12 Hz for α-helical conformations to -0.65 ± 0.14 Hz for β-sheets in ubiquitin (Cordier and Grzesiek, 1999). An exponential correlation between the coupling constant and the N•••O distance was described for protein G by the Bax group (Cornilescu et al., 1999b) where $^{h3}J_{NC'}$ is given as $-59*10^3$ Hz*exp$[-4*R_{NO}/Å]$. Neglecting again angular dependencies, it follows from this relation that for a typical α-helical or β-sheet conformation, a variation in the value of $^{h3}J_{NC'}$ by 0.1 Hz corresponds to a change in the N•••O distance of 0.05 or 0.07 Å, respectively.

7.2   H-Bond Angles

At present, the knowledge about the angular dependencies of the H-bond coupling constants is limited and systematic experimental data are missing. Density functional theory simulations (Scheurer and Brüschweiler, 1999) indicate that the $|^{h3}J_{NC'}|$ value has a maximum for a straight N-H•••O angle of 180° and it drops by about 20 to 30 % for a decrease in this angle to 140°. Density functional theory calculations of the H•••O=C angle dependency in the formamide dimer (Bagno, 2000) yield similar results that a more straight conformation yields larger values for $|^{h3}J_{NC'}|$. Depending on other parameters of the H-bond geometry, the maximum of the $|^{h3}J_{NC'}|$ was found at an H•••O=C angle of 150 to 170° (Bagno, 2000). This seems to be in line with experimental findings that the $^{h3}J_{NC'}$ couplings in the nucleic acid G-tetrad are weaker than the average of $^{h3}J_{NC'}$ couplings observed in proteins (Dingley et al., 2000). The principal difference in the geometry of the two systems is that the H•••O=C angle in the G-tetrad is only 125 ± 5° whereas this angle has typical values around 150° in proteins (see above). Similar observations were made for $^{h3}J_{NP}$ couplings in N-H•••O-P H-bonds (Mishima et al., 2000). In this study a $^{h3}J_{NP}$ coupling of 4.6 Hz was observed for an H•••O-P angle of 173°, but the values of $^{h3}J_{NP}$ dropped below 0.35 Hz for conformations where the H•••O-P angle was smaller than 126°.

8.   THEORY

The phenomenon of scalar couplings across H-bonds has been the subject of a large number of quantum-chemical studies aimed at an explanation of the experimental findings and at elucidating the correlations with H-bond geometry and energy as well as other NMR parameters (Bagno, 2000; Barfield et al., 2001; Benedict et al., 2000; Case, 2000; Dannenberg et al., 1999; Del Bene, 2000; Del Bene and Bartlett, 2000; Del Bene and Jordan, 2000; Del Bene et al., 2000; Dingley et al., 1999; Pecul et al., 2000; Perera and Bartlett, 2000; Scheurer and Brüschweiler, 1999; Shenderovich et al., 1998). In general, reasonable to very good agreement is found between experimentally determined H-bond scalar coupling constants and the theoretical work. Besides the computer simulations (mainly density functional theory calculations), a simple three orbital model was found to be sufficient to explain the main features of the $^{h2}J_{NN}$, $^{h1}J_{NH}$, and $^{1}J_{NH}$ couplings in the N-H•••N bond (Barfield et $al.$, 2001).

These findings clearly establish that the H-bond spin-spin coupling is caused by the same mechanism as other covalent spin-spin couplings, i.e. first a magnetic nucleus on one side of the H-bond polarizes the electron cloud in the hydrogen brigde, then this polarized electron cloud magnetically polarizes a second nucleus on the other side of the H-bond. In biomolecular H-bonds, the Fermi contact term dominates this scalar magnetization transfer (Scheurer and Brüschweiler, 1999). Therefore, the observation of the H-bond couplings implies that the movement of s-electrons on both sides of the H-bond must be correlated. As mentioned before such a correlation does not necessarily indicate an attractive contribution to the total H-bond energy and a covalent character of the hydrogen bond. A number of studies have addressed this question recently (Arnold and Oldfield, 2000; Benedict et $al.$, 2000; Ghanty et $al.$, 2000). Whereas the results of Limbach and coworkers (Benedict et $al.$, 2000) show that various hydrogen bond properties in simple A-H•••B systems can be approximated by products of valence bond orders, a recent analysis by Bader's theory of Atoms In Molecules indicates that transition from a non-covalent to a covalent character is continuous as the donor-acceptor distance decreases (Arnold and Oldfield, 2000).

9. CONCLUSION

Clearly, many of the possible applications of the H-bond couplings will be directed to the establishment of assignment as well as secondary or tertiary structure information. Although information about H-bonds in the backbone of proteins can often be derived from NOE data, the observation of an H-bond coupling yields completely unambiguous evidence about the existence of the H-bond. The exponential dependence of the coupling constant on the H-bond distance should yield very strong distance constraints in structure calculations. In contrast to the backbone H-bonds, the establishment of H-bonds involving the sidechains of glutamic and aspartic acids (Cornilescu et $al.$, 1999b), histidines (Hennig and Geierstanger, 1999) or arginines (Liu et $al.$, 2000c) by means of long-range NOEs is considerably more difficult. In all cases, the good spectral dispersion of $^{1}H_N$, ^{15}N, ^{13}C-carbonyl or ^{13}C-carboxylate resonances makes the identification of the donor and acceptor groups by H-bond J-couplings rather easy. Particular important applications may be the unambiguous establishment of H-bond networks in the active site of enzymes or ribozymes, during protein folding, as well as the detection of intermolecular H-bonds in macromolecular interactions.

Compared to proteins, nucleic acids have a lower abundance of protons. As a consequence structural information from NOEs is often more limited. Since the few long-range interactions in folded nucleic acids generally involve hydrogen bonds, the detection of the H-bond couplings not only for Watson-Crick base pairs but also for a wide variety of non-standard base pairs and to phosphates yields tertiary structure information that might not be obtainable by any other NMR parameter. In this context, it is remarkable that the $^{h2}J_{NN}$, $^{h2}J_{HP}$, and $^{h3}J_{NP}$ couplings are large and that sensitivity issues are only a minor problem. In particular, easily obtainable information from $^{h2}J_{NN}$-correlations has been used in recent nucleic acid structure determinations of DNA/RNA (Collin et al., 2000) and RNA/RNA (Kim and Tinoco, 2000) kissing complexes and of a GAAA tetraloop (Rüdisser and Tinoco, 2000) as well as for the detection of novel structural elements in DNA triads, tetrads, and hexads (Kettani et al., 2000a; Kettani et al., 2000b).

A number of applications can be imagined from a quantitative analysis of the H-bond couplings: (1) very strong couplings should be detected in systems that have been proposed as low barrier hydrogen bonds; (2) The good correlation with the proton chemical shift is particularly striking. A quantitative understanding of this phenomenon might lead to an understanding of the proton shift itself; (3) At present, it is unclear how the size of the couplings relates to the energy of the hydrogen bond. If a simple correlation exists, the J-values could be used as a direct indicator of the H-bond stability; (4) The correlation between the size of the H-bond couplings and the donor-acceptor distances is well established. A good understanding of the angular dependencies is missing at present. Changes in the size of the coupling constants should yield a measure for changes in H-bond geometry when macromolecules are subjected to different physico-chemical conditions. Indeed recent studies have detected such changes during variations of temperature in nucleic acids (Kojima et al., 2000), and pressure in proteins (Li et al., 2000), upon ligand binding (Cordier et al., 2000), and during protein folding (Jaravine et al., 2001).

9.1 Acknowledgments

We thank Profs. Juli Feigon, Michael Barfield, Linda K. Nicholson and Masatsune Kainosho for many stimulating discussions as well as Chunyu Wang, James Masse and Robert D. Peterson for the preparation of the c-Src SH3 and DNA triplex samples. A.J.D. acknowledges funding by the Australian National Health and Medical Research Council C.J. Martin

Fellowship (Regkey 987074) and by DFG grant RI 252/17-1. F.C. is a recipient of an A. v. Humboldt fellowship. This work was supported by SNF grant 31-61'757.00 to S.G.

10. REFERENCES

Abragam, A., 1961, *The Principles of Nuclear Magnetism*, Clarendon Press, Oxford.

Altman, L. J., Laungani, D., Gunnarsson, G., Wennerström, H., and Forsen, S., 1978, *J. Am. Chem. Soc.* **100**: 8264–8266.

Arnold, W. D., and Oldfield, E., 2000, *J. Am. Chem. Soc.* **122**: 12835-12841.

Asakawa, N., Kuroki, S., Kuroso, H., Ando, I., Shoji, A., and Ozaki, T., 1992, *J. Am. Chem. Soc.* **114**: 3261-3265.

Bagno, A., 2000, *Chem. Eur. J.* **6**: 2925-2930.

Baker, E. N., and Hubbard, R. E., 1984, *Prog. Biophys. Molec. Biol.* **44**: 97-179.

Barfield, M., Dingley, A. J., Feigon, J., and Grzesiek, S., 2001, *J. Am. Chem. Soc.* **123**: 4014-4022.

Bax, A., Vuister, G. W., Grzesiek, S., Delaglio, F., Wang, A. C., Tschudin, R., and Zhu, G., 1994, *Methods Enzymol.* **239**: 79-105.

Becker, E. D., (1996), Hydrogen bonding. In *Encyclopedia of Nuclear Magnetic Resonance* (Grant, D. M., and Harris, R. K., eds.), Vol. **4**, pp. 2409–2415. John Wiley, New York.

Benedict, H., Shenderovich, I. G., Malkina, O. L., Malkin, V. G., Denisov, G. S., Golubev, N. S., and Limbach, H.-H., 2000, *J. Am. Chem. Soc.* **122**: 1979-1988.

Blake, P. R., Lee, B., Summers, M. F., Adams, M. W., Park, J. B., Zhou, Z. H., and Bax, A., 1992a, *J. Biomol. NMR.* **2**: 527-533.

Blake, P. R., Park, J. B., Adams, M. W., and Summers, M. F., 1992b, *J. Am. Chem. Soc.* **114**: 4931-4933.

Bowers, P. and Klevitt, R., 2000, *J. Am. Chem. Soc.* **122**: 1030-1033.

Boyd, J., Mal, T. K., Soffe, N., and Campbell, I. D., 1997, *J. Magn. Reson.* **124**: 61–71.

Case, D. A., 2000, *Curr. Opinion Struct. Biol.* **10**: 197-203.

Collin, D., van Heijenoort, C., Boiziau, C., J.-J., T. and Guittet, E., 2000, *Nucleic Acids Res.* **28**: 3386-3391.

Cordier, F., and Grzesiek, S., 1999, *J. Am. Chem. Soc.* **121**: 1601–1602.

Cordier, F., Rogowski, M., Grzesiek, S., and Bax, A., 1999, *J. Magn. Reson.* **140**: 510-512.

Cordier, F., Wang, C., Grzesiek, S. and Nicholson, L. K., 2000, *J. Mol. Biol.* **304**: 497-505.

Cornilescu, G., Hu, J.-S., and Bax, A., 1999a, *J. Am. Chem. Soc.* **121**: 2949-2950.

Cornilescu, G., Ramirez, B. E., Frank, M. K., Clore, G. M., Gronenborn, A. M., and Bax, A., 1999b, *J. Am. Chem. Soc.* **121**: 6275-6279.

Crabtree, R., Siegbahn, P., Eisenstein, O., Rheingold, A., and Koetzle, T., 1996, *Acc. Chem. Res.* **29**: 348-354.

Dannenberg, J. J., Haskamp, L., and Masunov, A., 1999, *J. Phys. Chem. A* **103**: 7083-7086.

Del Bene, J. E., 2000, *J. Am. Chem. Soc.* **122**: 3560-3561.

Del Bene, J. E., and Bartlett, R. J., 2000, *J. Am. Chem. Soc.* **122**: 10480-10481.

Del Bene, J. E., and Jordan, M. J. T., 2000, *J. Am. Chem. Soc.* **122**: 4794-4797.

Del Bene, J. E., Perera, S. A., Bartlett, R. J., Alkorta, I., and Elguero, J., 2000, *J. Phys. Chem. A* **104**: 7165-7166.

Delaglio, F., Grzesiek, S., Vuister, G. W., Zhu, G., Pfeifer, J. and Bax, A., 1995, *J. Biomol. NMR.* **6**: 277-293.

Dingley, A., and Grzesiek, S., 1998, *J. Am. Chem. Soc.* **120**: 8293-8297.

Dingley, A., Masse, J., Feigon, F., and Grzesiek, S., 2000, *J. Biomol. NMR.* **16**: 279-289.

Dingley, A. J., Masse, J. E., Peterson, R. D., Barfield, M., Feigon, J., and Grzesiek, S., 1999, *J. Am. Chem. Soc.* **121**: 6019-6027.

Dunger, A., Limbach, H.-H., and Weisz, K., 2000, *J. Am. Chem. Soc.* **122**: 10109-10114.

Fersht, A., 1985, *Enzyme Structure and Mechanism.* Second Edition edit, W. H. Freeman, New York.

Ghanty, T. K., Staroverov, V. N., Koren, P. R., and Davidson, E. R., 2000, *J. Am. Chem. Soc.* **122**: 1210-1214.

Golubev, N. S., Shenderovich, I. G., Smirnov, S. N., Denisov, G. S., and Limbach, H.-H., 1999, *Chem. Eur. J.* **5**: 492-497.

Grzesiek, S., and Bax, A., 1993, *J. Am. Chem. Soc.* **115**: 12593-12594.

Gunnarsson, G., Wennerström, H., Ega, W., and Forsen, S., 1976, *Chem. Phys. Lett.* **38**: 96-99.

Hennig, M., and Geierstanger, B., 1999, *J. Am. Chem. Soc.* **121**: 5123-5126.

Hennig, M., and Williamson, J., 2000, *Nucleic Acids Res.* **28**: 1585-1593.

Hu, J., Facelli, J., Alderman, D., Pugmire, R., and Grant, D., 1998, *J. Am. Chem. Soc.* **120**: 9863-9869.

Huggins, M., 1919, Thesis, University of California.

Hvidt, A., and Nielsen, S., 1966, *Adv. Prot. Chem.* **21**: 287-385.

Ippel, J. H., Wijmenga, S. S., de Jong, R., Heus, H. A., Hilbers, C. W., de Vroom, E., van der Marcel, G. A. and van Boom, J. H., 1996, *Magn. Reson. Chem.* **34**: S156-S176.

Jaravine, V. A., Alexandrescu, A. T., and Grzesiek, S., 2001, *Protein Sci.* **10**: 943-50.

Jeffrey, G. A., and Saenger, W., 1991, *Hydrogen Bonding in Biological Structures*, Springer, New York.

Juranic, N., Ilich, P. K., and Macura, S., 1995, *J. Am. Chem. Soc.* **117**: 405-10.

Juranic, N., Likic, V. A., Prendergast, F. G., and Macura, S., 1996, *J. Am. Chem. Soc.* **118**: 7859-7860.

Kainosho, M. and Nakamura, A., 1969, *Tetrahedron* **25**: 4071-4081.

Kettani, A., Basu, G., Gorin, A., Majumdar, A., Skripkin, E., and Patel, D. J., 2000a, *J. Mol. Biol.* **301**: 129-146.

Kettani, A., Gorin, A., Majumdar, A., Hermann, T., Skripkin, E., Zhao, H., Jones, R., and Patel, D. J., 2000b, *J. Mol. Biol.* **297**: 627-644.

Kim, C.-H., and Tinoco, I., 2000, *Proc. Nat'l. Acad. Sci. USA.* **97**: 9396-9401.

Kojima, C., Ono, A., and Kainosho, M., 2000, *J. Biomol. NMR.* **18**: 269-77.

Kontaxis, G., Clore, G. M. and Bax, A., 2000, *J. Magn. Reson.* **143**: 184-196.

Kuntz, I. D., Kosen, P. A., and Craig, E. C., 1991, *J. Am. Chem. Soc.* **113**: 1406-1408.

Kwon, O., and Danishefsky, S., 1998, *J. Am. Chem. Soc.* **120**: 1588-1599.

Latimer, W. M., and Rodebush, W. H., 1920, *J. Am. Chem. Soc.* **42**: 1419-1433.

Li, H., Yamada, H., Akasaka, K., and Gronenborn, A. M., 2000, *J. Biomol. NMR* **18**: 207-16.

Liu, A., Hu, W., Qamar, S., and Majumdar, A., 2000a, *J. Biomol. NMR.* **17**: 55-61.

Liu, A., Majumdar, A., Hu, W., Kettani, A., Skripkin, E., and Patel, D. J., 2000b, *J. Am. Chem. Soc.* **122**: 3206-3210.

Liu, A., Majumdar, A., Jiang, F., Chernichenko, N., Skripin, E., and Patel, D., 2000c, *J. Am. Chem. Soc.* **122**:11226-11227.

LiWang, A. C., and Bax, A., 1996, *J. Am. Chem. Soc.* **118**: 12864-12865.

LiWang, A. C., and Bax, A., 1997, *J. Magn. Reson.* **127**: 54-64.

Loh, S., and Markley, J., 1994, *Biochemistry* **33**: 1029-1036.

Löhr, F., Mayhew, S. G., and Rüterjans, H., 2000, *J. Am. Chem. Soc.* **122**: 9289-9295.

Luy, B., and Marino, J., 2000, *J. Am. Chem. Soc.* **122**: 8095-8096.

Majumdar, A., Kettani, A., and Skripkin, E., 1999a, *J. Biomol. NMR* **14**: 67-70.

Majumdar, A., Kettani, A., Skripkin, E., and Patel, D. J., 1999b, *J. Biomol. NMR* **15**: 207-11.

Markowski, V., Sullivan, G. R., and Roberts, J. D., 1977, *J. Am. Chem. Soc.* **99**: 714–718.

McDermott, A., and Ridenour, C. I., (1996), In *Encyclopedia of Nuclear Magnetic Resonance* (Grant, D. M. and Harris, R. K., eds.), pp. 3820-3825.

Meissner, A., and Sørensen, O. W., 2000a, *J. Magn. Reson.* **143**: 431-434.

Meissner, A., and Sørensen, O. W., 2000b, *J. Magn. Reson.* **143**: 387-390.

Mishima, M., Hatanaka, M., Yokoyama, S., Ikegami, T., Wälchli, M., Ito, Y. and Shirakawa, M., 2000, *J. Am. Chem. Soc.* **122**: 5883–5884.

Niimura, N., 1999, *Curr. Opin. Struct. Biol.* **9**: 602-8.

Pauling, L., 1960, The Nature of the Chemical Bond, Cornell University Press.

Pecul, M., Leszczynski, J., and Sadlej, J., 2000, *J. Phys. Chem. A* **104**: 8105-8113.

Perera, S., and Bartlett, R., 2000, *J. Am. Chem. Soc.* **122**: 1231-1232.

Pervushin, K., Fernandez, C., Riek, R., Ono, A., Kainosho, M., and Wuthrich, K., 2000, *J. Biomol. NMR.* **16**: 39-46.

Pervushin, K., Ono, A., Fernandez, C., Szyperski, T., Kainosho, M., and Wuethrich, K., 1998, *Proc. Natl. Acad. Sci U.S.A.* **95:** 14147-14151.

Pervushin, K., Riek, R., Wider, G. and Wüthrich, K., 1997, *Proc. Natl. Acad. Sci U.S.A.* **94**: 12366–12371.

Petrakis, L., and Sederholm, C., 1961, *J. Chem. Phys.* **35**: 1243-1248.

Piotto, M., Saudek, V., and Sklenar, V., 1992, *J. Biomol. NMR* **2**: 661-665.

Rüdisser, S., and Tinoco, I., 2000, *J. Mol. Biol.* **295**: 1211-1223.

Salzmann, M., Pervushin, K., Wider, G., Senn, H., and Wuthrich, K., 1998, *Proc. Natl. Acad. Sci. USA.* **95**: 13585-90.

Scheurer, C., and Brüschweiler, R., 1999, *J. Am. Chem. Soc.* **121**: 8661-8662.

Shenderovich, I. G., Smirnov, S. N., Denisov, G. S., Gindin, V. A., Golubev, N. S., Dunger, A., Reibke, R., Kirpekar, S., Malkina, O. L., and Limbach, H.-H., 1998, *Ber. Bunsenges. Phys. Chem.* **102**: 422 428.

Shoup, R. R., Miles, H. T., and Becker, E. D., 1966, *Biochem. Biophys. Res. Commun.* **23**. 194-201.

Takahashi, A., Kuroki, S., Ando, I., Ozaki, T., and Shoji, A., 1998, *J. Mol. Struct.* **442**: 195-199.

Tessari, M., Vis, H., Boelens, R., Kaptein, R., and Vuister, G. W., 1997, *J. Am. Chem. Soc.* **119**: 8985–8990.

Tjandra, N., and Bax, A., 1997, *J. Am. Chem. Soc.* **119**: 8076–8082.

Vuister, G., Tessari, M., Karimi-Nejad, Y., and Whitehead, B., (1998), *Modern Techniques in Protein NMR.* In *Biological Magnetic Resonance* (Krishna, N. R., and Berliner, L. J, eds.), Vol. **16**, pp. 195-257. Kluwer Academic.

Wagner, G., 1983, *Q. Rev. Biophys.* **16**: 1–57.

Wagner, G., Pardi, A. and Wuthrich, K., 1983, *J. Am. Chem. Soc.* **105**: 5948-9.

Wang, Y. X., Jacob, J., Cordier, F., Wingfield, P., Stahl, S. J., Lee-Huang, S., Torchia, D., Grzesiek, S., and Bax, A., 1999, *J. Biomol. NMR* **14**: 181-184.

Wishart, D. S., Sykes, B. D., and Richards, F. M., 1991, *J. Mol. Biol.* **222**: 311-333.

Wöhnert, J., Dingley, A. J., Stoldt, M., Görlach, M., Grzesiek, S., and Brown, L. R., 1999, *Nucleic Acids Res.* **27**: 3104-10.

Yan, X., Kong, X., Xia, Y., Sze, K. and Zhu, G., 2000, *J. Magn. Reson.* **147**: 357-360.

Zhou, N. E., Zhu, B.-Y., Sykes, B. D., and Hodges, R. S., 1992, *J. Am. Chem. Soc.* **114**: 4320-4326.

III

NMR Methods for Screening Bioactive Ligands

Chapter 10

NMR Methods for Screening the Binding of Ligands to Proteins – Identification and Characterization of Bioactive Ligands

Thomas Peters, Thorsten Biet and Lars Herfurth
Institute of Chemistry, Medical University of Luebeck, Luebeck, Germany

Abstract   It has been shown during recent years that NMR provides powerful tools for the screening and identification of bioactive ligand molecules. Depending on the experimental parameters employed, these novel techniques detect and characterize the binding of ligands to proteins via the observation of either protein resonances, or the ligand signals. Obviously, an intelligent combination of different experiments greatly enhances the options for screening compound libraries. This chapter summarizes the recent developments in the field.

1.   INTRODUCTION

NMR experiments have a large potential to study binding or recognition reactions that occur between molecules. Certainly, the detection of binding of low-molecular weight compounds to large receptor molecules, mostly proteins, plays a prominent role. The design of new drugs initially relies on the detection of binding activity of substances for target proteins. The intention of this chapter is to survey and illustrate NMR techniques that allow to identify and characterize the binding of low-molecular weight compounds to receptor proteins. Topics such as protein-protein interactions though clearly important will not be covered in detail here. Several articles are available which review the use of NMR experiments for the detection of binding activity highlighting different experimental approaches (Stockman, 1998; Moore, 1999; Roberts, 1999; Roberts, 2000; Lepre, 2001). Here, we would like to emphasize more practical aspects of the different approaches by discussing selected examples.

Compared to other screening techniques such as RIA, EIA or fluorescence based assays two main advantages of NMR are that (1) the assay system is essentially undisturbed, i.e. no immobilization or attachment of chemical labels is necessary and that (2) not only binding is detected but also atomic details of the binding process become available. In general, NMR may be used at different levels of complexity when investigating the binding of ligands to proteins. Depending on the particular type of experiment, NMR delivers, e.g. bound conformations of ligands, topologies of receptor sites, or binding epitopes of ligand molecules. Therefore, in comparison to brute-force HTS protocols NMR issues valuable information that significantly can speed up the drug-discovery process, as this will be shown for several examples below.

To detect and characterize binding of ligands to proteins any NMR parameter that changes upon binding is useful to observe. E.g., chemical shifts, NOEs, saturation transfer, scalar and dipolar coupling constants, spin-lattice and spin-spin relaxation times, and cross-correlated relaxation have been employed to follow the binding of ligands to proteins. Two major experimental approaches can be identified, in which either ligand resonances or receptor protein resonances are observed. The information obtained from the two strategies is complementary, and leads to a detailed picture of the binding mechanism.

2.   METHODS BASED ON THE OBSERVATION OF PROTEIN SIGNALS

Upon binding of a ligand to a receptor protein, chemical shifts of nuclei located in the protein binding pocket will be affected. Of course, the most sensitive nuclei to observe are protons. This has been used in the past to follow the binding of a ligand to a protein via observation of well separated His residues in the protein binding pocket (Feeney *et al.*, 1979; Birdsall *et al.*, 1984; Lian and Roberts, 1993). Theoretically, a complete ^{1}H chemical shift map can be obtained from a 2D homonuclear NMR experiment such as a TOCSY experiment, and spectra with and without ligand present may be compared. In practice, an unambiguous assignment of corresponding chemical shift changes upon addition of a ligand would be very difficult because of severe signal overlap, and moreover it would be impossible to identify the amino acids involved in protein-ligand interactions. It follows, that in order assign ^{1}H chemical shift changes that occur upon binding it is of advantage to employ heteronuclear correlation NMR experiments that lead to a deconvolution of the ^{1}H NMR signals.

2.1 ^{1}H/^{15}N-HSQC Experiments

In principle, ^{1}H/^{15}N-HSQC NMR experiments have been known to be useful to study the binding of ligands to proteins. For instance, ^{1}H/^{15}N-HSQC experiments were known to provide an effective method to investigate interactions of arginine guanidino groups with charged groups placed on the ligand (Pascal et al., 1995; Yamazaki et al., 1995; Feng et al., 1996; Gargaro et al., 1996; Nieto et al., 1997; Morgan et al., 1999). However, the value of this chemical shift perturbation technique for the screening of compound libraries was only discovered more recently. The first application of ^{1}H/^{15}N-HSQC experiments to screen ligands for binding activity was demonstrated for the FK506-binding protein FKBP (Shuker et al., 1996). In this study a linked-fragment approach was introduced, termed SAR by NMR, which will be explained below. When complexed to FK506 (Fig. 1), FKBP binds tightly to calcineurin that plays a role in the regulation of a variety of transcription factors. As a result of the binding to calcineurin significant immuno suppression is observed. Therefore, FK506 is an important drug that for instance suppresses rejection reactions after organ transplants. Many current attempts aim at the development of drugs that maintain the FKBP binding activity but at the same time lack the strong toxicity of FK506.

First, a compound library of approximately 1000 substances was screened for binding activity. A solution containing uniformly ^{15}N,^{13}C-labeled FKBP at a concentration of 2 mM was used for the HSQC-based binding experiments. From the library, a candidate (compound **2** in Fig. 1) with a K_D value of 2 μM was identified. In order to improve the binding activity, the authors applied a linked-fragment approach. In a second screen, a complex of FKBP and the pipecolinic acid derivative **2**, both at a concentration of 2 mM, was used to screen for ligands that would bind to a second site nearby to the first binding site. A benzanilide derivative **3** was identified that bound to FKBP with a K_D of 100 μM. The ^{1}H/^{15}N-HSQC spectra of FKBP in the presence of saturating amounts of **2** and in the absence and presence of compound **3** are shown in Fig. 2. The chemical shift changes indicate that compound **3** binds to a second site not competing with the binding of **2** to the first site. In order to generate a higher affinity ligand, the two fragments **2** and **3** were chemically linked. Five different linked molecules were synthesized from which compound **10** had the highest binding affinity for FKBP (K_D = 19 nM). The complex of FKBP and **10** was then subjected to a detailed structural analysis using NMR.

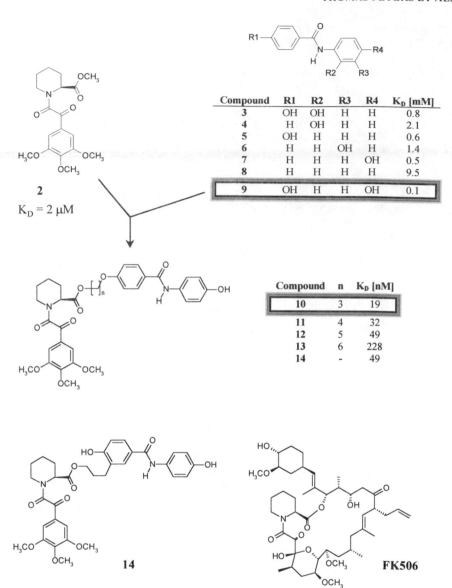

Compound	R1	R2	R3	R4	K_D [mM]
3	OH	OH	H	H	0.8
4	H	OH	H	H	2.1
5	OH	H	H	H	0.6
6	H	H	OH	H	1.4
7	H	H	H	OH	0.5
8	H	H	H	H	9.5
9	OH	H	H	OH	0.1

2

$K_D = 2\ \mu M$

Compound	n	K_D [nM]
10	3	19
11	4	32
12	5	49
13	6	228
14	-	49

14

FK506

Figure 1. Compounds used in $^1H/^{15}N$ -HSQC binding experiments by Shuker *et al.*, 1996 (adapted with permission from Shuker, S.B., Hajduk, P.J., Meadows, R.P., and Fesik, S.W., 1996, *Science* **274**, 1531-1534. Copyright 1996 American Association for the Advancement of Science).

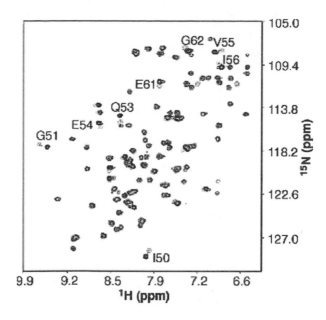

Figure 2. A superposition of $^1H/^{15}N$-HSQC spectra for FKBP in the absence (grey contours) and presence (black contours) of compound **3**. Both spectra were acquired in the presence of saturating amounts of **2** (2.0 mM). Significant chemical shifts changes are observed for labeled residues (reprinted with permission from Shuker, S.B., Hajduk, P.J., Meadows, R.P., and Fesik, S.W., 1996, *Science* 274, 1531-1534. Copyright 1996 American Association for the Advancement of Science).

SAR by NMR follows five main steps: 1. Identification of ligands with high binding affinity from a library of compounds utilizing $^1H/^{15}N$-HSQC experiments; 2. Optimization of ligands by chemical modification; 3. Identification of ligands binding in the presence of saturating amounts of optimized ligands from step 2 utilizing $^1H/^{15}N$-HSQC experiments; 4. Optimization of ligands for the second site; 5. Linking the two ligands for the primary and the secondary binding sites. This process is illustrated in Fig. 3. In general, SAR by NMR cannot uncover the identity of a compound with binding activity, if a mixture of compounds is present. Therefore, after having tested positive, a mixture has to be deconvoluted. This can be achieved either by testing each individual component of a sample – if the components are known – or by applying one of the techniques that are based on the observation of ligand instead of protein signals, as this will be shown below.

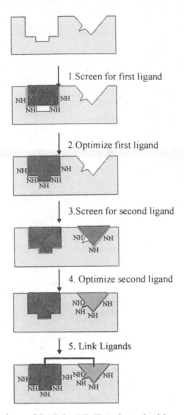

Figure 3. Schematic overview of SAR by NMR (adapted with permission from Shuker, S.B., Hajduk, P.J., Meadows, R.P., and Fesik, S.W., 1996, *Science* **274**, 1531-1534. Copyright 1996 American Association for the Advancement of Science).

If enough labeled protein is available $^{1}H/^{15}N$-HSQC spectra can be obtained in a few minutes. Therefore, for samples that contain receptor protein in the presence of e.g. ten different compounds it is readily estimated that 100 to 200 compounds could be screened per hour. Increasing the number of potential ligands per sample clearly increases the amount of compounds that can be tested for binding activity per time unit. On the other hand, apart from potential solubility problems adding more compounds to each sample makes it more time consuming to identify the compound that has binding activity after a positive answer, i.e. a shift of $^{15}N/^{1}H$ cross peaks, has been identified. It is obvious, that the sample size is dependent on individual criteria that have to be evaluated before a screening process is initiated. It is important to realize the amounts of protein and respective ligands that are required for SAR by NMR. If, e.g. a protein of molecular weight of 15 kDa is subjected to screening, 7.5 mg of protein will be necessary to produce 0.5 mL of a 1 mM NMR sample. It follows, that for

100 samples 750 mg of protein has to be available. Certainly, protein can be recovered via dialysis after having completed the measurement but in order to generate considerable throughput, or in order to get into the limits of HTS with more than 5.000 compounds tested per day it is necessary to have the ^{15}N labeled protein available in gram quantities. As will be shown below, the sensitivity of SAR by NMR can be increased utilizing several improvements.

Since its first application to FKBP, SAR by NMR has been used for the design of high-affinity ligands for several other proteins of therapeutic relevance as this is compiled in Table 1.

Table 1: Examples where SAR by NMR has been used to identify ligands with binding affinity.

Protein	Ligand	Reference
Stromelysin	nonpeptide inhibitors	Hajduk et al., 1997a
Stromelysin	Optimization of biaryl derivatives	Olejniczak et al., 1997
E2	Biphenyl ether derivatives	Hajduk et al., 1997b
FKBP12	C24-Deoxyascomycin	Wiedemann et al., 1999
ErmAM, ErmC'	2-Amino-4-piperidinyl-1,3,5-triazines	Hajduk et al., 1999b
Urokinase	2-Aminobenzimidazoles	Hajduk et al., 2000a
Adenosine Kinase	Optimization of pyridopyrimidine	Hajduk et al., 2000b

Methodologically, several improvements of the method were made in the mean time. Before the introduction of TROSY (Pervushin et al., 1997; see also the chapter by Pervushin) ^{1}H/^{15}N-HSQC screening was limited to proteins with a molecular weight of less than ca. 20 kDa in size. With TROSY much larger receptor proteins may be targeted (Hajduk et al., 2000c) which will be discussed in 2.2 below. Although larger proteins become tractable by utilizing TROSY experiments problems remain with the tremendous number of cross peaks that have to be assigned. Strategies to achieve spectroscopic assignment as well as structural analysis of large proteins are known but require significantly more effort and time.

As a simple and practical alternative, individual ^{15}N labeled amino acids can be incorporated into a protein (Wikstroem, 2000; Kainosho, 1982). By labeling e.g. Ala residues only, assignment problems are substantially reduced. In order to completely map a binding site on the protein surface this approach requires the preparation of several different ^{15}N-labeled proteins.

A significant increase in sensitivity is achieved by applying cryo-probe technology (Hajduk et al., 1999a). It has been shown that in this case the concentration of the ^{15}N labeled protein target can be reduced to 50 μM thus allowing an increase in the number of potential ligands that can be tested simultaneously. Using stromelysin as a target protein it was demonstrated that testing 100 compounds in one sample resulting in a 5 mM overall ligand concentration is possible without difficulties. Deconvolution of samples that display binding activity is achieved by first testing samples of ten compounds each, and then testing each individual component in the active sample of ten. With conventional probe technology, ca. 10 compounds can be tested simultaneously. It follows, that a 10 fold increase in speed can be achieved by using cryoprobes and it has been estimated that ca. 200.000 compounds can be tested per month including deconvolution of active libraries (Hajduk, 1999a).

Docking algorithms may be used to suggest a structure of a protein-ligand complex, but a subsequent NMR or X-ray structural analysis is necessary to unambiguously orient the ligand in the binding site. An interesting alternative to this approach is provided by HSQC-based screening of libraries of structurally related ligands binding to the same protein (Medek et al., 2000). It is argued that if a variety of structurally similar ligands is available, it should be possible to define the orientation of the ligands in the binding pocket because differences in the HSQC chemical shift maps can be directly related to the portion of the ligand that has been altered. Utilizing this principle it was possible to analyze the orientation of ascomycin, a FK506 analogue, when bound to FKBP (Medek et al., 2000). Similarity between ligands is not only useful for the analysis of ligand orientations in protein binding pockets. For instance, a statistical analysis of binding data on 11 protein targets has identified molecular motifs that are preferred for protein binding (Hajduk et al., 2000d). Following this study, the biphenyl element is one such structural building block that delivers binding activity whilst at the same time achieving significant binding specificity.

2.2 $^1H/^{13}C$-HSQC Experiments

In general, labeling of proteins with ^{13}C is more expensive than with ^{15}N. Also, $^1H/^{13}C$-HSQC spectra are more complex than $^1H/^{15}N$-HSQC spectra. Therefore, $^1H/^{13}C$-HSQC experiments have not widely been employed to test for binding. To overcome the shortcomings of $^1H/^{13}C$-HSQC for screening a labeling technique was suggested that allows to selectively label the δ1 protons of the methyl groups of valine, leucine, and isoleucine only (Hajduk et al., 2000c). This has one advantage that the complexity of resulting HSQC spectra is significantly reduced. At the same time the presence of three protons in methyl groups vs. a single proton in an NH group increases the sensitivity, and the favorable relaxation properties of methyl groups also allows to apply HSQC-based screening to larger proteins also. The labeling protocol was based on a previous labeling strategy where fully ^{13}C-labeled α-ketobutyrate and α-ketoisovalerate was used for the incorporation of protonated methyl groups for valine, leucine, and isoleucine into perdeuterated, uniformly [13]C- and ^{15}N-labeled proteins (Gardner, *et al.*, 1997; Gardner *et al.*, 1998; Goto *et al.*, 1999). A synthetic route was devised to produce [3-^{13}C]-α–ketobutyrate and [3,3'-^{13}C]-α-ketoisovalerate utilizing ^{13}C-methyl iodide as ^{13}C source. Proteins with molecular weights ranging from 12 to 110 kDa were subjected to ^{1}H/ ^{13}C-HSQC screening.

For FKBP (12 kDa) and Bcl-xL (19 kDa) it was found that the use of ^{1}H/^{13}C-HSQC experiments vs ^{1}H/^{15}N-HSQC experiments results in a nearly 3-fold higher sensitivity which is understandable given the relative sensitivity of the nuclei. Therefore, throughput rates are possible that for ^{15}N-labeled samples can only be achieved using cryo-probe technology (Hajduk *et al.*, 1999a). In Fig. 4 ^{1}H/^{15}N- and ^{1}H/^{13}C-HSQC spectra of a 50μM sample of [13]C(methyl)/U-^{15}N-labeled Bcl-xL are compared. It is obvious that the ^{1}H/^{13}C-HSQC spectrum is of superior quality.

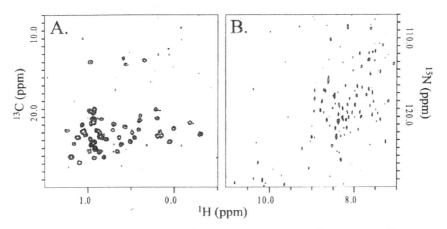

Figure 4. (A) ^{1}H/^{13}C-HSQC and (B) ^{1}H/^{15}N-HSQC spectra of [13]C(methyl)/U-^{15}N-labeled Bcl-xL (19 kDa) at 500 MHz. Reprinted with permission from Hajduk *et al.*, 2000c. Copyright 2000 American Chemical Society.

For larger proteins (MW > 40 kDa) MBP (42 kDa) and DHNA (110 kDa) served as examples. It was shown that acceptable throughput rates can only be achieved if the protein is perdeuterated. Compared to ^{1}H/^{15}N-TROSY [13]C(methyl)-HSQC turned out to be much more sensitive, as this is evident from Fig. 5 that compares a ^{1}H/^{13}C-HSQC to a ^{1}H/^{15}N-TROSY spectrum of a 50 μM sample of [13]C(methyl)/U-^{15}N,[2]H-labeled MBP. It can be envisioned that the use of ^{1}H/^{13}C-TROSY experiments will yield even better results. One general advantage of using ^{1}H/^{13}C-HSQC is that deuterated buffers are not necessary because no H/D exchange can occur.

The applicability of the ^{13}C(methyl)-labeling strategy clearly depends on the presence of valine, leucine, or isoleucine residues in the active site of the protein under investigation. A statistical analysis of 191 crystal structures of proteins bound to ligand molecules shows that 92% of the ligands were found to have a heavy atom within 6 Å of at least one methyl carbon of valine, leucine, or isoleucine as compared to 82% of the ligands within 6 Å of at least one backbone nitrogen.

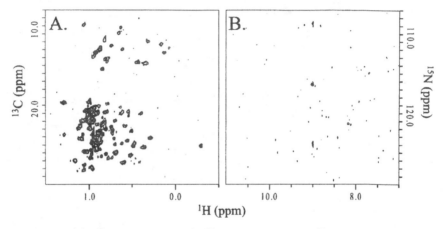

Figure 5. (A) ^{1}H/^{13}C-HSQC and (B) ^{1}H/^{15}N-TROSY spectra of ^{13}C(methyl)/U-^{15}N-labeled MBP (42kDa) at 800 MHz. Reprinted with permission from Hajduk *et al.*, 2000c. Copyright 2000 American Chemical Society.

2.3   Cross-Saturation Experiments

Instead of monitoring differences in chemical shifts upon binding of ligands to a receptor protein it is possible to use experiments that are based on saturation transfer (Mayer and Meyer, 1999a; see also 3.1). It was demonstrated that this method is also useful for mapping the sites of interaction in protein-protein complexes (Takahashi *et al.*, 2000; Ramos *et al.*, 2000). This cross-saturation mapping of protein-protein interfaces utilizes the steady-state NOE-difference experiment (Neuhaus and Williamson, 2000) where ^{1}H resonances of a nondeuterated component (target resonances) are saturated leading to the observation of magnetization of amide protons of a second deuterated and uniformly ^{15}N-labeled protein (reporter resonances). The reason for deuteration is twofold: (1) a spectral window has to be present to selectively irradiate target resonances, and (2) spin-diffusion on the side of the reporter protein is reduced.

Since deuteration of a protein is costly and not always feasible, for RNA-protein complexes an elegant extension of this method was suggested (Ramos *et al.*, 2000). By using the spectral window of the H1' sugar resonances and H5 base resonances of the nucleic acid centered at ~ 6 ppm, or the imino protons with chemical shifts in excess of 12 ppm it was possible to generate cross-saturation ^{1}H/^{15}N-HSQC spectra. For a complex of a multidomain RNA-binding protein, Nova1 KH3, cross-saturation experiments were performed in the presence of a 15 nucleotide RNA target. The binding site for the RNA was mapped, and the mapping was compared to the chemical shift perturbation method using ^{1}H/^{15}N-HSQC spectra and to X-ray data for the protein-RNA complex. The result from cross-saturation mapping is very similar to the one from the chemical shift perturbation experiments with the latter giving a slightly too large binding site as compared to the X-ray data. This suggests that the combined use of chemical shift perturbation and cross-saturation data allows the identification of interaction surfaces with greater confidence. The RNA-binding sites as identified by three different methods are depicted in Fig. 6.

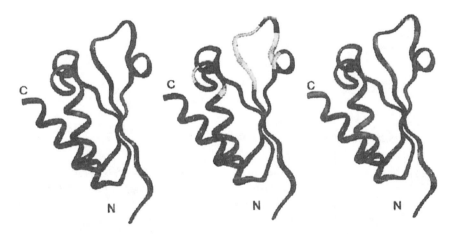

Figure 6. Structural homology model of Nova1 KH3 protein (Swissprot) showing the protein backbone as a ribbon. Three different methods of identifying surface residues are compared. Left: residues reported to make direct contact in the crystal structure of a closely related complex (grey). Center: residues identified using cross saturation experiments (light grey). Right: chemical shift perturbation (grey). Reprinted with permission from Ramos *et al.*, 2000. Copyright 2000 American Chemical Society.

3.   METHODS BASED ON THE OBSERVATION OF LIGAND SIGNALS

Methods that detect changes in ligand spectral properties upon binding to a receptor protein are complementary to the techniques discussed above. In principle, almost any spectral property can be utilized to screen libraries for binding activity, and to further characterize the binding process. Some methods are more practical than others, of course, and the information obtained can vary greatly. Before describing the details of individual techniques we would like to summarize the most important criteria that determine the choice of the method. One of the most important questions certainly is: How much protein is required for the measurement? The time for an individual measurement is another substantial criterion, especially since some biologically very interesting proteins are rather labile. One also needs to know how many ligands can be screened at the same time in one sample, and which kind of information is delivered by the experiment. Almost all techniques work without the requirement of isotope labeling, neither for the protein nor for the ligands. Also, most methods are not limited by the size of the protein, but especially for larger receptor proteins it becomes of prime importance that the experiment delivers unambiguous results that are independent from boundary conditions such as concentration, temperature etc.. Taking all factors into consideration, saturation transfer difference (STD) NMR experiments are by far the most robust and reliable experiments that can be performed to detect binding. Therefore, we will describe these experiments first.

3.1   Saturation Transfer Difference (STD) NMR Experiments

Saturation transfer experiments had been used for many years to study for example the kinetics of exchange reactions. Also, saturation transfer was applied to investigate the binding of carbohydrates to carbohydrate binding proteins (Akasaka, 1979; Poppe *et al.*, 1997), but the true potential of this approach was not realized until recently when it was shown that saturation transfer difference (STD) NMR is well suited to screen compound libraries for binding activity (Mayer and Meyer, 1999a). In addition, it was shown that the STD principle may be combined with any NMR pulse sequence generating a whole suite of STD NMR experiments such as STD TOCSY or STD HSQC (Mayer and Meyer, 1999a; Vogtherr and Peters, 2000).

In its most basic form the experiment applies the steady state NOE difference pulse sequence, and therefore is identical with the cross-saturation NMR experiments described above. The experiment was first applied to screen a library of carbohydrate molecules for binding activity towards a carbohydrate binding protein, wheat germ agglutinin (WGA). At the same time it was shown that STD NMR is an excellent technique to also determine the binding epitope of the ligand, information that is of prime importance for the directed development of drugs (Mayer and Meyer, 1999a).

The experiment is performed as follows. Ligands are added to a solution of the receptor protein and one ^{1}H NMR experiment is performed where a spectral region is selectively irradiated that contains no ligand but only protein signals. Usually such regions are easily identified depending on the chemical nature of the ligands. In the present case of a mixture of carbohydrate molecules the region of aromatic protons around 7 ppm would be suitable for selective irradiation but it is also possible to utilize the fact that for large receptor proteins the lines are very broad, and irradiation at a frequency of e.g. -3 ppm still yields sufficient saturation of the protein. The irradiation is performed during the relaxation time of the NMR experiments, and via spin diffusion leads to saturation of all protein signals that increases with time. If a ligand binds to the protein it will also be saturated. The degree of saturation depends on the residence time of the ligand in the protein binding pocket. In a second experiment the irradiation frequency is set at a value that is far from any signal, e.g. 40 ppm. Subtraction of these so called on- and off-resonance experiments leads to a difference spectrum, in which only signals are visible that result from signal attenuation via saturation transfer. Clearly, if the binding is very tight, and consequentially off rates are in the range of several seconds, saturation transfer of ligand molecules is not very efficient. This is usually the case for nanomolar and smaller K_D values. If the K_D values are in the range of several 100 nanomolar or larger fast exchange of free and bound ligands leads to a very efficient transfer of saturation onto the free ligand molecules. Consequentially, in the difference spectra only signals of those ligands are observed that bind to the protein. The STD principle is visualized in Fig. 7.

The intensity of the STD signals depends on the irradiation time and on the excess of ligand molecules used. Fig. 8 shows the dependence of STD signals as a function of the irradiation time and as a function of excess of ligand used. The more ligand is used and the longer the irradiation time the stronger is the STD signal. It is seen that both curves in Fig. 8 asymptotically approach a maximum value. In general, an irradiation time of 2 s and a 100-fold excess of ligand gives good results. From the high ligand to protein ratio it is clear that the amount of protein required for the measurements is very small. With a probe of 800:1 S/N at 500 MHz an amount of 1 to 10 nmol of protein is sufficient to record STD spectra. At a molecular weight of 50 kDa this translates into 50 to 100 µg. This clearly underlines one of the main strengths of the method. A complete relaxation and conformational exchange matrix (CORCEMA) theory for the quantitative analysis of STD intensities has been presented recently (Jayalakshmi and Krishna, 2002).

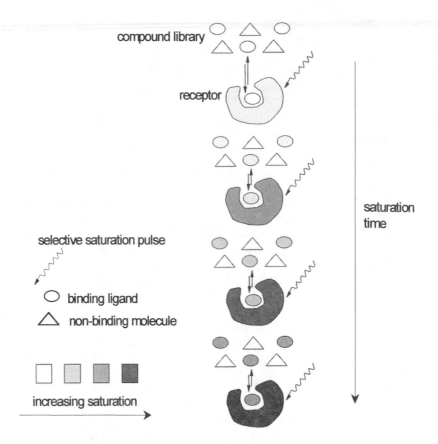

Figure 7. Principle of STD NMR. The receptor protein is saturated with a selective saturation pulse. In general, the saturation pulse consists of a cascade of gaussian shaped pulses. The duration of saturation times ranges from 1 to 2 s. The ligand is used in a ca. 100 fold molar excess over the protein, allowing to work with ☐M protein concentrations. Adapted with permission from Mayer and Meyer, 1999a. Copyright 1999 Wiley-VCH.

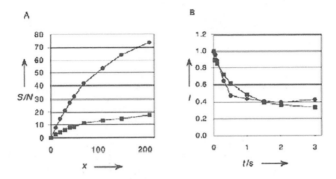

Figure 8. (A) Signal-to-noise ratio (S/N) of STD difference NMR spectra at 500 MHz as a function of molar excess (x) of ligand (GlcNAc, circles: *N*-acetyl group, squares: H1) over protein (WGA). effect as a function of excess ligand. (B) Integral values (I) of selected ^{1}H-NMR signals (circles: *O*-methyl group of FucOMe, squares: H6-methyl group of FucOMe) as a function of saturation time (t). The ligand was *O*-methyl-α-L-fucose (FucOMe), the protein was *Aleuria aurantia* agglutinin (AAA). FucOMe was used with 30 fold molar excess over AAA. It is obvious that the decrease of intensity levels out at about 60% of the original intensity. Reprinted with permission from Mayer and Meyer, 1999a. Copyright 1999 Wiley-VCH.

Using 1D STD spectra the compound *N*-acetylglucosamine (GlcNAc) was identified as the only one with binding affinity for WGA. All other molecules showed no response in the STD spectra. As mentioned above the STD principle can be combined with any NMR pulse sequence, and one rather powerful experiment is the STD TOCSY experiment. Especially in cases where the library is more complex the additional deconvolution of signals brought about by the second dimension is very helpful. The STD TOCSY experiment was also shown to be ideally suited for mapping the binding epitope of ligands (Mayer and Meyer, 1999; Maaheimo *et al.*, 2000; Mayer and Meyer, 2001). For instance the binding epitope of the sialyl Lewis[x] tetrasaccharide when bound to the L-fucose-recognizing lectin *Aleuria aurantia* agglutinin (AAA) can be readily mapped by employing STD TOCSY (Haselhorst, 1999; Haselhorst *et al.,* 2001). A normal TOCSY spectrum of sialyl Lewis[x] compared to the STD TOCSY spectrum, both in the presence of AAA, discloses the L-fucose residue as the major binding epitope, since only the cross-peak pattern of the L-fucose residue is left in the STD TOCSY spectrum (Fig. 9).

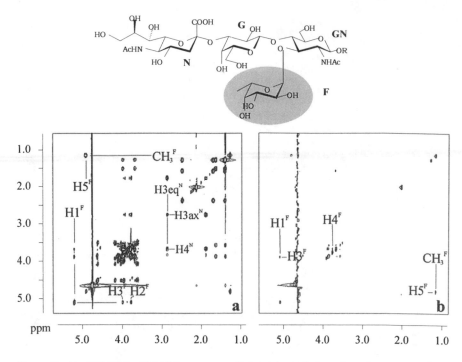

Figure 9. (a) 500 MHz TOCSY spectrum of sialyl Lewis[x] (see formula on top of the spectra): the complete spin systems of all pyranose rings are clearly visible. (b) STD TOCSY spectrum of sialyl Lewis[x] in the presence of AAA (molar ratio of ligand to protein was 100:1): only the spin system of the L-fucose residue (F, grey circle in the formula) is visible indicating that only this part of the tetrasaccharide makes contact with the protein. For the TOCSY spin lock field a MLEV17 sequence was used. The mixing time was 60 ms. For the saturation of the protein (on-resonance 8.80 ppm, off-resonance 40.00 ppm) a cascade of 40 selective Gaussian pulses (50 ms each) were applied resulting in a total saturation time of 2s.

In a practical example a library of 20 D-galactose derivatives was tested for binding towards a lectin from elderberry, *Sambucus nigra* agglutinin (SNA) (Vogtherr and Peters, 2000). Although all compounds had the same molecular weights and almost identical polarity, the stereochemistry being the only criterion to tell them apart, a clear identification of binding components was straightforward using STD TOCSY spectra (Fig. 10). With conventional screening technology, identification of the compound with binding activity directly from the mixture would have been inherently difficult, because of the requirement of a clean separation of all different compounds. Therefore, all 20 components of the library would have had to be synthesized individually as compared to a one-step random methylation used in the STD NMR approach.

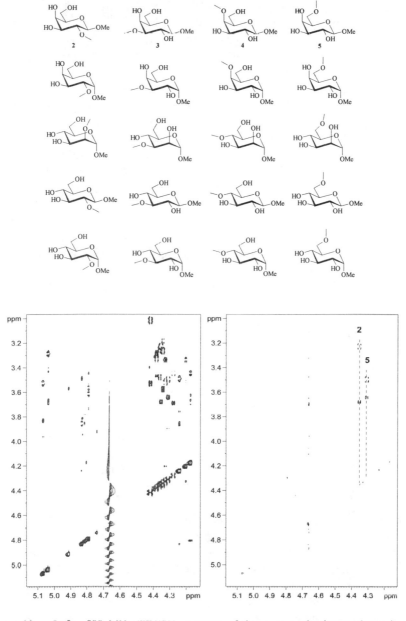

Figure 10:   Left: 500 MHz TOCSY spectrum of the compound mixture shown in the scheme above. Only the portion of the anomeric protons of the monosaccharide derivatives is shown. Right: STD TOCSY spectrum of the mixture in the presence of the lectin SNA (ca. 50 fold molar excess of ligands) showing that only the two β-D-galactose derivatives 2 and 5 bind to the protein. Reprinted with permission from Vogtherr and Peters, 2000. Copyright 2000, American Chemical Society.

Also, it was shown that with little chemical effort, a random O-methylated library of O-methyl-β-D-galactoside, containing ^{13}C labeled O-methyl groups only, can be obtained and subjected to HMQC and HMBC STD experiments readily revealing those hydroxyl groups of the sugar that are essential for binding to the protein. Using standard binding assays this task would normally require considerable synthetic effort.

Membrane bound proteins are inherently difficult to investigate. When devoid of their natural membrane environment they usually loose their structure and hence their functionality. The only way to solve this problem is to study these proteins in a biological membrane. For high-resolution NMR spectroscopy this poses many problems, and therefore structural work on membrane-bound proteins has been confined to MAS NMR spectroscopy so far. STD NMR has the significant advantage that binding to membrane bound proteins can be studied even within the natural membrane environment without causing any problems. In a pilot experiment, wheat germ agglutinin was immobilized to controlled pore glass particles and the binding of carbohydrate derivatives was followed by STD NMR utilizing HRMAS (Klein et $al.$, 1999). The resulting STD spectra clearly indicated the binding of only one of the compounds in the mixture.

Finally, it should be mentioned that with STD NMR it is possible to screen compound libraries at HTS rates (50,000 compounds per day).

WaterLOGSY

A variant of STD NMR utilizes bound water at protein-ligand interfaces. It is well documented that NMR is well suited to study this bound water (Otting, 1997). The observation of negative intermolecular water-ligand NOEs may be explained either by bound water squeezed in between ligand and protein or by a water shell surrounding the ligand (Dalvit et $al.$, 1999). Based on these observations experiments were developed that use the bulk water to detect the binding of ligands to proteins (Dalvit et $al.$, 2000). The experimental setup either utilizes the steady state NOE experiment, where on-resonance saturation is applied to the water chemical shift, or an experimental scheme is applied where the water resonance is selectively inverted such as in the NOE-ePHOGSY scheme (Dalvit and Hommel, 1995). In order to generate best sensitivity, the experiments are performed in H_2O containing only small amounts of D_2O. In order to avoid problems associated with radiation damping and to eliminate artifacts, pulsed field gradients were employed for proper water suppression giving the method the name WaterLOGSY (Water-Ligand Observation with Gradient Spectroscopy).

The technique has successfully been applied to study the binding of ten low molecular weight ligands (at a concentration of 100 μM each) to cyclin-dependent kinase 2 (cdk2, MW ca. 34 kDa, concentration 10 μM). A clear discrimination between the binding ligand, an indole derivative, and the non-binding ligands was straightforward and the resulting spectrum is shown in Fig. 11. It appears that this approach is especially useful for complexes where bound water plays a major role, e.g. for RNA-ligand interactions.

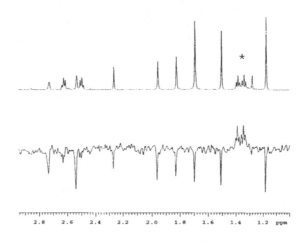

Figure 11. WaterLOGSY spectrum with NOE-ePHOGSY of a library of ten compounds in the presence of cdk2 (lower spectrum). Signals from the indole derivative with binding affinity for cdk2 are easily identified (positive signals). The upper spectrum displays the ¹H-NMR reference. The signals belonging to the binding ligand (α-(ethoxycarbonyl)-3-indoleacrylate) are marked with an asterisk. Reprinted with permission from Dalvit *et al.*, 2000. Copyright 2000 Kluwer/Escom.

3.2 Detection of binding from trNOE experiments

The transferred NOE experiment was described almost three decades ago (Balaram *et al.*, 1972a; Balaram *et al.*, 1972b), and since then has been continuously applied to study the conformation of low molecular weight ligands bound to much larger proteins (Clore and Gronenborn, 1982; Clore and Gronenborn, 1983). With the advent of modern multidimensional NMR experiments and considerable improvement of spectrometer hardware, this class of experiments has seen a steadily growing number of applications, and for cases where 3D structures of the protein receptor are known algorithms have now been developed that allow to calculate the conformation of the bound ligand from trNOEs taking into account the relaxation environment of the protein binding pocket and the binding kinetics (Ni, 1994, Moseley *et al.*, 1995; Moseley *et al.*, 1997; Krishna and Moseley, 1999). The suitability of trNOEs to screen compound libraries for binding activity was demonstrated for a carbohydrate library in the presence of a carbohydrate binding protein, *Aleuria aurantia* agglutinin (AAA) (Meyer *et al.*, 1997), and a second example with a slightly more complex library of carbohydrate derivatives with potential binding affinity towards E-selectin was published shortly thereafter (Henrichsen *et al.*, 1999).

The principle of utilizing trNOE experiments for screening purposes is simple. Once a low molecular weight ligand binds to a protein receptor it adopts the motional properties of the much larger proteins. Exchange between free and bound state of the ligand leads to a superposition of NOEs from the free and the bound ligands on the signals of the free ligand, generating so called trNOEs. For ligands with a molecular weight less than ca. 1 kDa this brings about a change of the sign of the observed NOE. Whereas non-binding ligands display small positive NOEs, trNOEs for the ligands with binding activity are strong and negative. This protocol is applicable if the binding kinetics is such that exchange between free and bound states is fast on the relaxation and preferably also on the chemical shift time scale. Usually, this is the case for K_D values ranging from μM to mM. The ligand should preferably be of such a size that positive NOEs are observed in the free state, since this allows for the easy discrimination between free and bound ligands.

As an example we show the results for a carbohydrate library that had been tested for binding activity for E-selectin (Henrichsen *et al.*, 1999), a carbohydrate binding protein that plays a key role in the initiation of inflammatory cascades. NOESY spectra of a library of 10 structurally similar carbohydrate derivatives were acquired in the presence of E-selectin. The protein concentration was approximately 20 μM, resulting in a 40 μM concentration of binding sites. The ligand concentrations were varied from 5 to 20 fold molar excess with respect to the binding site concentration. From a comparison of the corresponding NOESY and trNOESY spectra the assignment of the component with binding activity was straightforward (Fig. 12). For clean base planes in trNOESY spectra the application of a T1$_\rho$ filter after the 90 degree excitation pulse in the NOESY pulse sequence is essential (Scherf and Anglister, 1993). Another experimental improvement is the application of z-gradient pulses during the mixing period of the NOESY experiment.

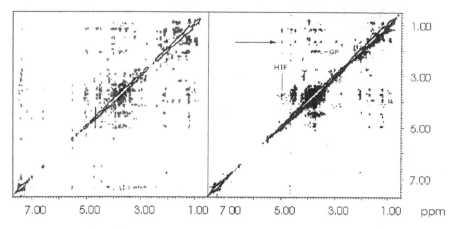

Figure 12. Left: NOESY spectrum of a carbohydrate library at 500 MHz. The mixing time was 900 ms, and all NOEs are positive. Right: NOESY spectrum of the library in the presence of E-selectin (molar ratio of ligands to protein ca. 15:1 for each individual compound) with a mixing time of 300 ms. Only negative contours are shown, highlighting the negative cross peaks originating from the compound with binding affinity towards E-selectin. Reprinted with permission from Henrichsen *et al.*, 1999. Copyright 1999 Wiley-VCH.

Other examples of the utilization of trNOEs for the detection of binding activity have been published in the mean time substantiating the practical value of this approach (Mayer and Meyer, 1999b; Moore, 1999b). It has been shown for instance that sequence information of oligosaccharides can be obtained directly from a mixture of carbohydrate molecules in the presence of a carbohydrate binding protein by applying a homonuclear 3D TOCSY trNOESY experiment (Herfurth *et al.*, 2000). Compared to STD NMR experiments trNOE experiments are less robust in the experimental setup, mainly because trNOEs are optimum only for a relatively small range of ligand excess, whereas in STD NMR experiments a larger excess of ligand does not disturb the experiment. For the observation of trNOEs ligand to protein ratios of about 10:1 to 30:1 are usually optimal, for STD NMR experiments best results are expected for more than ca. 100 fold excess of ligand. This also explains why for STD NMR up to an order of magnitude less protein is required to detect binding activity. Nevertheless, trNOE experiments are a useful complement of STD NMR experiments when testing for binding activity. For instance, STD NMR gives unique information on the binding epitope, whereas trNOE experiments deliver valuable data for the analysis of the bound conformation of the ligand.

3.3   NOE Pumping

When binding to a protein a ligand not only alters its tumbling time but also changes its relaxation environment. Protons located in the binding pocket of the protein contribute to the relaxation of the ligand protons and vice versa. In the original NOE pumping experiment it was proposed to suppress ligand signals by way of diffusion filtering followed by a NOESY type mixing time in order to make relaxation pathways from the protein to the ligand visible by way of subtraction from a suitable reference spectrum (Chen and Shapiro, 1998). In an improved version the reverse experimental setup was chosen (Chen and Shapiro, 2000). In a first experiment, after excitation a T2 filter is applied followed by a NOESY type mixing period in order to suppress protein signals and to allow "NOE pumping" from the ligand to the protein only. In a second reference experiment the T2 filter is applied after the mixing period so that during the mixing time both processes "NOE pumping" from the ligand to the protein and vice versa cancel each other. The difference of the two experiments delivers signals only from those compounds that were in contact with the protein binding site. An example for this reverse NOE pumping is shown in Fig. 13 showing the binding of octanoic acid to human serum albumin. The method provides an alternative to the more robust STD NMR in cases where selective irradiation of the protein is not possible.

3.4   Methods Based on Relaxation Filtering

Line broadening upon binding has long been used to study the binding of ligands to proteins. Certainly, the observation and quantification of such effects requires well separated signals, especially if compound mixtures are to be investigated (Moore, 1999b). Therefore, line broadening effects itself are of limited value for the detection of binding of component of complex mixtures. Other experimental approaches rely on the separate acquisition of spectra of ligand molecules with and without protein receptor present (Hajduk *et al.*, 1997c). Subsequent subtraction leads to spectra that only contain contributions from binding ligands. Unfortunately this requires a rather careful, and therefore more time consuming experimental setup, and the resulting difference spectra are prone to artifacts.

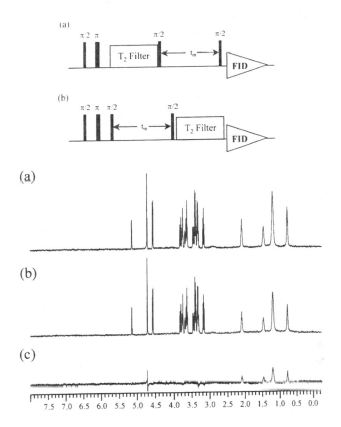

Figure 13. Example of a reverse NOE pumping experiment (RNP). Pulse sequence (a) generates RNP, whereas sequence (b) is used for referencing. The spectra below are obtained for a sample of 1 mM octanoic acid and 1 mM glucose in the presence of 20 µM HSA. (a) Reference spectrum, (b) RNP spectrum, (c) difference spectrum showing only the signals of octanoic acid that binds to HSA. Reprinted with permission from Chen and Shapiro 2000. Copyright 2000 American Chemical Society.

In a recent paper it was proposed to apply a spin labeled first ligand that binds to a main binding site of a protein, and subsequently to screen for so called second binding sites (Jahnke *et al.*, 2000). The presence of a spin label constitutes a T2 relaxation sink for ligands that bind to the second binding site of the protein. Therefore, in spin-lock filtered NMR spectra ligands that bind to this second site are not visible because of fast T2 relaxation, and therefore corresponding signal disappear from the spectrum. An example for this type of relaxation editing is shown in Fig. 14.

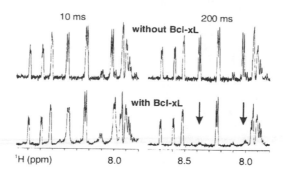

Figure 14. Spectra of a mixture of 8 aromatic compounds in the presence of a spin labeled biphenyl derivative, and in the absence (upper trace) and presence (lower trace) of Bcl-xL. The spectra on the left were recorded with a 10 ms spin-lock filter, the spectra on the right with a 200 ms spin-lock filter. Signals ligand that binds to the second site vanish in the relaxation-filtered spectrum (arrows). Reprinted with permission from Jahnke *et al.*, 2000. Copyright 2000 American Chemical Society.

3.5 Methods Based on Diffusion Editing

With the development of reliable gradient technology for high resolution NMR spectroscopy the investigation of diffusion processes with NMR became practicable. A variety of NMR pulse sequences now exist that allow to precisely determine diffusion properties of molecules in solution. Diffusion editing has been extensively used to analyze compound mixtures, and especially to detect molecular association processes (Lin and Shapiro, 1996; Lin *et al.*, 1997a; Lin *et al.*, 1997b; Hajduk *et al.*, 1997c; Bleicher *et al.*, 1998; Anderson *et al.*, 1999). For small and intermediate size molecules this works well, and in principle, diffusion is also applicable to the detection of the binding of low molecular weight compounds to large protein receptors. In one approach it was suggested to prepare different samples (Hajduk *et al.*, 1997c, cf. also 3.5) of ligand mixtures with and without protein present and then to subtract corresponding diffusion edited spectra. Again this method is prone to artifacts and therefore of limited practical value. The main problem is that for the detection of changes in the diffusion constant of a ligand that binds to a protein, it is required that a significant amount of the ligand is required to be bound by the protein on the time average. Therefore, no large excess of ligand molecules would be allowed if binding was to be detected directly. Also, the size of the protein would be rather limited because at low ligand to protein ratios no individual signals of the ligand could be identified because of severe line broadening. Therefore, diffusion editing so far have been proven to be a valuable tool where molecular interactions of small to intermediate size molecules is to be characterized. For the detection of binding of small molecules to large protein receptors diffusion editing does not provide a practical alternative

3.6 Transfer of Cross-Correlated Relaxation and Residual Dipolar Coupling Constants

It has been shown that a phenomenon related to the transfer NOE occurs when observing so called cross-correlation processes (Reif *et al.*, 1998). If a small molecular weight ligand binds to a large protein, the change in the overall molecular tumbling time causes cross-correlation to become effective. In analogy to trNOEs transferred cross-correlation is now observable (Blommers *et al.*, 1999; Carlomagno *et al.*, 1999). The observation of transferred cross-correlation effects that are dependent on the 3D structure of the ligand in the bound state allows to determine the bound conformation independently from trNOE experiments. Unfortunately, the observation of transfer of cross-correlation between e.g. dipolar relaxation processes of CH vectors in a bound nucleotide requires labeling of the ligand with ^{13}C. This certainly limits the approach for the application to compound libraries.

For the determination of 3D structures of proteins the development of methods that allow to utilize residual dipolar couplings in partly oriented media, usually in the presence of bicelles or phages, constitutes a major milestone (Tjandra and Bax, 1997). The measurement of residual dipolar couplings has also been applied to study the binding of ligands to proteins (Shimizu *et al.*, 1999; Olejniczak *et al.*, 1999; Thompson *et al.*, 2000). The measurement first requires the preparation of a suitable medium that allows the formation of bicelles in which the solute molecules partially orient themselves to the magnetic field. This causes problems if compound libraries are utilized. In order to make the effects of binding visible a large concentration of the protein is required, and the ligands need to be ^{13}C and ^{15}N labeled. Therefore, the measurement of changes in residual dipolar coupling constants of a ligand upon binding to a protein certainly has its value for the analysis of the bound conformation but so far no practical approach has been suggested to make use of this technique for the screening of compound mixtures for binding activity.

4. PERSPECTIVES

As this has already been emphasized by many authors, and highlighted by the examples above NMR as a screening tool provides more information than other methods. NMR screening technology is ideally suited to detect so called lead compounds, i.e. compounds with micro- or even only millimolar binding activity but at the same time allows to explore the region of strong binding with nanomolar dissociation constants.

The use of brute force HTS protocols for the development of novel drugs is not very efficient given 10^{40} molecules that approximately constitute the low molecular weight compound space. The extra information that is available from NMR based screening certainly is of great help to speed up drug development, and therefore will play an increasing role in the future. With the development of cryo probes and higher fields the sensitivity of NMR experiments has been increased almost by an order of magnitude during the recent years. This already had a great impact on the use of NMR based screening. The combination with novel concepts that aim at the design of more intelligent compound libraries, and especially at the design of compound libraries that are tailor made for NMR screening protocols considerably strengthens the technology. One such approach is comprised by the SHAPES strategy that is based on libraries that contain scaffolds of potentially bioactive molecules that after detection of binding activity are further modified to yield high affinity ligands (Bemis and Murcko, 1996; Bemis and Murcko, 1999; Fejzo et al., 1999). It is certainly not exaggerated to say that NMR already plays a key role in screening for novel leads and optimized drugs.

4.1   Acknowledgments

This work was supported by the BMBF, the DFG, and the VW foundation. Financial support from the Verband der Chemischen Industrie is also gratefully acknowledged. We thank Dr. T. Keller and Dr. G. Wolff (Bruker Analytik GmbH) for excellent support. We thank Dr. Andrew Benie for his comments on this article.

5.   REFERENCES

Akasaka, K., 1979, J. Magn. Reson. 36, 135-140.
Anderson, R. C., Lin, M., and Shapiro, M. J., 1999, J. Comb. Chem. 1, 69-72.
Balaram, P., Bothner-By, A. A., and Dadok, J., 1972a, J. Am. Chem. Soc. 94, 4015-4017.
Balaram, P., Bothner-By, A. A., and Breslow, E., 1972b, J. Am. Chem. Soc. 94, 4017-4018.
Bemis, G. W., and Murcko, M. A., 1996, J. Med. Chem. 39, 2887-2893.
Bemis, G. W., and Murcko, M. A., 1999, J. Med. Chem. 42, 5095-5099.
Birdsall, B., Feeney, J., Pascual, C., Roberts, G. C. K., Kompis, I., Then, R. L., Muller, K., and Kroehn, A., 1984, J. Med. Chem. 23, 1672-1676.
Bleicher, K., Lin, M., Shapiro, M. J., and Wareing, J. R., 1998, J. Org. Chem. 63, 8486-8490.

Blommers, M. J. J., Stark, W., Jones, C. E., Head, D., Owen, C. E., and Jahnke, W., 1999, *J. Am. Chem. Soc.* **121**, 1949-1953.

Carlomagno, T., Felli, I. C., Czech, M., Fischer, R., Sprinzl, M., and Griesinger, C., (1999), *J. Am. Chem. Soc.* **121**, 1945-1948.

Chen, A., and Shapiro, M.J., 1998, *J. Am. Chem. Soc.* **120**, 10258-10259.

Chen, A., and Shapiro, M.J., 2000, *J. Am. Chem. Soc.* **122**, 414-415.

Clore, G. M., and Gronenborn, A. M., 1982, *J. Magn. Reson.* **48**, 402-417.

Clore, G. M., and Gronenborn, A. M., 1983, *J. Magn. Reson.* **53**, 423-442.

Dalvit, C., and Hommel, U., 1995, *J. Magn. Reson.* **109**, 334-338.

Dalvit, C., Cottens, S., Ramage, P., and Hommel, U., 1999, *J. Biomol. NMR.* **13**, 43-50.

Dalvit, C., Pevarello, P., Tatò, M., Veronesi, M., Vulpetti, A., and Sundström, M., 2000, *J. Biomol. NMR* **18**, 65-68.

Feeney, J., Batchelor, G., Albrand, J. P., and Roberts, G. C. K., 1979, *J. Magn. Reson.* **33**, 519-529.

Fejzo, J., Lepre, C. A., Peng, J. W., Bemis, G. W., Ajay, Murcko, M. A., and Moore, J. M, 1999, *Chem. Biol.* **6**, 755-769.

Feng, M. -H., Philippopoulos, M., McKerell, A. D., Jr., and Lim, C., 1996, *J. Am. Chem. Soc.* **118**, 11265-11277.

Gardner, K. H., and Kay, L. E., 1997, *J. Am. Chem. Soc.* **119**, 7599-7600.

Gardner, K. H., Zhang, X. C., Gehring, K., and Kay, L. E., 1998, *J. Am. Chem. Soc.* **120**, 11738-11748.

Gargaro, A. R., Frenkiel, T.A., Nieto, P. M., Birdsall, B., Polshakov, V. I., Morgan, W. D., and Feeney, J., 1996, *Eur. J. Biochem.* **238**, 435-439.

Goto, N. K., Gardner, K. H., Mueller, G. A., Willis, R. C., and Kay, L. E., 1999, *J. Biomol. NMR* **13**, 369-374.

Hajduk, P. J., Sheppard, G., Nettesheim, D. G., Olejniczak, E. T., Shuker, S. B., Meadows, R. P., Steinman, D. H., Carrera, G. M., Marcotte, P. A., Severin, J., Walter, K., Smith, H., Gubbins, E., Simmer, R., Holzman, T. F., Morgan, D. W., Davidsen, S. K., and Summ, J. B., 1997a, *J. Am. Chem. Soc.* **119**, 5818-5827.

Hajduk, P. J., Dinges, J., Miknis, G. F., Merlock, M., Middleton, T., Kempf, D. J., Egan, D. A., Walter, K. A., Robins, T. S., Shuker, S. B., Holzman, T. F., and Fesik, S. W., 1997b, *J. Med. Chem.* **40**, 3144-3150.

Hajduk, P. J., Olejniczak, E. T., and Fesik, S. W., 1997c, *J. Am. Chem. Soc.* **119**, 12257-12261.

Hajduk, P. J., Gerfin, T., Boehlen, J. M., Häberli, M., Marek, D., and Fesik, S. W., 1999a, *J. Med. Chem.* **42**, 2315-2317.

Hajduk, P. J., Dinges, J., Schkeryantz, J. M., Janowick, D., Kaminski, M., Tufano, M., Augeri, D. J., Petros, A., Nienaber, V., Zhong, P., Hammond, R., Coen, M., Beutel, B., Katz, L., and Fesik S. W., 1999b, *J. Med. Chem.* **42**, 3852-3859.

Hajduk, P. J., Boyd, S., Nettesheim, D., Nienaber, V., Severin, J., Smith, R., Davidson, D., Rockway, T., and Fesik, S. W., 2000a, *J. Med. Chem.* **43**, 3862-3866.

Hajduk, P. J., Gomtsyan, A., Didomenico, S., Cowart, M., Bayburt, E. K., Solomon, L., Severin, J., Smith, R., Walter, K., Holzman, T. F., Stewart, A., McGaraughty, S., Jarvis, M.F., Kowaluk, E. A., and Fesik, S. W., 2000b, *J. Med. Chem.* **43**, 4781-4786.

Hajduk, P.J., Augeri, D.J., Mack, J., Mendoza, R., Yang, J., Betz, S.F., and Fesik, S.W., 2000c, *J. Am. Chem. Soc.* **122**, 7898-7904.

Hajduk, P. J., Bures, M., Prestegard, J., and Fesik, S. W., 2000d, *J. Med. Chem.* **43**, 3443-3447.

Haselhorst, T., 1999, PhD thesis, Medical University of Luebeck, ISBN 3-89712-698-2, pp. 126-169.

Haselhorst, T., Weimar, T., and Peters, T., 2001, *J.Am. Chem. Soc.* **123**, 10705-10714.

Henrichsen, D., Ernst, B., Magnani, J. L., Wang, W. T., Meyer, B., and Peters, T., 1999, *Angew. Chem. Int. Ed.* **38**, 98-102.

Herfurth, L., Weimar, T., and Peters, T., 2000, *Angew. Chem. Int. Ed. Engl.* **39**, 2097-2099.

Jahnke, W., Perez, L. B., Paris, C. G., Strauss, A., Fendrich, G., and Nalin, C. M., 2000, *J. Am. Chem. Soc.* **122**, 7394-7395.

Jayalakshmi, V., and Krishna, N. R., 2002, *J. Magn. Reson.*, **155**, 106-118.

Kainosho, M., and Tsuji, T., 1982, *Biochemistry* **21**, 6273-6279.

Klein, J., Meinecke, R., Mayer, M. and Meyer, B., 1999, *J. Am. Chem. Soc.* **121**, 5336-5337.

Krishna, N. R., and Moseley, H. N. B., 1999, in *Biol. Magn. Reson.*, **17**, (N. R. Krishna and L. J. Berliner, eds), Kluwer Academic / Plenum Press, New York, pp. 223-307.

Lepre, C. A., 2001, *Drug Discovery Today* **6**, 133-140.

Lian, L.Y., and Roberts, G. C. K., 1993, in *NMR of Macromolecules*, (Roberts, G. C. K., ed.), Oxford University Press, Oxford, pp. 153-182.

Lin, M., and Shapiro, M. J., 1996, *J. Org. Chem.* **61**, 7617-7619.

Lin, M., Shapiro, M. J., and Wareing, J. R., 1997a, *J. Am. Chem. Soc.* **119**, 5249-5350.

Lin, M., Shapiro, M. J., and Wareing, J. R., 1997b, *J. Org. Chem.* **62**, 8930-8931.

Maaheimo, H., Kosma, P., Brade, L., Brade, H., and Peters, T., 2000, *Biochemistry* **39**, 12778-12788.

Mayer, M., and Meyer, B., 1999a, *Angew. Chem. Int. Ed. Engl.* **35**, 1784-1788.

Mayer, M., and Meyer, B., 1999b, *J. Med. Chem.* **43**, 2093-2099.

Mayer, M., and Meyer, B., 2001, *J. Am. Chem.Soc.* **123**, 6108-6117.

Meyer, B., Weimar, T., and Peters, T., 1997, *Eur. J. Biochem.* **246**, 705-709.

Medek, A., Hajduk, P. J., Mack, J., and Fesik, S. W., 2000, *J. Am. Chem. Soc.* **122**, 1241-1242.

Moore, J. M., 1999, *Curr. Opin. Biotech.* **10**, 54-58.

Moore, J. M., 1999b, *Biopolymers* **51**, 221-243.

Morgan, W. D., Birdsall, B., Nieto, P. M., Gargaro, A. R., and Feeney, J., 1999, *Biochemistry* **38**, 2127-2134.

Moseley, H. N. B., Curto, E. V., and Krishna, N. R., 1995, *J. Magn. Reson.* **108B**, 243-261.

Moseley, H. N. B., Lee, W., Arrowsmith, C., and Krishna, N. R., 1997, *Biochemistry* **36**, 5239-5299.

Neuhaus, D., and Williamson, M. P., 2000, *The Nuclear Overhauser Effect in Structural and Conformational Analysis*, Wiley-VCH, New York.

Ni, F., 1994, *Prog. NMR Spectr.* **26**, 517-606.

Nieto, P. M., Birdsall, B., Morgan, W. D., Frenkiel, T. A., Gargaro, A. R., and Feeney, J., 1997, *FEBS Lett.* **405**, 16-20.

Olejniczak, E. T., Hajduk, P. J., Marcotte, P. A., Nettesheim, D. G., Meadows, R. P., Edalji, R., Holzman, T. F., and Fesik, S. W., 1997, *J. Am. Chem. Soc.* **119**, 5828-5832.

Olejniczak, E. T., Meadows, R. P., Wang, H., Cai, M., Nettesheim, D. G., and Fesik, S. W., 1999, *J. Am. Chem. Soc.* **121**, 9249-9250.

Otting, G., 1997, *Progr. NMR Spectrosc.* **31**, 259-285.

Pascal, S. M., Yamazaki, T., Singer, A. U., Kay, L. E., and Forman-Kay, J. D., 1995, *Biochemistry* **34**, 11353-11362.

Pervushin, K., Riek, R., Wider, G., and Wüthrich, K., 1997, *Proc. Natl. Acad. Sci. USA* **23**, 12366-12371.

Poppe, L., Brown, G. S., Philo, J. S., Nikrad, P. V., and Shah, B. H., 1997, *J. Am. Chem. Soc.* **119**, 1727-1736.

Ramos, A., Kelly, G., Hollingworth, D., Pastore, A., and Frenkiel, T., 2000, *J. Am. Chem. Soc.* **122**, 11311-11314.

Reif, B., Hennig, M., and Griesinger, C., 1997, *Science* **276**, 1230-1233.

Roberts, G. C. K., 1999, *Curr. Opin. Biotech.* **10**, 42-47.

Roberts, G. C. K., 2000, *Drug Discovery Today* **5**, 230-240.

Scherf, T., and Anglister, J., 1993, *Biophys. J.* **64**, 754-761.

Shimizu, H., Donohue-Rolfe, A., and Homans, S. W., 1999, *J. Am. Chem. Soc.* **121**, 5815-5816.

Shuker, S. B., Hajduk, P. J., Meadows, R. P., and Fesik, S. W., 1996, *Science* **274**, 1531-1534.

Stockman, B. J., 1998, *Progr. NMR Spectrosc.* **33**, 109-151.

Takahashi, H., Nakanishi, T., Kami, K., Arata, Y., and Shimada, I., 2000, *Nat. Struct. Biol.* **7**, 220-223.

Thompson, G. S., Shimizu, H., Homans, S. W., and Donohue-Rolfe, A., 2000, *Biochemistry* **39**, 13153-13156.

Tjandra, N., and Bax, A., 1997, *Science* **278**, 1111-1114.

Vogtherr, M., and Peters, T., 2000, *J. Am. Chem. Soc.* **122**, 6093-6099.

Wiedemann, P. E., Fesik, S. W., Petros, A. M., Nettesheim, D. G., Mollison, K. W., Lane, B. C., Or, Y. S., and Luly, J. R., 1999, *J. Med. Chem.* **42**, 4456-4461.

Wikstroem, M., 2000, Lecture at the Conference *NMR: Drug Discovery* and *Design*, Oct. 24-26, McLean, Virginia, USA.

Yamazaki, T., Pascal. S. M., Singer, A. U., Forman-Kay, J. D., and Kay, L. E., 1995, *J. Am. Chem. Soc.* **117**, 3556-3564.

Contents of Previous Volumes

Index